Constrained Nonlinear Control: Performance and Robustness

Ruihang Ji · Shuzhi Sam Ge · Dongyu Li

Constrained Nonlinear Control: Performance and Robustness

Springer

Ruihang Ji
Department of Electrical and Computer Engineering, College of Design and Engineering
National University of Singapore
Singapore, Singapore

Shuzhi Sam Ge
Department of Electrical and Computer Engineering, College of Design and Engineering
National University of Singapore
Singapore, Singapore

Dongyu Li
School of Cyber Science and Technology
Beihang University
Beijing, China

ISBN 978-981-96-2970-1 ISBN 978-981-96-2971-8 (eBook)
https://doi.org/10.1007/978-981-96-2971-8

This Springer imprint is published by the registered company Springer Nature Singapore Pte Ltd.
The registered company address is: 152 Beach Road, #21-01/04 Gateway East, Singapore 189721, Singapore

Preface

The development of stable and robust nonlinear control has attracted significant research interest within the control community in recent years due to its theoretical challenges and practical demands. It is well known that the controlled systems encounter various constraints in engineering, stemming not only from the physical limitations but also from application-specific or user-defined performance requirements. These constraints, whether related to system capabilities (e.g., input saturation, communication bandwidth, and measurable information) or specific performance criteria (e.g., transient response, steady-state accuracy, convergence speed, and overshoot performance), bring immediate challenges in the control design, as they can potentially degrade system performance or even compromise closed-loop stability, especially under uncertain and complex environments. Consequently, the study of constrained nonlinear control is both meaningful and necessary for achieving high tracking performance of the controlled systems, which motivates the research presented in this book.

There is a vast amount of literature on constrained control for nonlinear systems under various conditions, for example, external disturbances, actuator faults, time delay, and unknown control directions. Nevertheless, there are some interesting problems that remain unsolved and deserve more in-depth studies. (a) If multiple constraints are considered at the same time, most studies adopt an implicit assumption that these constraints should be satisfied always, even under serious conditions. Actually, serious conditions, such as actuator faults, time delays, and external disturbances, often lead to the violation of performance-related constraints due to the limited tracking ability, and it is frequently encountered in both numerical simulation and practical engineering. There is a lack of comprehensive discussion on the conflict between different constraints or their interconnections in the existing studies. (b) Secondly, note that many existing works are tailored to specific conditions, which may not be applicable to other scenarios. For example, the commonly used Prescribed Performance Control is only applicable for the initial conditions that are within the allowable performance boundaries, but suffers from singularity problem if the initial conditions are outside performance boundaries. How to design a unified control design that can guarantee the system to reach the prescribed performance region within finite time, regardless of initial conditions is still an open problem. (c) Practical systems often operate under limited computational loads but need to maintain certain performance constraints. As systems grow in complexity, how to save communication resources and guarantee system performance at the same time deserve more in-depth studies. Although some efforts on Event-Triggered Control have been made to reduce unnecessary signal transmissions but the signal encoding-decoding process is always neglected. This could aggravate communication delays and congestion issues due to the limited bandwidth, particularly when communication bits are critical.

This book is motivated by the above discussion and aims to systematically investigate and address the constrained nonlinear control problems under various conditions. With this in mind, this book is carefully tailored to (i) propose Tunnel Prescribed Performance and Finite-Time Tunnel Prescribed Performance to improve the tracking performance in terms of overshoot performance and settling time; (ii) define a new Entry Capture Problem which relaxes the restriction on the initial conditions, making control design more flexible and generalized; (iii) establish a balance between input saturation and prescribed performance by introducing Saturation-Tolerant Prescribed Performance; and (iv) introduce a novel coded event-triggered scheme to further reduce the communication burden and enhance the communication security at the same time. For completeness, rigorous proofs of the closed-loop stability are also provided, and numerical simulation examples are conducted to verify the effectiveness of the proposed control strategies.

This book starts with a brief summary of some frequently encountered constraints in the controlled systems, and basic mathematical concepts and stability theorems are introduced in Chap. 2, which is the basis of the subsequent chapters.

In Chap. 3, several structured constrained control strategies are proposed for Euler-Lagrange systems with input saturation and uncertainties. Auxiliary systems are designed to compensate for the nonlinearities from input saturation. In order to alleviate the chattering phenomenon, a novel concept of second-order dynamic surface control is introduced, which extends the control input to its time derivative.

In Chap. 4, the tracking control problem of more general nonlinear systems with input saturation and performance constraints is addressed. Tunnel prescribed performance and finite-time tunnel prescribed performance are proposed to improve the tracking performance. Two tunnel prescribed control schemes are designed for nonlinear systems with known and unknown control directions, respectively.

In Chap. 5, a new Entry Capture Problem is defined, that is, for any initial tracking error, it can be regulated into the prescribed performance constraints within a user-given time. An error self-tuning function and self-adjustable prescribed performance provide two different solutions for the underlying problem. The resulted tunnel prescribed control can not only guarantee finite-time convergence but also be applicable for any initial conditions.

In Chap. 6, more complex Multi-input Multi-output nonlinear systems with input saturation and prescribed performance constraints are considered. We establish a bridge between input saturation and prescribed performance instead of neglecting their contradictions. Two auxiliary systems are specifically designed for systems with known and unknown control directions. The proposed saturation-tolerant prescribed control can always guarantee both constraints even under serious conditions, such as actuator faults and external disturbances.

In summary, this book provides a comprehensive study of constrained nonlinear control for various systems under different conditions. The book will be of interest to researchers, academics, engineers, and research students in the fields of nonlinear systems, constrained control, and control engineering who would like to find out about the performance, robustness, and applications of constrained nonlinear systems.

Singapore — Ruihang Ji
Singapore — Shuzhi Sam Ge
Beijing, China — Dongyu Li
November 2024

Acknowledgements

The creation of this book would not have been possible without the support, encouragement, and guidance of many individuals and institutions. We would like to express my sincere gratitude to those who have contributed to this book in various ways, including invaluable insights, continuous guidance, and constructive feedback throughout the research and writing process. We would like to thank (in alphabetical order of last names)

Mou Chen of Nanjing University of Aeronautics and Astronautics;
Prokhorov Danil of Toyota Technical Center;
Xin Huo of Harbin Institute of Technology;
Qinglong Han of Swinburne University of Technology;
Wanyue Jiang of Qingdao University;
Zhen Kan of University of Science and Technology of China;
Haizhou Li of Nation University of Singapore;
Yibei Li of Chinese Academy of Sciences;
Jiang Long of Northwestern Polytechnical University;
Xiaoling Liang of Nation University of Singapore;
Yong Liu of Agency for Science, Technology and Research;
Kun Li of Chongqing University;
Qiong Liu of Tsinghua University;
Jiafeng Li of Beijing Jiaotong University;
Xiaorui Liu of Qingdao University;
Jie Ma of Harbin Institute of Technology;
Kemao Ma of Harbin Institute of Technology;
Rong Su of Nanyang Technological University;
Haining Sun of Agency for Science, Technology and Research;
Yongduan Song of Chongqing University;
Lihan Sun of Harbin Institute of Technology;
Zheng Mike Shou of National University of Singapore;
Ruochen Tai of Nanyang Technological University;
Pey Yuen Tao of Agency for Science, Technology and Research;

Yujuan Wang of Chongqing University;
Fenglan Wang of Nation University of Singapore;
Wentao Wu of Shanghai Jiao Tong University;
Lantao Xing of Shandong University;
Bin Yao of Purdue University;
Moyu Yao of Tsinghua University;
Ze Yang of Harbin Institute of Technology;
Baoqing Yang of Harbin Institute of Technology;
Kai Zhao of Chongqing University;
Weidong Zhang of Shanghai Jiao Tong University;
Hui Zhao of Harbin Institute of Technology;
Xinyu Zhang of Tsinghua University;
Zhirong Zhang of University of Macau;
Pengyu Zhang of Nation University of Singapore;
Yuxiang Zhang of Nation University of Singapore;
Yu Zhang of Harbin Institute of Technology;
Jinxi Zhang of Northeastern University;
Liangli Zhen of Agency for Science, Technology and Research; among others.

This research is supported by Singapore RIE2025 Manufacturing, Trade and Connectivity Industry Alignment Fund – Pre-Positioning (IAF-PP) under Grant M24N2a0039 and the National Research Foundation Singapore under its AI Singapore Programme AISG2-GC-2023-007. Special thanks towards the Springer editorial team for their support and guidance throughout the publication process, and towards IEEE for granting us the permission to use the materials from our publications. Finally, we would like to thank all the readers for their interest in this book. We hope that the book will be useful to researchers, academics, and engineers in the field of constrained nonlinear control.

Singapore — Ruihang Ji
Singapore — Shuzhi Sam Ge
Beijing, China — Dongyu Li
November 2024

Contents

Chapter 1
Introduction

The control problem for nonlinear systems, particularly those with constraints, has garnered substantial attention due to its practical and theoretical significant [4, 33, 40, 41, 45–47]. The constraints frequently encountered in engineering systems could influence the tracking performance, stability, and robustness of the control system. Typically, these constraints can be categorized into two types: system capability-related constraints [27, 32] (i.e., input saturation, communication resources, measurable information); and performance-related constraints [39, 42] (e.g., transient response, convergence rate, and steady-state behavior). In this book, we will discuss the control designs of Euler–Lagrange Systems, more general nonlinear systems and Multi-input Multi-output (MIMO) systems under single or multiple constraints. Hereafter, we will provide an overview of constrained nonlinear control designs under different scenarios.

1.1 Input Saturation Constraint and Control Design

In practical applications, due to the physical limitation, input saturation on hardware indicates that the magnitude of the control signal is always constrained within a feasible region [37, 70, 77, 81, 95]. However, the introduced nonlinearities by input saturation may severely limit the tracking performance and even cause system instability [12, 75, 97], which makes the control design more complex and challenging. The development of nonlinear control schemes for uncertain systems with input saturation has been a task of major practical interest as well as theoretical significance. There are two strategies in handling input saturation: direct design and anti-saturation design.

The direct design methods consider the input saturation in the control design stage and aim to achieve a stable closed-loop systems [10, 11, 20]. Its design is

R. Ji et al., *Constrained Nonlinear Control: Performance and Robustness*,
https://doi.org/10.1007/978-981-96-2971-8_1

often conservative and may not achieve high-performance control systems. The anti-saturation design methods, also called two-step methods, first design a control box under unconstrained conditions, then construct a virtual auxiliary system to compensate for nonlinearity caused by input saturation. Inspired by these two strategies, many studies have been investigated to develop control schemes for nonlinear systems under various scenarios. To compensate for the nonlinearities caused by input saturations, an adaptive tracking control is proposed in [12] for MIMO nonlinear systems, where an auxiliary system is first introduced to analyze the constraints effect. In [86], the finite-time tracking control problem is addressed for the Single-input Single-Output (SISO) nonlinear systems with asymmetric input saturations. And smooth hyperbolic tangent functions are utilized to approximate the nonlinear saturation functions. In [76], a novel auxiliary dynamic system is constructed to tackle saturations where additional auxiliary states can be integrated into the control scheme, and a nonlinear finite-time observer is developed to estimate the unknown disturbance. In [8, 77], the problem of actuator faults including loss of effectiveness, bias faults and stuck conditions are further addressed for the constrained systems. Note that the above approaches are strictly confined within the state feedback control field. Namely, all the system states should be measurable, which may not always be met in the practical engineering. Output feedback control provides a strong tool by developing various observers to estimate unmeasurable system states. In [54], the tracking control problem for a class of stochastic nonlinear time-delay systems is addressed under input saturation and unmeasurable states, in which the growth assumptions on the drift and diffusion terms have been relaxed. In [33], a finite-time output feedback control is developed for MIMO nonlinear systems, where fuzzy logic systems based observer is constructed to estimate unmeasurable system states despite saturations and actuator faults simultaneously occur.

1.2 Prescribed Performance Constraint and Control Design

Prescribed Performance Control (PPC) pioneered in [3] is proposed for handling the prescribed performance constraints by defining Prescribed Performance Functions (PPFs) and Error Transformation Function (ETF). With aid of the universal approximation abilities of Fuzzy Logic Systems (FLSs) and Neural Networks (NNs) [48, 59], PPC is greatly promoted the development of uncertain nonlinear systems. More specifically, a novel fuzzy observer-based PPC is proposed in [43], where semi-globally uniformly ultimately bounded is guaranteed. To obtain low-complexity control scheme, the developed PPC in [7] avoids any approximations or adaptive techniques. In [44], a NN output feedback optimized control is developed, together with a barrier-type optimal cost functions and actor-critic structure, where the system states are guaranteed within the constrained sets. Unfortunately, all these studies need the prior knowledge of the control direction, which is not usually available in engineering. The Nussbaum-type function as an effective technique has been utilized to handle such problem [14, 25, 51, 57, 73, 80, 82, 90]. Together with this technique,

PPC can be subsequently applied to various nonlinear systems under unknown control directions. In [90], a simple fault-tolerant control is designed for nonlinear systems, and the predefined performance requirements are also guaranteed. In [51], the consensus problem for multiagent systems is addressed, where prescribed performance is imposed on the consensus errors in switching networks.

Note that the above results mainly focus on infinite-time control performance, whereas finite-time performance is much more expected in most practical systems [15, 67, 68]. An overview of the finite/fixed-time tracking methods and their applications are fully presented in [50], where some recent approaches together with their performance are discussed. In [43], finite-time NN control tracking method is investigated for a class of nonlinear systems having powers of odd rational numbers, where the closed-loop system is semi-global practical finite-time stability. However, the following problems have not been well addressed in the current literature. (i) The common PPFs in [7, 43, 59] are distributed on different sides, leading to a loose funnel-shape allowable set. Under this condition, such PPFs are difficult to fulfill the increasing tracking requirements (i.e., for the flight tracking systems, the tracking error is required within a tight allowable set due to the limited perspective region and smaller overshoot performance is preferred to avoid losing tracking object). (ii) The existing studies in [15, 67, 68] can guarantee finite-time tracking performance by integrating PPC with finite-time control. Notice that their settling times are dependent on the system initial conditions and unable to be determined in advance. Moreover, the additional introduced fractional power items increase the complexity level of the control implementation. Therefore, how to develop a unified yet concise prescribed performance to provide better transient performance is an interesting yet challenging problem.

(Prescribed Performance Control under Input Saturation) As discussed above, the input saturation is a common phenomenon in practice, which may severely degrade the tracking ability. Many PPC methods simultaneously considering input saturation problem have evolved as an exciting research area due to their theoretical significance and practical interest (see [13, 61, 66, 72, 83, 84, 88, 94, 96] and references therein). However, there are two issues that are worth noting: (i) There exists an implicit assumption that both input and performance constraints are never violated in these studies even under complex conditions and uncertainties. Unfortunately, in practice, the uncertain nonlinearities could seriously degrade the tracking performance since the input saturation limits the tracking capability. It may result in performance-constraint violation, and make the classical PPC methods inapplicable since they are undefined for the tracking error outside the allowable set. In this way, PPC suffers from singularity problem, rendering these current methods infeasible; and (ii) Most PPFs are prespecified by users, whose expressions keep fixed from the beginning of system operation. In this way, the robustness cannot be well guaranteed since the nonlinearities are always time varying and unknown. Sometime, the control design should be conservative to ensure the feasibility. The doubt whether the redeveloped ones are practicable for any initial condition and unknown nonlinearities still remains open and needs further research. This problem stems from that the input saturation and performance constraints are always analyzed independently without comprehensively considering their interconnections and conflicts. It is therefore a promising direction by developing flexible performance boundaries to bridge a balance between input and performance constraints. Inspired by [34, 85], where auxiliary systems are investigated for known control directions, and by [30] in which case nonlinear time-delay systems are discussed, how to achieve the aforementioned trade-off strategy for more general nonlinear systems under unknown control directions and other uncertainties (i.e., actuator faults, external disturbances, time delay, etc.) deserves further studies.

(Prescribed Performance Control for any Initial Conditions) During the past decades, Prescribed Performance Control (PPC) has attracted much attention since it can preassign performance attributes [3, 16, 45, 52, 90], such as overshoot, convergence rate and steady-state error. Driven by its practical and theoretical significance, this technique has been applied to various nonlinear systems [31, 65, 91, 93]. In [58], an arbitrary convergence time is integrated in PPC for the tracking control of high-order systems. In [63], the PPC is developed for the formation control, in which some system states are constrained in allowable sets. An observer-based control is studied for manipulators in [64] using only generalized coordinates as measurements. Moreover, flight experiments are implemented in [36] and a low-complexity control scheme is developed in [62] to reduce the computation burden in practice. Despite these remarkable results, some challenging issues still exist in the classical PPC.

(i) All these established results are based on an implicit assumption that the tracking error $e_i(t)$ should satisfy the prescribed performance from the beginning of system operation, i.e., $|e_i(0)| < \rho_i(0)$ with $\rho_i(t)$ being the performance function. If the system is rerun or the reference signal is modified, it is critical to reselect $\rho_i(0)$ correspondingly, or control singularity problem may occur since $e_i(t)$ outside the allowable set is not defined. Recently, funnel control (FC) [5, 6, 23, 24], viewed as an adaptive high-gain control scheme, provides an effective solution to handle such initial condition restriction by allowing the funnel boundary to infinity at the initial time. Note that these results only focus on the case of relative degree one. In [22, 56, 92], a class of novel global performance functions are then developed for uncertain nonlinear systems with arbitrary relative degree, making the control design more applicable. However, these methodologies are based on a sequence of specific performance functions, which are challenging to extend to more general performance cases. Moreover, the calculations of derivatives of virtual control signals are inevitable during the backstepping design. Hence, how to develop a simple yet generalized mechanism for the PPC design is interesting although challenging.

(ii) It can be also observed that both funnel boundaries and global performance functions are always symmetric or distributed on different sides, resulting in funnel-shape allowable sets. In this case, loose performance constraints are imposed on error signals which may cause obvious overshoot performance. In practice, more tight allowable set is expected in most applications from the considerations of system operation and safety. One representative example is a flight tracking system which needs to be within a small region around the target due to the limited perception range of sensors. Therefore, we have proposed an asymmetric Tunnel Prescribed Performance (TPP) in [34] capable of providing more strict constraints on the tracking error, and extended its results to the case of unknown control directions [28, 30]. We, in this book, consider a new Entry Capture Problem (ECP) for a class of uncertain nonlinear systems such that the tracking error can be regulated into the asymmetric prescribed performance within a given time regardless of any initial condition (i.e., initial-condition violation). Such ECP is frequently encountered but have not been fully addressed in these existing works.

1.3 Communication Resource Constraint and Control Design

The above control schemes are executed periodically which is not preferable particularly when the system remains at a satisfactory level. As the controlled systems grow in complexity and size, efficient resource usage and network congestion avoidance become essential. The event-triggered control proposed in [1, 2, 29] has been an effective tool to reduce communication burden, by which the control input is updated only upon the occurrence of a predefined event. Since then, event-triggered control has aroused widespread research interest, leading to developments of various alternative approaches, see, [19, 26, 35, 49, 60, 69, 78, 89] and references therein. Inspired by [69], various triggering conditions have been proposed, such as state-independent threshold [38, 74], dynamic threshold [19], self-trigger threshold [17], and rate-coded event-triggered scheme [27]. In [55], the typically encountered denial-of-service attacks is further investigated, and a novel attack detection is introduced under event-case condition. In [21], a distributed dynamic event-based consensus is proposed for both leaderless and leader-following networks, respectively. In [71], denial-of-service attacks are further considered in the development of a dynamic memory event-triggered strategy. Most these event-triggered control methods concentrate on reducing the frequency of signal update to save communication resources, but the transmission encoding-decoding process is ignored. Typically, upon the occurrence of the predefined event, the control signal is encoded into a fixed-length string, such as 16-bit string, before being transmitted over a communication channel. Once received by the actuator box, a decoding process is employed before it being applied to the actuator. Note that using such a number of bits for each communication might be redundant and inefficient, especially for a small control input. Moreover, if the communication channel can only handle a limited number of bits at a time, multiple transmissions between control and actuator may lead to transmission delays and packet losses [18, 53]. Recently, some excellent works in [9, 79, 87] have investigated a class of event-triggered mechanisms from an encoding-decoding viewpoint, by which only 1-bit or 2-bit string is required for each communication. However, to our best knowledge, there is no result on investigating the ECP with event-triggered control.

References

1. Åarzén, K.E.: A simple event-based pid controller. IFAC Proceedings Volumes **32**(2), 8687–8692 (1999)
2. Åström, K.J., Bernhardsson, B.: Comparison of periodic and event based sampling for first-order stochastic systems. IFAC Proceedings Volumes **32**(2), 5006–5011 (1999)
3. Bechlioulis, C.P., Rovithakis, G.A.: Robust adaptive control of feedback linearizable mimo nonlinear systems with prescribed performance. IEEE Transactions on Automatic Control **53**(9), 2090–2099 (2008)
4. Bechlioulis, C.P., Rovithakis, G.A.: Robust partial-state feedback prescribed performance control of cascade systems with unknown nonlinearities. IEEE Trans. Autom. Control **56**(9), 2224–2230 (2011)
5. Berger, T., Dennstädt, D., Ilchmann, A., Worthmann, K.: Funnel model predictive control for nonlinear systems with relative degree one. SIAM Journal on Control and Optimization **60**(6), 3358–3383 (2022)
6. Berger, T., Rauert, A.L.: Funnel cruise control. Automatica **119**, 109061 (2020)
7. Bikas, L.N., Rovithakis, G.A.: Tracking performance guarantees in the presence of quantization for uncertain nonlinear systems. IEEE Transactions on Automatic Control (2020)
8. Cao, L., Li, H., Dong, G., Lu, R.: Event-triggered control for multiagent systems with sensor faults and input saturation. IEEE Trans. Syst., Man, Cybern.: Syst. (2019)
9. Cao, L., Ren, H., Li, H., Lu, R.: Event-triggered output-feedback control for large-scale systems with unknown hysteresis. IEEE Transactions on Cybernetics **51**(11), 5236–5247 (2020)
10. Chen, B.M., Lee, T.H., Peng, K., Venkataramanan, V.: Composite nonlinear feedback control for linear systems with input saturation: theory and an application. IEEE Transactions on automatic control **48**(3), 427–439 (2003)
11. Chen, B.S., Lu, H.C.: State estimation of large-scale systems. International Journal of Control **47**(6), 1613–1632 (1988)
12. Chen, M., Ge, S.S., Ren, B.: Adaptive tracking control of uncertain mimo nonlinear systems with input constraints. Automatica **47**(3), 452–465 (2011)
13. Chen, M., Shao, S., Jiang, B.: Adaptive neural control of uncertain nonlinear systems using disturbance observer. IEEE Transactions on Cybernetics **47**(10), 3110–3123 (2017)
14. Chen, W., Li, X., Ren, W., Wen, C.: Adaptive consensus of multi-agent systems with unknown identical control directions based on a novel nussbaum-type function. IEEE Transactions on Automatic Control **59**(7), 1887–1892 (2013)
15. Cui, G., Yu, J., Shi, P.: Observer-based finite-time adaptive fuzzy control with prescribed performance for nonstrict-feedback nonlinear systems. IEEE Transactions on Fuzzy Systems (2020)
16. Dimanidis, I.S., Bechlioulis, C.P., Rovithakis, G.A.: Output feedback approximation-free prescribed performance tracking control for uncertain mimo nonlinear systems. IEEE Trans. Autom. Control **65**(12), 5058–5069 (2020)
17. Dimarogonas, D.V., Frazzoli, E., Johansson, K.H.: Distributed event-triggered control for multi-agent systems. IEEE Transactions on Automatic Control **57**(5), 1291–1297 (2011)
18. Ding, L., Han, Q.L., Ge, X., Zhang, X.M.: An overview of recent advances in event-triggered consensus of multiagent systems. IEEE transactions on cybernetics **48**(4), 1110–1123 (2017)
19. Girard, A.: Dynamic triggering mechanisms for event-triggered control. IEEE Transactions on Automatic Control **60**(7), 1992–1997 (2014)
20. Glattfelder, A., Schaufelberger, W.: Stability analysis of single loop control systems with saturation and antireset-windup circuits. IEEE Transactions on Automatic Control **28**(12), 1074–1081 (1983)
21. He, W., Xu, B., Han, Q.L., Qian, F.: Adaptive consensus control of linear multiagent systems with dynamic event-triggered strategies. IEEE transactions on cybernetics **50**(7), 2996–3008 (2019)

22. Hua, C., Li, H., Li, K., Ding, W.: Low-computation tracking control of nonlinear systems with asymmetric full-state constraints and unknown control directions. IEEE Transactions on Systems, Man, and Cybernetics: Systems **53**(2), 1051–1059 (2023)
23. Ilchmann, A., Ryan, E.P., Sangwin, C.J.: Tracking with prescribed transient behaviour. ESAIM: Control, Optimisation and Calculus of Variations **7**, 471–493 (2002)
24. Ilchmann, A., Schuster, H.: Pi-funnel control for two mass systems. IEEE Transactions on Automatic Control **54**(4), 918–923 (2009)
25. Ji, R., Ge, S.S.: Event-triggered tunnel prescribed control for nonlinear systems. IEEE Transactions on Fuzzy Systems (2023)
26. Ji, R., Ge, S.S.: Consensus for heterogeneous multiagent systems: Output rate-coded secure control. IEEE Transactions on Cybernetics (2024)
27. Ji, R., Ge, S.S.: Rate-coded secure control for multi-agent systems. IEEE Transactions on Automatic Control (2024)
28. Ji, R., Ge, S.S., Li, D.: Saturation-tolerant prescribed control for nonlinear systems with unknown control directions and external disturbances. IEEE Transactions on Cybernetics pp. 1–13 (2023). https://doi.org/10.1109/TCYB.2023.3244988
29. Ji, R., Ge, S.S., Zhao, K.: Coded event-triggered control for nonlinear systems. Automatica **167**, 111753 (2024)
30. Ji, R., Li, D., Ge, S.S.: Saturation-tolerant prescribed control for nonlinear time-delay systems. IEEE Transactions on Fuzzy Systems (2022)
31. Ji, R., Li, D., Ge, S.S., Li, H.: Tunnel prescribed control of nonlinear systems with unknown control directions. IEEE Transactions on Neural Networks and Learning Systems (2023)
32. Ji, R., Li, D., Ma, J., Ge, S.S.: Saturation-tolerant prescribed control of mimo systems with unknown control directions. IEEE Transactions on Fuzzy Systems pp. 1–1 (2022)
33. Ji, R., Ma, J., Li, D., Ge, S.S.: Finite-time adaptive output feedback control for mimo nonlinear systems with actuator faults and saturations. IEEE Transactions on Fuzzy Systems **29**(8), 2256–2270 (2020)
34. Ji, R., Yang, B., Ma, J., Ge, S.S.: Saturation-tolerant prescribed control for a class of mimo nonlinear systems. IEEE Transactions on Cybernetics (2021)
35. Ji, R., Zhang, Y., Ge, S.S.: Coded secure control for multi-agent systems. In: 2024 14th Asian Control Conference (ASCC), pp. 599–604. IEEE (2024)
36. Jiang, T., Huang, J., Li, B.: Composite adaptive finite-time control for quadrotors via prescribed performance. Journal of the Franklin Institute **357**(10), 5878–5901 (2020)
37. Jing, Y.H., Yang, G.H.: Fuzzy adaptive fault-tolerant control for uncertain nonlinear systems with unknown dead-zone and unmodeled dynamics. IEEE Transactions on Fuzzy Systems (2019)
38. Kumari, K., Bandyopadhyay, B., Reger, J., Behera, A.K.: Event-triggered sliding mode control for a high-order system via reduced-order model based design. Automatica **121**, 109163 (2020)
39. Li, D., Ge, S.S., Lee, T.H.: Fixed-time-synchronized consensus control of multiagent systems. IEEE transactions on control of network systems **8**(1), 89–98 (2020)
40. Li, D., Ge, S.S., Lee, T.H., Li, D., Ge, S.S., Lee, T.H.: Time-synchronized consensus of network systems. Time-Synchronized Control: Analysis and Design pp. 165–187 (2022)
41. Li, D., Tong, S., Yang, H., Hu, Q.: Time-synchronized control for spacecraft reorientation with time-varying constraints. IEEE/ASME Transactions on Mechatronics (2024)
42. Li, D., Yu, H., Tee, K.P., Wu, Y., Ge, S.S., Lee, T.H.: On time-synchronized stability and control. IEEE Transactions on Systems, Man, and Cybernetics: Systems **52**(4), 2450–2463 (2021)
43. Li, Y., Li, K., Tong, S.: Adaptive neural network finite-time control for multi-input and multi-output nonlinear systems with positive powers of odd rational numbers. IEEE Trans. Neural Netw. Learning Syst. (2019)
44. Li, Y., Liu, Y., Tong, S.: Observer-based neuro-adaptive optimized control of strict-feedback nonlinear systems with state constraints. IEEE Transactions on Neural Networks and Learning Systems **33**(7), 3131–3145 (2022). https://doi.org/10.1109/TNNLS.2021.3051030
45. Li, Y., Tong, S.: Adaptive neural networks prescribed performance control design for switched interconnected uncertain nonlinear systems. IEEE transactions on neural networks and learning systems **29**(7), 3059–3068 (2017)

46. Li, Y., Tong, S., Li, T.: Composite adaptive fuzzy output feedback control design for uncertain nonlinear strict-feedback systems with input saturation. IEEE Transactions on Cybernetics **45**(10), 2299–2308 (2014)
47. Liang, H., Zhang, Y., Huang, T., Ma, H.: Prescribed performance cooperative control for multiagent systems with input quantization. IEEE Transactions on cybernetics **50**(5), 1810–1819 (2019)
48. Liu, M., Ji, R., Ge, S.S., Fellow, I.: Adaptive neural control for a tilting quadcopter with finite-time convergence. Neural Computing and Applications **33**(23), 15987–16004 (2021)
49. Liu, T., Jiang, Z.P.: Event-based control of nonlinear systems with partial state and output feedback. Automatica **53**, 10–22 (2015)
50. Liu, Y., Li, H., Lu, R., Zuo, Z., Li, X.: An overview of finite/fixed-time control and its application in engineering systems. IEEE/CAA Journal of Automatica Sinica **9**(12), 2106–2120 (2022)
51. Liu, Y., Yang, G.H.: Prescribed performance-based consensus of nonlinear multiagent systems with unknown control directions and switching networks. IEEE Transactions on Systems, Man, and Cybernetics: Systems **50**(2), 609–616 (2017)
52. Lv, M., De Schutter, B., Cao, J., Baldi, S.: Adaptive prescribed performance asymptotic tracking for high-order odd-rational-power nonlinear systems. IEEE Transactions on Automatic Control **68**(2), 1047–1053 (2023)
53. Mazo Jr, M., Cao, M.: Asynchronous decentralized event-triggered control. Automatica **50**(12), 3197–3203 (2014)
54. Min, H., Xu, S., Zhang, B., Ma, Q.: Output-feedback control for stochastic nonlinear systems subject to input saturation and time-varying delay. IEEE transactions on Automatic Control **64**(1), 359–364 (2018)
55. Ni, J., Duan, F., Shi, P.: Fixed-time consensus tracking of multiagent system under dos attack with event-triggered mechanism. IEEE Transactions on Circuits and Systems I: Regular Papers **69**(12), 5286–5299 (2022)
56. Ning, P., Hua, C., Li, K., Meng, R.: A blow-up function approach to global event-triggered prescribed tracking output feedback control of nonlinear systems. IEEE Transactions on Circuits and Systems I: Regular Papers **69**(10), 4225–4236 (2022)
57. Nussbaum, R.D.: Some remarks on a conjecture in parameter adaptive control. Sys. Control Lett. **3**(5), 243–246 (1983)
58. Qi, X., Ma, H., Jing, Y.: Arbitrary settling time control with prescribed performance for high-order nonlinear systems. Asian Journal of Control (2022)
59. Qiu, J., Sun, K., Wang, T., Gao, H.: Observer-based fuzzy adaptive event-triggered control for pure-feedback nonlinear systems with prescribed performance. IEEE Trans. Fuzzy Syst. **27**(11), 2152–2162 (2019)
60. Sahoo, A., Xu, H., Jagannathan, S.: Adaptive neural network-based event-triggered control of single-input single-output nonlinear discrete-time systems. IEEE transactions on neural networks and learning systems **27**(1), 151–164 (2015)
61. Salmanpour, Y., Arefi, M.M., Khayatian, A., Kaynak, O.: Event-triggered fuzzy adaptive leader-following tracking control of non-affine multi-agent systems with finite-time output constraint and input saturation. IEEE Transactions on Fuzzy Systems (2021)
62. Shen, Z., Wang, Q., Dong, S., Yu, H.: Dynamic surface control for tracking of unmanned surface vessel with prescribed performance and asymmetric time-varying full state constraints. Ocean Engineering **253**, 111319 (2022)
63. Shojaei, K., Chatraei, A.: Robust platoon control of underactuated autonomous underwater vehicles subjected to nonlinearities, uncertainties and range and angle constraints. Applied Ocean Research **110**, 102594 (2021)
64. Shojaei, K., Kazemy, A., Chatraei, A.: An observer-based neural adaptive pid controller for robot manipulators including motor dynamics with a prescribed performance. IEEE/ASME Transactions on Mechatronics **26**(3), 1689–1699 (2020)
65. Sui, S., Tong, S.: Finite-time fuzzy adaptive ppc for nonstrict-feedback nonlinear mimo systems. IEEE Transactions on Cybernetics **53**(2), 732–742 (2023)

66. Sui, S., Tong, S., Li, Y.: Observer-based fuzzy adaptive prescribed performance tracking control for nonlinear stochastic systems with input saturation. Neurocomputing **158**, 100–108 (2015)
67. Sun, K., Qiu, J., Karimi, H.R., Fu, Y.: Event-triggered robust fuzzy adaptive finite-time control of nonlinear systems with prescribed performance. IEEE Transactions on Fuzzy Systems **29**(6), 1460–1471 (2020)
68. Sun, W., Wu, Y.Q., Sun, Z.Y.: Command filter-based finite-time adaptive fuzzy control for uncertain nonlinear systems with prescribed performance. IEEE Transactions on Fuzzy Systems **28**(12), 3161–3170 (2020)
69. Tabuada, P.: Event-triggered real-time scheduling of stabilizing control tasks. IEEE Transactions on Automatic Control **52**(9), 1680–1685 (2007)
70. Wang, H., Liu, P.X., Zhao, X., Liu, X.: Adaptive fuzzy finite-time control of nonlinear systems with actuator faults. IEEE Trans. Cybern. (2019)
71. Wang, J., He, W., Zhang, H., Xie, X., Wang, Y., Meng, Q.: Novel dynamic event-triggered consensus control of multiagent systems with markovian switching topologies under dos attacks. IEEE Transactions on Cybernetics (2024)
72. Wang, T., Qiu, J., Gao, H.: Adaptive neural control of stochastic nonlinear time-delay systems with multiple constraints. IEEE Transactions on Systems, Man, and Cybernetics: Systems **47**(8), 1875–1883 (2016)
73. Wang, W., Wang, D., Peng, Z., Li, T.: Prescribed performance consensus of uncertain nonlinear strict-feedback systems with unknown control directions. IEEE Transactions on Systems, Man, and Cybernetics: Systems **46**(9), 1279–1286 (2015)
74. Wang, W., Wen, C., Huang, J., Zhou, J.: Adaptive consensus of uncertain nonlinear systems with event triggered communication and intermittent actuator faults. Automatica **111**, 108667 (2020)
75. Wen, C., Zhou, J., Liu, Z., Su, H.: Robust adaptive control of uncertain nonlinear systems in the presence of input saturation and external disturbance. IEEE Transactions on Automatic Control **56**(7), 1672–1678 (2011)
76. Xia, G., Sun, C., Zhao, B., Xue, J.: Prescribed performance adaptive fuzzy containment control for with input saturation based on finite-time disturbance observer. International J. Contr., Autom. Syst. **17**(2), 370–379 (2019)
77. Xiao, B., Hu, Q., Zhang, Y.: Adaptive sliding mode fault tolerant attitude tracking control for flexible spacecraft under actuator saturation. IEEE Transactions on Control Systems Technology **20**(6), 1605–1612 (2011)
78. Xing, L., Wen, C., Liu, Z., Su, H., Cai, J.: Event-triggered adaptive control for a class of uncertain nonlinear systems. IEEE Transactions on Automatic Control **62**(4), 2071–2076 (2016)
79. Xing, L., Wen, C., Liu, Z., Su, H., Cai, J.: Event-triggered output feedback control for a class of uncertain nonlinear systems. IEEE Transactions on Automatic Control **64**(1), 290–297 (2018)
80. Xu, J.X., Yan, R.: Iterative learning control design without a priori knowledge of the control direction. Automatica **40**(10), 1803–1809 (2004)
81. Xu, P., Li, Y., Tong, S.: Fuzzy adaptive finite time fault-tolerant control for multi-input and multi-output nonlinear systems with actuator faults. International J. Contr., Autom. Syst. **17**(7), 1655–1665 (2019)
82. Xudong, Y.: Asymptotic regulation of time-varying uncertain nonlinear systems with unknown control directions. Automatica **35**(5), 929–935 (1999)
83. Yang, Y., Ge, C., Wang, H., Li, X., Hua, C.: Adaptive neural network based prescribed performance control for teleoperation system under input saturation. Journal of the Franklin Institute **352**(5), 1850–1866 (2015)
84. Yang, Y., Tan, J., Yue, D.: Prescribed performance tracking control of a class of uncertain pure-feedback nonlinear systems with input saturation. IEEE Transactions on Systems, Man, and Cybernetics: Systems **50**(5), 1733–1745 (2018)
85. Yong, K., Chen, M., Shi, Y., Wu, Q.: Flexible performance-based robust control for a class of nonlinear systems with input saturation. Automatica **122**, 109268 (2020)

86. Yu, J., Zhao, L., Yu, H., Lin, C., Dong, W.: Fuzzy finite-time command filtered control of nonlinear systems with input saturation. IEEE Trans. Cybern. **48**(8), 2378–2387 (2017)
87. Zhang, C., Chen, Z., Wang, J., Liu, Z., Chen, C.P.: Fuzzy adaptive two-bit-triggered control for a class of uncertain nonlinear systems with actuator failures and dead-zone constraint. IEEE Transactions on Cybernetics **51**(1), 210–221 (2020)
88. Zhang, C., Ma, G., Sun, Y., Li, C.: Observer-based prescribed performance attitude control for flexible spacecraft with actuator saturation. ISA Trans. **89**, 84–95 (2019)
89. Zhang, C.H., Yang, G.H.: Event-triggered adaptive output feedback control for a class of uncertain nonlinear systems with actuator failures. IEEE transactions on cybernetics **50**(1), 201–210 (2018)
90. Zhang, J.X., Yang, G.H.: Prescribed performance fault-tolerant control of uncertain nonlinear systems with unknown control directions. IEEE Trans. Autom. Control **62**(12), 6529–6535 (2017)
91. Zhang, L., Che, W.W., Deng, C., Wu, Z.G.: Prescribed performance control for multiagent systems via fuzzy adaptive event-triggered strategy. IEEE Transactions on Fuzzy Systems **30**(12), 5078–5090 (2022)
92. Zhao, K., Song, Y., Chen, C.P., Chen, L.: Adaptive asymptotic tracking with global performance for nonlinear systems with unknown control directions. IEEE Transactions on Automatic Control **67**(3), 1566–1573 (2021)
93. Zhao, K., Song, Y., Ma, T., He, L.: Prescribed performance control of uncertain euler–lagrange systems subject to full-state constraints. IEEE Transactions on Neural Networks and Learning Systems **29**(8), 3478–3489 (2018)
94. Zheng, Z., Feroskhan, M.: Path following of a surface vessel with prescribed performance in the presence of input saturation and external disturbances. IEEE/ASME Transactions on Mechatronics **22**(6), 2564–2575 (2017)
95. Zheng, Z., Xie, L.: Finite-time path following control for a stratospheric airship with input saturation and error constraint. International J. Contr. **92**(2), 368–393 (2019)
96. Zhou, Q., Li, H., Wang, L., Lu, R.: Prescribed performance observer-based adaptive fuzzy control for nonstrict-feedback stochastic nonlinear systems. IEEE Transactions on Systems, Man, and Cybernetics: Systems **48**(10), 1747–1758 (2017)
97. Zhu, Z., Xia, Y., Fu, M.: Adaptive sliding mode control for attitude stabilization with actuator saturation. IEEE Transactions on Industrial Electronics **58**(10), 4898–4907 (2011)

Chapter 2
Preliminaries

2.1 Lyapunov Stability Theorem

Definition 2.1 (*Class-$\mathcal{K}$ Function*) For a continuous function $\alpha(t) : [0, a) \to \mathbb{R}_+$, it is said to belong to class $\mathcal{K}$ if:

(i) it is strictly increasing; and
(ii) it is s.t. $\alpha(0) = 0$.

It is said to belong to class $\mathcal{K}_\infty$ if $a = \infty$ and $\alpha(r) \to \infty$ as $r \to \infty$.

Definition 2.2 (*Class-$\mathcal{KL}$ Function*) A continuous function $\alpha : [0, a) \to [0, \infty)$ is said to belong to class $\mathcal{KL}$ if:

(i) for each fixed t, the mapping $\beta(r, t)$ belongs to class $\mathcal{K}$ with respect to r and,
(ii) for each fixed r, the mapping $\beta(r, t)$ is decreasing with respect to t and $\beta(r, t) \to 0$ as $t \to \infty$.

Definition 2.3 Given a continuous function $V(x, t) : \mathbb{R}^n \times \mathbb{R}_+ \to \mathbb{R}$, it is said to be:

(i) *locally positive definite* if there exists a class $\mathcal{K}$ function $\alpha(\cdot)$ such that $V(x, t) \geq \alpha(\|x\|)$ for all $t \geq 0$ and in the neighborhood $\mathcal{N}$ of the origin $\mathbb{R}^n$;
(ii) *positive definite* if $\mathcal{N} = \mathbb{R}^n$;
(iii) *(locally) negative definite* if $-V(x, t)$ is (locally) positive definite; and
(iv) *(locally) decrescent* if there exists a class $\mathcal{K}$ function $\beta(\cdot)$ such that $V(x, t) \leq \beta(\|x\|)$ for all $t \geq 0$ and in (the neighborhood $\mathcal{N}$ of the origin) $\mathbb{R}^n$.

Definition 2.4 For a continuous differential function $V(x, t) : \mathbb{R}^n \times \mathbb{R}_+ \to \mathbb{R}$, together with a system of differential equations:

$$\dot{x} = f(x, t).$$

R. Ji et al., *Constrained Nonlinear Control: Performance and Robustness*,
https://doi.org/10.1007/978-981-96-2971-8_2

The derivative of $V(x,t)$ relative to the above system is therefore given by:

$$\dot{V}(x,t) = \frac{\partial V(x,t)}{\partial t} + \frac{\partial V(x,t)}{\partial x} f(x,t),$$

where

$$\frac{\partial V(x,t)}{\partial x} \triangleq \left[\frac{\partial V(x,t)}{\partial x_1}, \frac{\partial V(x,t)}{\partial x_2}, \cdots, \frac{\partial V(x,t)}{\partial x_n} \right]. \tag{2.1}$$

Theorem 2.1 (Lyapunov Stability Theorem) *Consider the following nonlinear dynamic system:*

$$\dot{x} = f(x,t), \; x(0) = x_0,$$

and the origin is an equilibrium point. Let $\mathcal{N}$ denote a neighborhood of the origin, that is, $\mathcal{N} = \{x : \|x\| \leq \epsilon\}$ for $\epsilon > 0$. Then, the origin is said to be:

(i) stable in the sense of Lyapunov, if there exists a scalar function $V(x,t)$ such that $V(x,t) > 0$ and $\dot{V}(x,t) \leq 0$ for all $x \in \mathcal{N}$;
(ii) uniformly stable, if there exists a scalar function $V(x,t)$ such that $V(x,t) > 0$ and decrescent, and $\dot{V}(x,t) \leq 0$ for all $x \in \mathcal{N}$;
(iii) asymptotically stable, if there exists a scalar function $V(x,t)$ such that $V(x,t) > 0$ and $\dot{V}(x,t) < 0$ for all $x \in \mathcal{N}$;
(iv) uniformly asymptotically stable, if there exists a scalar function $V(x,t)$ such that $V(x,t) > 0$ and decrescent, and $\dot{V}(x,t) < 0$ for all $x \in \mathbb{R}^n$ (i,e., $\mathcal{N} = \mathbb{R}^n$);
(v) global, uniformly, asymptotically stable, if there exists a scalar function $V(x,t)$ such that $V(x,t) > 0$ and decrescent and is radially unbounded, and $\dot{V}(x,t) < 0$ for all $x \in \mathbb{R}^n$;
(vi) exponentially stable, if there exist positive constants α, β, γ such that, for all $x \in \mathcal{N}$, we have $\alpha\|x\|^2 \leq V(x,t) \leq \beta\|x\|^2$ and $\dot{V}(x,t) \leq -\gamma\|x\|^2$; and
(vii) globally exponentially stable, if there exist positive constants α, β, γ such that, for all $x \in \mathbb{R}^n$, we have $\alpha\|x\|^2 \leq V(x,t) \leq \beta\|x\|^2$ and $\dot{V}(x,t) \leq -\gamma\|x\|^2$.

2.2 Useful Stability Theorems and Formula

Consider a class of nonlinear autonomous systems or time-invariant systems

$$\dot{x} = f(x), \; x \in \mathbb{D}, \tag{2.2}$$

where $\mathbb{D} \subseteq \mathbb{R}^n$ denotes the domain of the system state variable x, and the nonlinear function $f : \mathbb{D} \to \mathbb{R}^n$ represents a mapping of the system state x. Let $x(t)$ denote the solution of the system starting from $x(t_0)$.

Lemma 2.1 ([6]) *For the nonlinear system (2.2), if a continuous positive definite function $V(x)$ satisfies the following differential inequality:*

$$\dot{V}(x) \leq -c_1 V(x) - c_2 V^{\beta}(x) + \epsilon, \tag{2.3}$$

where c_1, c_2, and ϵ are positive numbers, and $0 < \beta < 1$, then the equilibrium point $x^* = 0$ is said to be semi-global practical finite-time stable, and the settling time can be expressed as

$$T = t_0 + \frac{1}{c_1(1-\beta)} \ln\left(\frac{c_2c_3 + c_1 V^{1-\beta}(x(t_0))}{c_2c_3}\right), \tag{2.4}$$

where $0 < c_3 < 1$.

Proof From (2.3), for $\forall 0 < c_3 < 1$, we have,

$$\dot{V}(x) \leq -c_1 V(x) - c_2c_3 V^{\beta} - (c_2(1-c_3)V^{\beta}(x) - \epsilon). \tag{2.5}$$

We consider the following two cases: $\Omega = \{x | V^{\beta}(x) \leq \frac{\epsilon}{c_2(1-c_3)}\}$ and $\bar{\Omega} = \{x | V^{\beta}(x) > \frac{\epsilon}{c_2(1-c_3)}\}$.

Case 1: If $x \in \bar{\Omega}$, we have,

$$\dot{V}(x) \leq -c_1 V(x) - c_2c_3 V^{\beta}(x) \tag{2.6}$$

By dividing both sides of the (2.6) to term $V^{\beta}(x)$ yields,

$$\int_{t_0}^{t} \frac{V^{-\beta}(x)}{c_2c_3 + c_1 V^{1-\beta}(x)} \mathrm{dV} \leq -\int_{t_0}^{t} \mathrm{dt}. \tag{2.7}$$

Then we have,

$$T = t_0 + \frac{1}{c_1(1-\beta)} \ln\left(\frac{c_2c_3 + c_1 V^{1-\beta}(x(t_0))}{c_2c_3}\right). \tag{2.8}$$

This means that $x \in \Omega$ for all $t \geq T$.

Case 2: If $x \in \Omega$, according to case 1, the system state x does not exceed the set Ω. □

Lemma 2.2 ([4]) *p, q, and r are positive constant, for any positive numbers x_1 and x_2,*

$$|x_1|^p |x_2|^q \leq \frac{p}{p+q} r |x_1|^{p+q} + \frac{q}{p+q} r^{-\frac{p}{q}} |x_2|^{p+q}. \tag{2.9}$$

Lemma 2.3 ([3]) *For any positive numbers x_i, $i = 1, \ldots, n$, and $0 < \alpha < 2$, there is*

$$\left(\sum_{i=1}^{n} x_i^2\right)^{\alpha} \leq \left(\sum_{i=1}^{n} x_i^{\alpha}\right)^{2} \leq n\left(\sum_{i=1}^{n} x_i^{2\alpha}\right) \tag{2.10}$$

Lemma 2.4 ([2]) *For any real numbers x_i, $(i = 1, \ldots, n)$ and $0 < \beta < 1$, the following inequalities hold:*

$$\left(\sum_{i=1}^{n} |x_i|\right)^{\beta} \leq \sum_{i=1}^{n} |x_i|^{\beta} \leq n^{1-\beta}\left(\sum_{i=1}^{n} |x_i|\right)^{\beta}. \tag{2.11}$$

Definition 2.5 ([7]) We call a continuous function $N(\cdot)$ as a Nussbaum-type function, if it has the properties given by:

$$\lim_{s \to +\infty} \sup \frac{1}{s} \int_0^s N(\zeta) \mathrm{d}\zeta = +\infty, \tag{2.12}$$

$$\lim_{s \to -\infty} \inf \frac{1}{s} \int_0^s N(\zeta) \mathrm{d}\zeta = -\infty. \tag{2.13}$$

Lemma 2.5 ([5]) *For the smooth functions $V(t)$ and $\zeta(t)$ defined on $[0, t)$ where $V(t) \geq 0$, $\forall t \in [0, t)$, bounded $V(0)$, and the Nussbaum-type function $N(\zeta)$, if the following inequality holds:*

$$\dot{V}(t) \leq -cV(t) + [gN(\zeta) + 1]\dot{\zeta} + \epsilon, \tag{2.14}$$

where $c > 0$, $\epsilon > 0$ and $g \neq 0$, then $[gN(\zeta) + 1]\dot{\zeta}$, $\zeta(t)$, and $V(t)$ are bounded on $[0, t)$.

Lemma 2.6 ([1]) *Consider the system $\dot{x}(t) = Kx(t) + \epsilon(t)$, where K is Metzler matrix and $\epsilon(t)$ is non-negative value. There exists a unique solution $x(t)$ if for any nonnegative $x(0)$ such that the inequality $x(t) \geq 0$ holds for $\forall t \geq 0$.*

References

1. Farina, L., Rinaldi, S.: Positive linear systems: theory and applications, vol. 50. John Wiley & Sons (2011)
2. Huang, X., Lin, W., Yang, B.: Global finite-time stabilization of a class of uncertain nonlinear systems. Automatica **41**(5), 881–888 (2005)
3. Ji, R., Ma, J., Sam Ge, S.: Modeling and control of a tilting quadcopter. IEEE Trans. Aerosp. Electro. Syst. **56**(4), 2823–2834 (2020). https://doi.org/10.1109/TAES.2019.2955525
4. Qian, C., Lin, W.: A continuous feedback approach to global strong stabilization of nonlinear systems. IEEE Transactions on Automatic Control **46**(7), 1061–1079 (2001)

5. Shi, W.: Adaptive fuzzy control for mimo nonlinear systems with nonsymmetric control gain matrix and unknown control direction. IEEE Transactions on Fuzzy Systems **22**(5), 1288–1300 (2013)
6. Yu, S., Yu, X., Shirinzadeh, B., Man, Z.: Continuous finite-time control for robotic manipulators with terminal sliding mode. Automatica **41**(11), 1957–1964 (2005)
7. Zhang, T., Ge, S.S.: Adaptive neural control of mimo nonlinear state time-varying delay systems with unknown dead-zones and gain signs. Automatica **43**(6), 1021–1033 (2007)

Chapter 3
Constrained Control of Euler–Lagrange Systems with Input Saturation Constraint

3.1 Introduction

In this chapter, we explore state-feedback and output-feedback control strategies for a class of Euler–Lagrange systems under input saturation, actuator faults, and measurement constraint. In practical applications, a large magnitude of control input is often required for achieving fast tracking convergence [9]. However, due to the limitations of physical mechanisms, the control input is inevitably constrained within an effective region. Due to the complex saturation nonlinearities, how to cope with these effects has become a challenging research topic. Moreover, most works are followed by the assumption that all the system states are available. The unmeasurable systems states make the traditional state feedback control in applicable. To address the abovementioned problems, we develop a series of nonlinear control designs for Euler–Lagrange systems and the main contributions are summarized as follows: To compensate for the saturation nonlinearity, a virtual auxiliary system is constructed, and a backstepping-based constrained control is proposed. A novel saturation constraint mechanism for the control input derivative is designed for alleviating the chattering phenomenon. Inspired by the second-order sliding mode control, a novel second-order dynamic surface control is developed, which not only ensures the finite-time convergence of the closed-loop system but also satisfies the input saturation constraints. Finally, we further consider actuator faults and unmeasurable states. A finite-time output feedback control is designed where fuzzy logic systems are utilized to estimate the unmeasurable states. The effectiveness of the proposed control designs is verified by numerical simulations.

R. Ji et al., *Constrained Nonlinear Control: Performance and Robustness*,
https://doi.org/10.1007/978-981-96-2971-8_3

3.2 Modeling and Analysis of Euler–Lagrange Systems

It is well known that second-order Euler–Lagrange systems can describe many practical systems [8, 10], such as satellites, unmanned aerial vehicles, robotic manipulators, etc. A general Euler–Lagrange systems can be represented by:

$$M(\xi)\ddot{\xi} + H(\xi, \dot{\xi})\dot{\xi} + D(\xi, \dot{\xi})\dot{\xi} + G_w + \Delta_w = \tau, \tag{3.1}$$

where $\xi,\ \dot{\xi},\ \ddot{\xi} \in \mathbb{R}^p$ represent the position, velocity, and acceleration vector, respectively. $M(\xi) \in \mathbb{R}^{p\times p}$ denotes the symmetric positive definite inertia matrix, $H(\xi, \dot{\xi})\dot{\xi} \in \mathbb{R}^p$ represents the centrifugal and Coriolis forces vector, $D(\xi, \dot{\xi})\dot{\xi} \in \mathbb{R}^p$ represents the damping vector, $G_w \in \mathbb{R}^p$ denotes the gravitational force vector, $\Delta_w \in \mathbb{R}^p$ represents the external disturbance vector, and $\tau \in \mathbb{R}^p$ represents the control input.

Lemma 3.1 ([13]) *For the Euler–Lagrange system (3.1), there exist positive real numbers $k_{\underline{m}}$ and $k_{\overline{m}}$ such that:*

1. *$M(\xi)$ is a positive definite symmetric matrix and satisfies the boundedness condition: $0 \le k_{\underline{m}} I_p \le M(\xi) \le k_{\overline{m}} I_p$, where $k_{\underline{m}}$ and $k_{\overline{m}}$ denote the minimum and maximum eigenvalues of the matrix $M(\xi)$.*
2. *The matrix $\dot{M}(\xi) - 2H(\xi, \dot{\xi})$ is a skew-symmetric matrix, satisfying: $x^T(\dot{M}(\xi) - 2H(\xi, \dot{\xi}))x = 0,\ x \in \mathbb{R}^p$.*
3. *The Euler–Lagrange system (3.1) can be parameterized as: $M(\xi)\ddot{\xi} + H(\xi, \dot{\xi})\dot{\xi} + D(\xi, \dot{\xi})\dot{\xi} = Y(\xi, \dot{\xi}, \ddot{\xi})\vartheta$, where $\vartheta \in \mathbb{R}^l$ represents the vector of uncertain parameters, and $Y(\xi, \dot{\xi}, \ddot{\xi}) = [y_{ij}] \in \mathbb{R}^{p\times l}$ represents the regressor matrix.*

As practical systems often operate in complex environments, the parameters in Euler–Lagrange systems (3.1) are difficult to obtain accurately. Therefore, system matrices suffer from uncertainties, which can be represented by:

$$\begin{aligned} &M(\xi) = M_0(\xi) + \Delta M(\xi),\ \ H(\xi, \dot{\xi}) = H_0(\xi, \dot{\xi}) + \Delta H(\xi, \dot{\xi}), \\ &D(\xi, \dot{\xi}) = D_0(\xi, \dot{\xi}) + \Delta D(\xi, \dot{\xi}),\ \ G_w = G_{w0} + \Delta G_w, \end{aligned} \tag{3.2}$$

where $M_0(\xi)$, $H_0(\xi, \dot{\xi})$, $D_0(\xi, \dot{\xi})$, and G_{w0} represent the nominal matrices, and $\Delta M(\xi)$, $\Delta H(\xi, \dot{\xi})$, $\Delta D(\xi, \dot{\xi})$, and ΔG_w are the uncertainty matrices. The Euler–Lagrange system in (3.1) can be rewritten as:

$$M_0(\xi)\ddot{\xi} + H_0(\xi, \dot{\xi})\dot{\xi} + D_0(\xi, \dot{\xi})\dot{\xi} + G_{w0} + \Delta = \tau, \tag{3.3}$$

where $\Delta = \Delta M_0(\xi)\ddot{\xi} + \Delta H_0(\xi, \dot{\xi})\dot{\xi} + \Delta D_0(\xi, \dot{\xi})\dot{\xi} + \Delta G_{w0} + \Delta_w$. For later control design, the following assumptions are made.

Assumption 3.1 According to (3.2) and (3.3), there exist positive constants $b_i,\ i = 1, \ldots, 5$ such that the following inequalities hold:

$$\begin{aligned}
&\|M_0\| \le b_1, \ \|H_0(\xi)\| \le b_2\|\dot{\xi}\|, \\
&\|D_0(\xi)\| \le b_3 + b_4\|\dot{\xi}\|, \ \|G_{w0}\| \le b_5.
\end{aligned} \tag{3.4}$$

Similarly, there exist positive constants $c_i, \ i = 1, \ldots, 5$ satisfying the following inequalities:

$$\begin{aligned}
&\|\Delta M_0\| \le c_1, \ \|\Delta H_0(\xi)\| \le c_2\|\dot{\xi}\|, \\
&\|\Delta D_0(\xi)\| \le c_3 + c_4\|\dot{\xi}\|, \ \|\Delta G_{w0}\| \le c_5.
\end{aligned} \tag{3.5}$$

Assumption 3.2 From the Euler–Lagrange systems in (3.3), there exists a positive constant δ such that $\|\Delta_w\| \le \delta$ holds.

Assumption 3.3 For the Euler–Lagrange systems in (3.1), there exist positive real numbers δ_1 and δ_2 such that the external disturbances and their derivatives satisfy $\|\Delta_w\| \le \delta_1$ and $\|\dot{\Delta}_w\| \le \delta_2$.

Assumption 3.4 Considering Euler–Lagrange systems in (3.1) and Assumption 3.1, we define $F_1(\xi, \dot{\xi}) = \Delta\dot{M}(\xi)\ddot{\xi} + \Delta\dot{H}(\xi, \dot{\xi})\dot{\xi} + \Delta\dot{D}(\xi, \dot{\xi})\dot{\xi} + \Delta\dot{G}_w, \ F_2(\xi, \dot{\xi}) = \Delta M(\xi)\frac{d}{dt}\ddot{\xi} + \Delta H(\xi, \dot{\xi})\ddot{\xi} + \Delta D(\xi, \dot{\xi})\ddot{\xi} + \dot{\Delta}_w + \Delta M(\xi)M_0^{-1}(\xi)\dot{\Delta}$. There exist positive real numbers $w_i, \ i = 1, \ldots, 14$ such that the following inequalities hold:

$$\begin{aligned}
&\|\frac{d}{dt}(M_0)^{-1}\| \le w_1 + w_2\|\dot{\xi}\|, \\
&\|F_1(\xi, \dot{\xi})\| \le w_3 + w_4\|\dot{\xi}\| + w_5\|\dot{\xi}\|^2 + w_6\|\dot{\xi}\|^3, \\
&\|F_2(\xi, \dot{\xi})\| \le w_7 + w_8\|\dot{\xi}\| + w_9\|\dot{\xi}\|^2 + w_{10}\|\dot{\xi}\|^3, \\
&\|\Delta\| \le w_{11} + w_{12}\|\dot{\xi}\| + w_{13}\|\dot{\xi}\|^2, \\
&\|M_0^{-1}\|\|(I + \Delta M(\xi)M_0^{-1}(\xi))^{-1}\| \le w_{14},
\end{aligned} \tag{3.6}$$

Based on the above assumptions, we can have:

$$\begin{aligned}
\Delta =& \Delta M_0\ddot{\xi} + \Delta H_0(\xi, \dot{\xi})\dot{\xi} + \Delta D_0(\xi, \dot{\xi})\dot{\xi} + \Delta G_{w0} + \Delta_w \\
=& \Delta M_0(\tau - M_0^{-1}(H_0(\xi, \dot{\xi})\dot{\xi} + D_0(\xi, \dot{\xi})\dot{\xi} + G_{w0} + \Delta)) + \Delta H_0(\xi, \dot{\xi})\dot{\xi} \\
&+ \Delta D_0(\xi, \dot{\xi})\dot{\xi} + \Delta G_{w0} + \Delta_w.
\end{aligned} \tag{3.7}$$

Referring to (3.4) and (3.5), it gives $\|\Delta M_0 M_0^{-1} + I_6\| \le d_1$ and $\|\Delta M_0 M_0^{-1}\| \le d_2$ where $d_i > 0, \ i = 1, 2$. Then, we obtain

$$\begin{aligned}
\|\Delta\| =& \|\Delta M_0 M_0^{-1} + I_6\|^{-1}\|\Delta M_0\tau - \Delta M_0 M_0^{-1}(H_0(\xi, \dot{\xi})\dot{\xi} + D_0(\xi, \dot{\xi})\dot{\xi} + G_{w0}) \\
&+ \Delta H_0(\xi, \dot{\xi})\dot{\xi} + \Delta D_0(\xi, \dot{\xi})\dot{\xi} + \Delta G_{w0} + \Delta_w\| \\
\le& d_1 c_1 \tau + (d_1 d_2 b_3 + d_1 c_3)\|\dot{\xi}\| + (d_1 d_2 b_2 + d_1 d_2 b_4 + d_1 c_2 + d_1 c_4)\|\dot{\xi}\|^2 \\
&+ d_1\delta + d_1 d_2 b_5 + d_1(c_5 + \delta).
\end{aligned} \tag{3.8}$$

It indicates that all uncertainties and external disturbances have been integrated, and their boundary information is related to control input and velocity signals. As the control input is always bounded in practical applications, there exists a positive constant d_3 such that $|\tau| \leq d_3$ holds. Therefore, (3.8) can be further rewritten as:

$$\|\Delta\| \leq \vartheta_1 + \vartheta_2\|\dot{\xi}\| + \vartheta_3\|\dot{\xi}\|^2, \tag{3.9}$$

where $\vartheta_1 = d_1c_1\tau + d_1d_2b_5 + d_1c_5 + d_1\delta + d_1\delta$, $\vartheta_2 = d_1d_2b_3 + d_1c_3$, $\vartheta_3 = d_1d_2b_2 + d_1d_2b_4 + d_1c_2 + d_1c_4$ are all unknown constants. The boundary information of uncertainties and external disturbances are integrated into a linear expression w.r.t. the parameters ϑ_i, $i = 1, 2, 3$, without the exact knowledge of ϑ_i. It significantly reduces the conservatism of control design.

3.3 Constrained State-Feedback Control Under Input Saturation

3.3.1 Problem Formulation

According to Euler–Lagrange systems in (3.1), its state-space representation is given by:

$$\begin{cases} \dot{x}_1 = x_2, \\ \dot{x}_2 = f_2(x_1, x_2) + d + u, \end{cases} \tag{3.10}$$

where $x_1 = \xi$, $x_2 = \dot{\xi}$, $f_2(x_1, x_2) = -M_0(\xi)^{-1}(H_0(\xi, \dot{\xi})\dot{\xi} + D_0(\xi, \dot{\xi})\dot{\xi} + G_{w0})$, $d = -M_0(\xi)^{-1}\Delta$, $f_2(x_1, x_2)$ is composed of the known nominal parameters, and $u = M_0(\xi)^{-1}\tau$ denotes the control input with considering saturation constraint:

$$\tau = \text{sat}(v) = \begin{cases} v_{\max}, & v > v_{\max} \\ v, & v_{\min} \leq v \leq v_{\max} \\ v_{\min}, & v < v_{\min} \end{cases} \tag{3.11}$$

where $v_{\max}$ and $v_{\min}$ represent the upper and lower bounds of the control input, respectively, v denotes the control input to be designed later, and d consists of all model uncertainties.

From Assumptions 3.1–3.2, we have:

$$\|d\| \leq \vartheta_1 + \vartheta_2\|\dot{\xi}\| + \vartheta_3\|\dot{\xi}\|^2, \tag{3.12}$$

where $\vartheta_i,\ i = 1, 2, 3$ are unknown positive constants. In this section, we aim to develop several constrained state-feedback control designs for Euler–Lagrange Systems under input saturation such that the tracking errors are guaranteed to converge to a small neighborhood around the origin within finite time.

3.3.2 Backstepping-Based Constrained Finite-Time Control

In order to compensate for nonlinearities from input saturation, we establish the following virtual auxiliary system:

$$\dot{\chi}_1 = -L_1\chi_1 + \chi_2 \tag{3.13}$$

$$\dot{\chi}_2 = -L_2\chi_2 + M_0(\xi)^{-1}\Delta\tau, \tag{3.14}$$

where $\Delta\tau = \tau - v$, L_1 and L_2 represent positive definite matrices, χ_1 and χ_2 are the virtual auxiliary states.

We are at the position to design our constrained backstepping-based finite-time control. First, we define the tracking error z_1 as:

$$z_1 = x_1 - x_{1d} - \chi_1, \tag{3.15}$$

where x_{1d} denotes the reference trajectory. The tracking error z_2 is given as:

$$z_2 = x_2 - \alpha_1 - \chi_2, \tag{3.16}$$

where α_1 is the virtual control input. From (3.10), the time derivative of z_1 is

$$\dot{z}_1 = \dot{x}_1 - \dot{x}_{1d} - \dot{\chi}_1 = x_2 - \dot{x}_{1d} + L_1\chi_1 - \chi_2 = z_2 + \alpha_1 - \dot{x}_{1d} + L_1\chi_1, \tag{3.17}$$

where α_1 is designed as

$$\alpha_1 = -K_1 z_1 - K_2 z_1^{\beta} + \dot{x}_{1d} - L_1\chi_1, \tag{3.18}$$

where $K_1 = \text{diag}(k_{1i})$ and $K_2 = \text{diag}(k_{2i})$ are positive definite matrices composed by positive real numbers k_{1i} and $k_{2i},\ i = 1, \ldots, p$. Taking the time derivative of z_2 gives

$$\dot{z}_2 = \dot{x}_2 - \dot{\alpha}_1 - \dot{\chi}_2 = f_2(x_1, x_2) + d + u - \dot{\alpha}_1 + L_2\chi_2 - M_0(\xi)^{-1}\Delta\tau, \tag{3.19}$$

where the control input v is designed as

$$v = v_1 + v_2, \tag{3.20}$$

$$v_1 = M_0(\xi)(-f_2(x_1, x_2) + \dot{\alpha}_1 - L_2\chi_2 - K_3 z_2 - K_4 z_2^{\beta} - z_1), \tag{3.21}$$

$$v_2 = -M_0(\xi)(\hat{\vartheta}_1 + \hat{\vartheta}_2\|x_2\| + \hat{\vartheta}_3\|x_2\|^2)\mathrm{sign}(z_2), \tag{3.22}$$

where $K_3 = \mathrm{diag}(k_{3i})$ and $K_4 = \mathrm{diag}(k_{4i})$ are positive definite matrices composed by positive real numbers k_{3i} and k_{4i}, $i = 1, \ldots, p$, $0 < \beta < 1$, $\hat{\vartheta}_i$, $i = 1, \ldots, p$ denotes the estimation of unknown parameter in the upper boundary of uncertainties. The adaptive law for $\hat{\vartheta}_i$ is designed by

$$\dot{\hat{\vartheta}}_1 = \Gamma_1(\|z_2\| - \hat{\vartheta}_1), \tag{3.23}$$

$$\dot{\hat{\vartheta}}_2 = \Gamma_2(\|z_2\|\|\dot{\xi}\| - \hat{\vartheta}_2), \tag{3.24}$$

$$\dot{\hat{\vartheta}}_3 = \Gamma_3(\|z_2\|\|\dot{\xi}\|^2 - \hat{\vartheta}_3), \tag{3.25}$$

where Γ_i, $i = 1, 2, 3$ are positive definite symmetric matrix. Now, we can have the following theorem of this section.

Theorem 3.1 *For the Euler–Lagrange systems in (3.10) with parameter uncertainties and external disturbances, with the aid of the control design (3.20) and adaptive laws (3.23)–(3.25), the closed-loop system is guaranteed to follow the reference trajectory, and the tracking errors z_1, z_2, and $\tilde{\vartheta} i$ converge to a small neighborhood around the origin within finite time, then remain within the compact set Ω_{z_1}, Ω_{z_2} and $\Omega_{\tilde{\vartheta}_i}$.*

$$\Omega_{z_1} := \left\{ z_1 \in \mathbb{R}^6 \mid \|z_1\| < \sqrt{\frac{2\varepsilon}{l_1(1 - l_3)}} \right\}, \tag{3.26}$$

$$\Omega_{z_2} := \left\{ z_2 \in \mathbb{R}^6 \mid \|z_2\| < \sqrt{\frac{2\varepsilon}{l_1(1 - l_3)}} \right\}, \tag{3.27}$$

$$\Omega_{\tilde{\vartheta}_i} := \left\{ \tilde{\vartheta}_i \in \mathbb{R} \mid \|\tilde{\vartheta}_i\| < \sqrt{\frac{2\Gamma_i\varepsilon}{l_1(1 - l_3)}} \right\}. \tag{3.28}$$

Proof We consider the following Lyapunov function candidate:

$$V_1 = \frac{1}{2} z_1^{\mathrm{T}} z_1 + \frac{1}{2} \chi_1^{\mathrm{T}} \chi_1. \tag{3.29}$$

Its time derivative along (3.13), (3.17) and (3.18) is

$$\begin{aligned} \dot{V}_1 =& z_1^T(z_2 + \alpha_1 - \dot{x}_{1d} + L_1\chi_1) + \chi_1^T(-L_1\chi_1 + \chi_2) \\ =& z_1^T(z_2 - K_1 z_1 - K_2 z_1^{\beta}) - \chi_1^T L_1 \chi_1 + \chi_1^T \chi_2. \end{aligned} \tag{3.30}$$

Based on the Young's inequality, we have

$$\chi_1^T \chi_2 \leq \frac{1}{2}\chi_1^T \chi_1 + \frac{1}{2}\chi_2^T \chi_2. \tag{3.31}$$

Substituting (3.31) into (3.30) yields

$$\begin{aligned}\dot{V}_1 =& z_1^T(z_2 - K_1 z_1 - K_2 z_1^\beta) - \chi_1^T L_1 \chi_1 + \frac{1}{2}\chi_1^T \chi_1 + \frac{1}{2}\chi_2^T \chi_2 \\ \leq& z_1^T z_2 - z_1^T K_1 z_1 - z_1^T K_2 z_1^\beta - (\lambda_{\min}(L_1) - \frac{1}{2})\chi_1^T \chi_1 + \frac{1}{2}\chi_2^T \chi_2, X\end{aligned} \tag{3.32}$$

where $\lambda_{\min}(L_1)$ denotes the minimum eigenvalue of the matrix L_1. By appropriately choosing parameters, let $\lambda_{\min}(L_1) > \frac{1}{2}$. Define the Lyapunov function V_2 as

$$V_2 = V_1 + \frac{1}{2}z_2^T z_2 + \frac{1}{2}\chi_2^T \chi_2 + \frac{1}{2}\sum_{i=1}^{3}\Gamma_i^{-1}\tilde{\vartheta}_i^2, \tag{3.33}$$

where $\tilde{\vartheta}_i = \vartheta_i - \hat{\vartheta}_i$. Upon (3.14), (3.19) and (3.20), the time derivative of V_2 is

$$\begin{aligned}\dot{V}_2 =& \dot{V}_1 + z_2^T \dot{z}_2 + \chi_2^T \dot{\chi}_2 - \sum_{i=1}^{3}\Gamma_i^{-1}\tilde{\vartheta}_i\dot{\hat{\vartheta}}_i \\ \leq& - z_1^T K_1 z_1 - z_1^T K_2 z_1^\beta - (\lambda_{\min}(L_1) - \frac{1}{2})\chi_1^T \chi_1 + \chi_2^T \chi_2 - z_2^T K_3 z_2 - z^T K_4 z_2^\beta \\ &+ \|z_2\|\|d\| - (\hat{\vartheta}_1 + \hat{\vartheta}_2\|x_2\| + \hat{\vartheta}_3\|x_2\|^2)\|z_2\| - \chi_2^T L_2 \chi_2 + \frac{1}{2}\|M_0^{-1}(\xi)\Delta\tau\|^2 \\ &- (\tilde{\vartheta}_1 + \tilde{\vartheta}_2\|x_2\| + \tilde{\vartheta}_3\|x_2\|^3)\|z_2\| + \sum_{i=1}^{3}\tilde{\vartheta}_i\hat{\vartheta}_i.\end{aligned} \tag{3.34}$$

Based on $\tilde{\vartheta}_i = \vartheta_i - \hat{\vartheta}_i$ and Lemma 2.2, it gives

$$\begin{aligned}\tilde{\vartheta}_i\hat{\vartheta}_i = \tilde{\vartheta}_i(\vartheta_i - \tilde{\vartheta}_i) \leq& -\frac{1}{2}\tilde{\vartheta}_i^2 + \frac{1}{2}\vartheta_i^2 \\ \leq& -\frac{1}{2}\tilde{\vartheta}_i^2 - \frac{1}{2}\tilde{\vartheta}_i^{\beta+1} + \frac{1}{2}\tilde{\vartheta}_i^{\beta+1} + \frac{1}{2}\vartheta_i^2 \\ \leq& -\frac{1-\beta}{4}\tilde{\vartheta}_i^2 - \frac{1}{2}\tilde{\vartheta}_i^{\beta+1} + \frac{1}{2}\vartheta_i^2 + \frac{1-\beta}{4}.\end{aligned} \tag{3.35}$$

Let $c_1 = \lambda_{\min}(L_1) - \frac{1}{2}$ and $c_2 = \lambda_{\min}(L_2) - 1$. For $\chi_1^T \chi_1$ and $\chi_2^T \chi_2$, they follow the similar process in (3.35), and therefore, (3.34) can be rewritten as

$$
\begin{aligned}
\dot{V}_2 \leq & - z_1^T K_1 z_1 - z_1^T K_2 z_1^\beta - c_1 \chi_1^T \chi_1 - z_2^T K_3 z_2 - z^T K_4 z_2^\beta \\
& - c_2 \chi_2^T \chi_2 + \frac{1}{2}\|M_0(\xi)^{-1}\Delta\tau\|^2 + \sum_{i=1}^{3} \tilde{\vartheta}_i \hat{\vartheta}_i \\
\leq & - z_1^T K_1 z_1 - z_1^T K_2 z_1^\beta - \frac{c_1(1-\beta)}{4}\chi_1^T \chi_1 - \frac{c_1}{2}\chi_1^T \chi_1^\beta - z_2^T K_3 z_2 - z^T K_4 z_2^\beta \\
& - \frac{c_2(1-\beta)}{4}\chi_2^T \chi_2 - \frac{c_2}{2}\chi_2^T \chi_2^\beta + \frac{(c_1+c_2+3)(1-\beta)}{4} + \frac{1}{2}\|M_0(\xi)^{-1}\Delta\tau\|^2 \\
& - \sum_{i=1}^{3} \frac{1-\beta}{4}\tilde{\vartheta}_i^2 - \sum_{i=1}^{3} \frac{1}{2}\tilde{\vartheta}_i^{\beta+1} + \sum_{i=1}^{3} \frac{1}{2}\vartheta_i^2.
\end{aligned}
\tag{3.36}
$$

Upon Lemma 2.3, it gives

$$
\begin{aligned}
\dot{V}_2 \leq & - z_1^T K_1 z_1 - z_1^T K_2 z_1^\beta - \frac{c_1(1-\beta)}{4}\chi_1^T \chi_1 - \frac{c_1}{2}\chi_1^T \chi_1^\beta - z_2^T K_3 z_2 - z^T K_4 z_2^\beta \\
& - \frac{c_2(1-\beta)}{4}\chi_2^T \chi_2 - \frac{c_2}{2}\chi_2^T \chi_2^\beta + \frac{1}{2}\|M_0(\xi)^{-1}\Delta\tau\|^2 + \frac{(c_1+c_2+3)(1-\beta)}{4} \\
& - \sum_{i=1}^{3} \frac{1-\beta}{4}\tilde{\vartheta}_i^2 - \sum_{i=1}^{3} \frac{1}{2}\tilde{\vartheta}_i^{\beta+1} + \sum_{i=1}^{3} \frac{1}{2}\vartheta_i^2 \\
\leq & - l_1\left(\frac{1}{2}z_1^T z_1 + \frac{1}{2}\chi_1^T \chi_1 + \frac{1}{2}z_2^T z_2 + \frac{1}{2}\chi_2^T \chi_2 + \frac{1}{2}\sum_{i=1}^{3} \Gamma_i^{-1}\tilde{\vartheta}_i\right) \\
& - l_2\left(\frac{1}{2}z_1^T z_1 + \frac{1}{2}\chi_1^T \chi_1 + \frac{1}{2}z_2^T z_2 + \frac{1}{2}\chi_2^T \chi_2 + \frac{1}{2}\sum_{i=1}^{3} \Gamma_i^{-1}\tilde{\vartheta}_i\right)^{\frac{1+\beta}{2}} + \varepsilon \\
\leq & - l_1 V_2 - l_2 V_2^{\frac{1+\beta}{2}} + \varepsilon.
\end{aligned}
\tag{3.37}
$$

where $l_1 = \min(2\lambda_{\min}(K_1), \frac{c_1(1-\beta)}{2}, 2\lambda_{\min}(K_3), \frac{c_2(1-\beta)}{2}, \frac{\Gamma_i(1-\beta)}{2})$, $l_2 = \min(2^\beta \lambda_{\min}(K_2), c_1 2^{\beta-1}, 2^\beta \lambda_{\min}(K_4), c_2 2^{\beta-1}, 2^{\beta-1})$, and $\varepsilon = \frac{1}{2}\|M_0(\xi)^{-1}\Delta\tau\|^2 + \frac{(c_1+c_2+3)(1-\beta)}{4} + \sum_{i=1}^{3}\frac{1}{2}\vartheta_i^2$. Based on Lemma 2.1, the closed-loop system is said to be semi-globally practical finite-time stable, and there exists a positive real number $0 < l_3 < 1$ such that

$$
\dot{V}_2 \leq -l_1 l_3 V_2 - l_2 V_2^{\frac{1+\beta}{2}} - l_1(1-l_3)V_2 + \varepsilon. \tag{3.38}
$$

Hence, if $V_2 \geq \frac{\varepsilon}{l_1(1-l_3)}$, $\dot{V}_2 < 0$ holds, which indicates that the system state z_1 converges to a small neighborhood around the origin within finite time, and then remains within the compact set Ω_{z_1}.

$$
\Omega_{z_1} := \left\{ z_1 \in \mathbb{R}^6 \mid \|z_1\| < \sqrt{\frac{2\varepsilon}{l_1(1-l_3)}} \right\}. \tag{3.39}
$$

For the error signals z_2 and $\tilde{\vartheta}_i$, the similar process can be conducted, and we omit here for clarity. This completes the proof. □

3.3.3 Second-Order Constrained Dynamic Surface Control

In the above section, for a class of Euler–Lagrange systems with uncertainties and external disturbances, a backstepping-based constrained control has been developed with considering input saturation. It is worthy noting that the control input v_2 in (3.22) includes a discontinuous sign function, which may lead to severe chattering problem when z_2 frequently crosses the origin. It could degrade the system's tracking performance. To address this issue, inspired by [6], we design a second-order constrained dynamic surface control under input saturation constraint, and establish a novel auxiliary system using the error of control input derivative as its input to compensate for the effects of nonlinear saturation. Additionally, dynamic surface control technique is employed to avoid tedious numerical computations in backstepping. It should be pointed out that our proposed second-order constrained dynamic surface control is not limited to the control design in this section; its design mechanism can be flexibly extended to other nonlinear control designs, greatly enhancing the generalization capability of our control design.

For Euler–Lagrange systems in (3.3), by taking the third derivative of the system state ξ, we have:

$$\begin{aligned}\frac{d}{dt}\ddot{\xi} =& M_0(\xi)^{-1}\dot{\tau} - M_0(\xi)^{-1}\frac{d}{dt}(H_0(\xi,\dot{\xi})\dot{\xi} + D_0(\xi,\dot{\xi})\dot{\xi} + G_{w0}) + \frac{d}{dt}M_0(\xi)^{-1}(\tau \\ &- H_0(\xi,\dot{\xi})\dot{\xi} - D_0(\xi,\dot{\xi})\dot{\xi} - G_{w0}) + \bar{\Delta},\end{aligned} \tag{3.40}$$

where $\bar{\Delta} = -M_0(\xi)^{-1}\dot{\Delta} - \frac{d}{dt}M_0(\xi)^{-1}\Delta$. According to Assumptions 3.1–3.4, we have

$$\|\bar{\Delta}\| \leq \vartheta_1 + \vartheta_2\|\dot{\xi}\| + \vartheta_3\|\dot{\xi}\|^2 + \vartheta_4\|\dot{\xi}\|^3, \tag{3.41}$$

where ϑ_i, $i = 1, \ldots, 4$ denotes unknown positive constants of the upper boundary of uncertainties. Let $y_1 = \xi$, $y_2 = \dot{\xi}$ and $y_3 = \ddot{\xi}$, then, Euler–Lagrange systems (3.3) can be rewritten as:

$$\dot{y}_1 = y_2 \tag{3.42}$$

$$\dot{y}_2 = y_3 \tag{3.43}$$

$$\begin{aligned}\dot{y}_3 =& M_0(\xi)^{-1}\dot{\tau} - M_0(\xi)^{-1}\frac{d}{dt}(H_0(\xi,\dot{\xi})\dot{\xi} + D_0(\xi,\dot{\xi})\dot{\xi} + G_{w0}) \\ &+ \frac{d}{dt}M_0(\xi)^{-1}(\tau - H_0(\xi,\dot{\xi})\dot{\xi} - D_0(\xi,\dot{\xi})\dot{\xi} - G_{w0}) + \bar{\Delta},\end{aligned} \tag{3.44}$$

To meet control input saturation, i.e., $\tau \in [\tau_{\min}, \tau_{\max}]$, we extend the control input saturation nonlinearity to the control input derivative, and design a novel constraint mechanism as follows:

$$\dot{\tau} = \begin{cases} \dot{\tau}_{\xi} \text{ (if } \tau_{\min} < \tau < \tau_{\max}), \\ 0 \text{ (others)}, \end{cases} \tag{3.45}$$

where τ_{ξ} is the actual control input to be designed later. With the aid of the above mechanism on the derivative of control input, we can ensure that the control input remains within the predefined reasonable range. To compensate for the effects of saturation nonlinearity, we establish a new virtual auxiliary system:

$$\dot{\chi}_1 = -L_1\chi_1 + \chi_2 \tag{3.46}$$

$$\dot{\chi}_2 = -L_2\chi_2 + \chi_3 \tag{3.47}$$

$$\dot{\chi}_3 = -L_3\chi_3 + M_0(\xi)^{-1}\Delta\dot{\tau}, \tag{3.48}$$

where $\chi_i,\ i = 1, 2, 3$ are the virtual auxiliary states, $L_i,\ i = 1, 2, 3$ are positive constants, and $\Delta\dot{\tau} = \dot{\tau} - \dot{\tau}_{\xi}$ denotes the error signal between the actual control input derivative and the designed control input derivative. For achieving finite-time convergence of Euler–Lagrange systems with alleviating chattering problem at the same time, we propose a second-order dynamic surface control. First, we define the tracking error z_1 as:

$$z_1 = y_1 - y_{1d} - \chi_1, \tag{3.49}$$

where y_{1d} denotes the desired trajectory. The time derivative of z_1 is

$$\dot{z}_1 = \dot{y}_1 - \dot{y}_{1d} - \dot{\chi}_1 = y_2 - \dot{y}_{1d} + L_1\chi_1 - \chi_2. \tag{3.50}$$

where the virtual control input α_1 is given by:

$$\alpha_1 = -K_1 z_1 - K_2 z_1^{\beta} + \dot{y}_{1d} - L_1\chi_1, \tag{3.51}$$

where $K_1 = \text{diag}(k_{1i}),\ K_2 = \text{diag}(k_{2i}),\ i = 1, \ldots, p$ are positive definite matrices composed by positive numbers k_{1i} and k_{2i}, and $0 < \beta < 1$. As backstepping design needs to repetitively calculate the time derivative of virtual control inputs which increases the computational burden, we, therefore, introduce a first-order filter to simplify its design process: $h_1\dot{\bar{y}}_2 + \bar{y}_2 = \alpha_1$ with h_1 being a positive constant. The tracking error z_2 is defined as:

$$z_2 = y_2 - \bar{y}_2 - \chi_2, \tag{3.52}$$

then, its derivative is

$$\dot{z}_2 = \dot{y}_2 - \dot{\bar{y}}_2 - \dot{\chi}_2 = y_3 - \frac{\alpha_1 - \bar{y}_2}{h_1} + L_2\chi_2 - \chi_3. \tag{3.53}$$

The virtual control input α_2 is designed as:

$$\alpha_2 = -z_1 - K_3 z_2 - K_4 z_2^{\beta} + \frac{\alpha_1 - \bar{y}_2}{h_1} - L_2\chi_2, \tag{3.54}$$

where $K_3 = \text{diag}(k_{3i})$, and $K_4 = \text{diag}(k_{4i}),\ i = 1, \ldots, p$ are positive definite matrices composed by positive numbers k_{3i} and k_{4i}. To avoid tedious numerical computations, we introduce a first-order filter: $h_2\dot{\bar{y}}_3 + \bar{y}_3 = \alpha_2$, where h_2 is a positive constant. The tracking error z_3 is defined by

$$z_3 = y_3 - \bar{y}_3 - \chi_3. \tag{3.55}$$

Taking time derivative of z_3 gives

$$\begin{aligned}\dot{z}_3 =& \dot{y}_3 - \dot{\bar{y}}_3 - \dot{\chi}_3\\ =& M_0(\xi)^{-1}\dot{\tau}_\xi - M_0(\xi)^{-1}\frac{d}{dt}(H_0(\xi,\dot{\xi})\dot{\xi} + D_0(\xi,\dot{\xi})\dot{\xi} + G_{w0})\\ &+ \frac{d}{dt}M_0(\xi)^{-1}(\tau - H_0(\xi,\dot{\xi})\dot{\xi} - D_0(\xi,\dot{\xi})\dot{\xi} - G_{w0}) + \bar{\Delta} - \frac{\alpha_2 - \bar{y}_3}{h_2} + L_3\chi_3.\end{aligned} \tag{3.56}$$

The time derivative of the actual control input $\tau_\xi = \tau_1 + \tau_2$ is designed as:

$$\dot{\tau}_\xi = \dot{\tau}_1 + \dot{\tau}_2, \tag{3.57}$$

$$\begin{aligned}\dot{\tau}_1 =& \frac{d}{dt}(H_0(\xi,\dot{\xi})\dot{\xi} + D_0(\xi,\dot{\xi})\dot{\xi} + G_{w0}) - M_0\Big(\frac{d}{dt}M_0(\xi)^{-1}(\tau - H_0(\xi,\dot{\xi})\dot{\xi} - G_{w0}\\ &- D_0(\xi,\dot{\xi})\dot{\xi}\Big) - \frac{\alpha_2 - \bar{y}_3}{h_2} + L_3\chi_3)\end{aligned} \tag{3.58}$$

$$\dot{\tau}_2 = -M_0(\xi)(z_2 + K_5 z_3 + K_6 z_3^{\beta} + (\hat{\vartheta}_1 + \hat{\vartheta}_2\|\dot{\xi}\| + \hat{\vartheta}_3\|\dot{\xi}\|^2 + \hat{\vartheta}_4\|\dot{\xi}\|^3 + \rho)\text{sign}(z_3)), \tag{3.59}$$

where $K_5 = \text{diag}(k_{5i})$ and $K_6 = \text{diag}(k_{6i}),\ i = 1, \ldots, p$ are positive definite matrices composed by positive numbers k_{5i} and k_{6i}, ρ is a positive constant, and $\hat{\vartheta}_i,\ i = 1, \ldots, p$ denotes the estimation of ϑ_i whose update laws are given by

$$\dot{\hat{\vartheta}}_1 = \Gamma_1(\|z_3\| - \hat{\vartheta}_1), \tag{3.60}$$

$$\dot{\hat{\vartheta}}_2 = \Gamma_2(\|z_3\|\|\dot{\xi}\| - \hat{\vartheta}_2), \tag{3.61}$$

$$\dot{\hat{\vartheta}}_3 = \Gamma_3(\|z_3\|\|\dot{\xi}\|^2 - \hat{\vartheta}_3), \tag{3.62}$$

$$\dot{\hat{\vartheta}}_4 = \Gamma_4(\|z_3\|\|\dot{\xi}\|^3 - \hat{\vartheta}_4), \tag{3.63}$$

where Γ_i, $i = 1, \ldots, 4$ are positive constants. Based on the above designed second-order dynamic surface control, we are now at the position to present the following theorem.

Theorem 3.2 *Considering Euler–Lagrange systems (3.3) subject to uncertainties and external disturbances and input saturation constraint, with the aid of the virtual auxiliary system (3.46)–(3.48), the second-order constrained dynamic surface control (3.57), and adaptive laws (3.60)–(3.63), all the closed-loop system states are bounded, and the tracking errors z_1, z_2, z_3, ϑ_i, and the filter errors Y_1 and Y_2, converge to a small neighborhood around the origin within finite time and remain within the regions Ωz_1, Ω_{z_2}, Ω_{z_3}, Ω_{ϑ_i}, Ω_{Y_1}, and Ω_{Y_2} thereafter.*

$$\Omega_{z_1} := \left\{ z_1 \in \mathbb{R}^p \mid \|z_1\| < \sqrt{\frac{2\varepsilon}{a(1-c)}} \right\}, \ \Omega_{z_2} := \left\{ z_2 \in \mathbb{R}^p \mid \|z_2\| < \sqrt{\frac{2\varepsilon}{a(1-c)}} \right\}, \tag{3.64}$$

$$\Omega_{z_3} := \left\{ z_3 \in \mathbb{R}^p \mid \|z_3\| < \sqrt{\frac{2\varepsilon}{a(1-c)}} \right\}, \ \Omega_{Y_1} := \left\{ Y_1 \in \mathbb{R}^p \mid \|Y_1\| < \sqrt{\frac{2\varepsilon}{a(1-c)}} \right\}, \tag{3.65}$$

$$\Omega_{\tilde{\vartheta}_i} := \left\{ \tilde{\vartheta}_i \in \mathbb{R} \mid \|\tilde{\vartheta}_i\| < \sqrt{\frac{2\Gamma_i\varepsilon}{a(1-c)}} \right\}, \ \Omega_{Y_1} := \left\{ Y_1 \in \mathbb{R}^p \mid \|Y_1\| < \sqrt{\frac{2\varepsilon}{a(1-c)}} \right\}. \tag{3.66}$$

where a, c, ε and the filter error Y_1 and Y_2 will be defined later.

Proof First, consider the following Lyapunov function candidate V_1 as

$$V_1 = \frac{1}{2} z_1^{\mathrm{T}} z_1 + \frac{1}{2} Y_1^{\mathrm{T}} Y_1 + \frac{1}{2} \chi_1^{\mathrm{T}} \chi_1, \tag{3.67}$$

where the filter error is defined as $Y_1 = \bar{y}_2 - \alpha_1$. The time derivative is

$$\dot{V}_1 = z_1^{\mathrm{T}} \dot{z}_1 + Y_1^{\mathrm{T}} \dot{Y}_1 + \chi_1^{\mathrm{T}} \dot{\chi}_1. \tag{3.68}$$

We have

$$\begin{aligned} z_1^{\mathrm{T}} \dot{z}_1 &= z_1^{\mathrm{T}} (y_2 - \dot{y}_{1d} + L_1 \chi_1 - \chi_2) \\ &\leq z_1^{\mathrm{T}} z_2 - (\lambda_{\min}(K_1) - \frac{1}{2}) z_1^{\mathrm{T}} z_1 - z_1^{\mathrm{T}} K_2 z_1^{\beta} + \frac{1}{2} Y_1^{\mathrm{T}} Y_1 \end{aligned} \tag{3.69}$$

and

$$\chi_1^{\mathrm{T}} \dot{\chi}_1 = -\chi_1^{\mathrm{T}} L_1 \chi_1 + \chi_1^{\mathrm{T}} \chi_2 \leq -(\lambda_{\min}(L_1) - \frac{1}{2}) \chi_1^{\mathrm{T}} \chi_1 + \frac{1}{2} \chi_2^{\mathrm{T}} \chi_2. \tag{3.70}$$

Substituting (3.69) and (3.70) into (3.68) gives

$$
\begin{aligned}
\dot{V}_1 \leq & z_1^{\mathrm{T}} z_2 - (\lambda_{\min}(K_1) - \frac{1}{2}) z_1^{\mathrm{T}} z_1 - z_1^{\mathrm{T}} K_2 z_1^{\beta} + \frac{1}{2} Y_1^{\mathrm{T}} Y_1 - \frac{1}{h_1} Y_1^{\mathrm{T}} Y_1 \\
& - Y_1^{\mathrm{T}} \dot{\alpha}_1 - (\lambda_{\min}(L_1) - \frac{1}{2}) \chi_1^{\mathrm{T}} \chi_1 + \frac{1}{2} \chi_2^{\mathrm{T}} \chi_2 \\
\leq & z_1^{\mathrm{T}} z_2 - (\lambda_{\min}(K_1) - \frac{1}{2}) z_1^{\mathrm{T}} z_1 - z_1^{\mathrm{T}} K_2 z_1^{\beta} + \frac{1}{2} \dot{\alpha}_1^{\mathrm{T}} \dot{\alpha}_1 - \left(\frac{1}{h_1} - 1\right) Y_1^{\mathrm{T}} Y_1 \\
& - (\lambda_{\min}(L_1) - \frac{1}{2}) \chi_1^{\mathrm{T}} \chi_1 + \frac{1}{2} \chi_2^{\mathrm{T}} \chi_2,
\end{aligned} \tag{3.71}
$$

where $0 < h_1 < 1$, $\lambda_{\min}(K_1) > \frac{1}{2}$, and $\lambda_{\min}(L_1) > \frac{1}{2}$. Based on Lemma 2.2, we have

$$
- Y_1^{\mathrm{T}} Y_1 \leq -Y_1^{\mathrm{T}} Y_1 - Y_1^{\mathrm{T}} Y_1^{\beta} + Y_1^{\mathrm{T}} Y_1^{\beta} \leq -\frac{1-\beta}{2} Y_1^{\mathrm{T}} Y_1 - Y_1^{\mathrm{T}} Y_1^{\beta} + \frac{1-\beta}{2}. \tag{3.72}
$$

Suppose that the derivative of the virtual control input satisfies $\|\dot{\alpha}_1\| \leq \delta_1 I_6$, where δ_1 is a positive constant. For $\chi_1^T \chi_1$, it follows the similar process in (3.72), it yields

$$
\begin{aligned}
\dot{V}_1 \leq & z_1^{\mathrm{T}} z_2 - (\lambda_{\min}(K_1) - \frac{1}{2}) z_1^{\mathrm{T}} z_1 - z_1^{\mathrm{T}} K_2 z_1^{\beta} - \left(\frac{1}{h_1} - 1\right) Y_1^{\mathrm{T}} Y_1^{\beta} + \frac{1}{2} \chi_2^{\mathrm{T}} \chi_2 + \frac{1}{2} \delta_1^2 \\
& - \frac{(1-h_1)(1-\beta)}{2h_1} Y_1^{\mathrm{T}} Y_1 - (\lambda_{\min}(L_1) - \frac{1}{2}) \chi_1^{\mathrm{T}} \chi_1^{\beta} + \frac{(1-h_1)(1-\beta)}{2h_1} \\
& - \frac{(2\lambda_{\min}(L_1) - 1)(1-\beta)}{4} \chi_1^{\mathrm{T}} \chi_1 + \frac{(2\lambda_{\min}(L_1) - 1)(1-\beta)}{4} \\
\leq & z_1^{\mathrm{T}} z_2 - c_1 z_1^{\mathrm{T}} z_1 - z_1^{\mathrm{T}} K_2 z_1^{\beta} - c_2 Y_1^{\mathrm{T}} Y_1^{\beta} - c_3 Y_1^{\mathrm{T}} Y_1 - c_4 \chi_1^{\mathrm{T}} \chi_1^{\beta} - c_5 \chi_1^{\mathrm{T}} \chi_1 + \frac{1}{2} \chi_2^{\mathrm{T}} \chi_2 + \varepsilon_1,
\end{aligned} \tag{3.73}
$$

where $c_1 = (\lambda_{\min}(K_1) - \frac{1}{2})$, $c_2 = (\frac{1}{h_1} - 1)$, $c_3 = \frac{(1-h_1)(1-\beta)}{2h_1}$, $c_4 = (\lambda_{\min}(L_1) - \frac{1}{2})$, $c_5 = \frac{(2\lambda_{\min}(L_1)-1)(1-\beta)}{4}$, and $\varepsilon_1 = \frac{1}{2}\delta_1^2 + \frac{(1-h_1)(1-\beta)}{2h_1} + \frac{(2\lambda_{\min}(L_1)-1)(1-\beta)}{4}$. Consider the Lyapunov function V_2

$$
V_2 = V_1 + \frac{1}{2} z_2^{\mathrm{T}} z_2 + \frac{1}{2} Y_2^{\mathrm{T}} Y_2 + \frac{1}{2} \chi_2^{\mathrm{T}} \chi_2. \tag{3.74}
$$

where $Y_2 = \bar{y}_3 - \alpha_2$. From (3.53), we can have

$$
\begin{aligned}
\dot{V}_2 = & \dot{V}_1 + z_2^{\mathrm{T}} \dot{z}_2 + Y_2^{\mathrm{T}} \dot{Y}_2 + \chi_2^{\mathrm{T}} \dot{\chi}_2 \\
\leq & \dot{V}_1 + z_2^{\mathrm{T}} (z_3 + Y_2 + \alpha_2 + \chi_3 - \frac{\alpha_1 - \bar{y}_2}{h_1} + L_2 \chi_2 - \chi_3) - \frac{1}{h_2} Y_2^{\mathrm{T}} Y_2 \\
& + \frac{1}{2} Y_2^{\mathrm{T}} Y_2 + \frac{1}{2} \dot{\alpha}_2^{\mathrm{T}} \dot{\alpha}_2 - \chi_2^{\mathrm{T}} L_2 \chi_2 + \frac{1}{2} \chi_2^{\mathrm{T}} \chi_2 + \frac{1}{2} \chi_3^{\mathrm{T}} \chi_3
\end{aligned}
$$

$$
\begin{aligned}
&\leq \dot{V}_1 + z_2^{\mathrm{T}}(z_3 + Y_2 - z_1 - K_3 z_2 - K_4 z_2^{\beta} + \frac{\alpha_1 - \bar{y}_2}{h_1} - L_2\chi_2 - \frac{\alpha_1 - \bar{y}_2}{h_1} \\
&\quad + L_2\chi_2) - \left(\frac{1}{h_2} - \frac{1}{2}\right) Y_2^{\mathrm{T}} Y_2 - (\lambda_{\min}(L_2) - \frac{1}{2})\chi_2^{\mathrm{T}}\chi_2 + \frac{1}{2}\chi_3^{\mathrm{T}}\chi_3 + \frac{1}{2}\delta_2^2,
\end{aligned} \tag{3.75}
$$

where $\|\dot{\alpha}_2\| \leq \delta_2 I_6$ holds, and δ_2 is a positive constant. Taking (3.73) into the above inequality, we have

$$
\begin{aligned}
\dot{V}_2 \leq & - c_1 z_1^{\mathrm{T}} z_1 - z_1^{\mathrm{T}} K_2 z_1^{\beta} - c_2 Y_1^{\mathrm{T}} Y_1^{\beta} - c_3 Y_1^{\mathrm{T}} Y_1 - c_4 \chi_1^{\mathrm{T}} \chi_1^{\beta} - c_5 \chi_1^{\mathrm{T}} \chi_1 \\
& - (\lambda_{\min}(K_3) - \frac{1}{2}) z_2^{\mathrm{T}} z_2 - z_2^{\mathrm{T}} K_4 z_2^{\beta} - \left(\frac{1}{h_2} - 1\right) Y_2^{\mathrm{T}} Y_2 \\
& - (\lambda_{\min}(L_2) - 1)\chi_2^{\mathrm{T}}\chi_2 + \frac{1}{2}\chi_3^{\mathrm{T}}\chi_3 + \frac{1}{2}\delta_2^2 + \varepsilon_1 + z_2^{\mathrm{T}} z_3 \\
\leq & - c_1 z_1^{\mathrm{T}} z_1 - z_1^{\mathrm{T}} K_2 z_1^{\beta} - c_2 Y_1^{\mathrm{T}} Y_1^{\beta} - c_3 Y_1^{\mathrm{T}} Y_1 - c_4 \chi_1^{\mathrm{T}} \chi_1^{\beta} - c_5 \chi_1^{\mathrm{T}} \chi_1 \\
& - c_6 z_2^{\mathrm{T}} z_2 - z_2^{\mathrm{T}} K_4 z_2^{\beta} - c_7 Y_2^{\mathrm{T}} Y_2^{\beta} - c_8 Y_2^{\mathrm{T}} Y_2 - c_9 \chi_2^{\mathrm{T}} \chi_2^{\beta} \\
& - c_{10} \chi_2^{\mathrm{T}} \chi_2 + \frac{1}{2}\chi_3^{\mathrm{T}}\chi_3 + z_2^{\mathrm{T}} z_3 + \varepsilon_2,
\end{aligned} \tag{3.76}
$$

where $c_6 = (\lambda_{\min}(K_3) - \frac{1}{2}) > 0$, $c_7 = (\frac{1}{h_2} - 1) > 0$, $c_9 = (-1 + \lambda_{\min}(L_2)) > 0$, $c_8 = \frac{(1-h_2)(1-\beta)}{2h_2}$, $c_{10} = \frac{(\lambda_{\min}(L_2)-1)(1-\beta)}{2}$, and $\varepsilon_2 = \varepsilon_1 + \frac{1}{2}\delta_2^2 + \frac{(1-\beta)(1-h_2)}{2h_2} + \frac{(1-\beta)(\lambda_{\min}(L_2)-1)}{2}$. Consider Lyapunov function V_3:

$$
V_3 = V_2 + \frac{1}{2} z_3^{\mathrm{T}} z_3 + \frac{1}{2}\chi_3^{\mathrm{T}}\chi_3 + \frac{1}{2}\sum_{i=1}^{4} \Gamma_i^{-1}\tilde{\theta}_i^2, \tag{3.77}
$$

where $\tilde{\vartheta}_i = \vartheta_i - \hat{\vartheta}_i$ denotes the estimation error. Referring to (3.56)–(3.59), it gives

$$
\begin{aligned}
\dot{V}_3 \leq & \dot{V}_2 + z_3^{\mathrm{T}}(M_0(\xi)^{-1}\dot{\tau}_{\xi} - M_0(\xi)^{-1}\frac{d}{dt}(H_0(\xi,\dot{\xi})\dot{\xi} + D_0(\xi,\dot{\xi})\dot{\xi} + G_{w0}) \\
& + \frac{d}{dt} M_0(\xi)^{-1}(\tau - H_0(\xi,\dot{\xi})\dot{\xi} - D_0(\xi,\dot{\xi})\dot{\xi} - G_{w0}) + \bar{\Delta} - \frac{\alpha_2 - \bar{y}_3}{h_2} \\
& + L_3\chi_3) - \chi_3^{\mathrm{T}} L_3 \chi_3 + \chi_3^{\mathrm{T}} M_0^{-1}\Delta\dot{\tau} - \sum_{i=1}^{4}\Gamma_i^{-1}\tilde{\vartheta}_i\dot{\hat{\vartheta}}_i \\
\leq & \dot{V}_2 - z_2^{\mathrm{T}} z_3 - z_3^{\mathrm{T}} K_5 z_3 - z_3^{\mathrm{T}} K_6 z_3^{\beta} + z_3^{\mathrm{T}}(\bar{\Delta} - (\hat{\vartheta}_1 + \hat{\vartheta}_2\|\xi\| + \hat{\vartheta}_3\|\xi\|^2 \\
& + \hat{\vartheta}_4\|\xi\|^3 + \rho)\mathrm{sign}(z_3)) - \chi_3^{\mathrm{T}}\mathrm{L}_3\chi_3 + \frac{1}{2}\chi_3^{\mathrm{T}}\chi_3 + \frac{1}{2}\Delta\mathrm{P}^{\mathrm{T}}\mathrm{M}_0(\xi)^{-\mathrm{T}}\mathrm{M}_0^{-1}\Delta\mathrm{P} \\
& - \|z_3\|(\tilde{\vartheta}_1 + \tilde{\vartheta}_2\|\dot{\xi}\| + \tilde{\vartheta}_3\|\dot{\xi}\|^2 + \tilde{\vartheta}_4\|\dot{\xi}\|^3) + \sum_{i=1}^{4}\tilde{\vartheta}_i\hat{\vartheta}_i,
\end{aligned} \tag{3.78}
$$

Upon Lemma 2.2, we obtain

$$\begin{aligned}\tilde{\vartheta}_i\hat{\vartheta}_i \leq & -\frac{1}{2}\tilde{\vartheta}_i^2-\frac{1}{2}\tilde{\vartheta}_i^{\beta+1}+\frac{1}{2}\tilde{\vartheta}_i^{\beta+1}+\frac{1}{2}\vartheta_i^2 \\ \leq & -\frac{1-\beta}{4}\tilde{\vartheta}_i^2-\frac{1}{2}\tilde{\vartheta}_i^{\beta+1}+\frac{1}{2}\vartheta_i^2+\frac{1-\beta}{4}\end{aligned} \tag{3.79}$$

By substituting (3.76) and (3.79) into (3.78), one has

$$\begin{aligned}\dot{V}_3 \leq & -c_1 z_1^{\mathrm{T}} z_1 - z_1^{\mathrm{T}} K_2 z_1^{\beta} - c_2 Y_1^{\mathrm{T}} Y_1^{\beta} - c_3 Y_1^{\mathrm{T}} Y_1 - c_4 \chi_1^{\mathrm{T}} \chi_1^{\beta} - c_5 \chi_1^{\mathrm{T}} \chi_1 \\ & - c_6 z_2^{\mathrm{T}} z_2 - z_2^{\mathrm{T}} K_4 z_2^{\beta} - c_7 Y_2^{\mathrm{T}} Y_2^{\beta} - c_8 Y_2^{\mathrm{T}} Y_2 - c_9 \chi_2^{\mathrm{T}} \chi_2^{\beta} - c_{10} \chi_2^{\mathrm{T}} \chi_2 \\ & - z_3^{\mathrm{T}} K_5 z_3 - z_3^{\mathrm{T}} K_6 z_3^{\beta} + \|z_3^{\mathrm{T}}\|(\|\bar{\Delta}\| - (\vartheta_1 + \vartheta_2\|\dot{\xi}\| + \vartheta_3\|\dot{\xi}\|^2 + \vartheta_4\|\dot{\xi}\|^3 \\ & + \rho)) - (\lambda_{\min}(L_3) - 1)\chi_3^{\mathrm{T}}\chi_3 - \sum_{i=1}^{4}\frac{1-\beta}{4}\tilde{\vartheta}_i^2 - \sum_{i=1}^{4}\frac{1}{2}\tilde{\vartheta}_i^{\beta+1} \\ & + \sum_{i=1}^{4}\frac{1}{2}\theta_i^2 + \sum_{i=1}^{4}\frac{1-\beta}{4} + \frac{1}{2}\Delta\dot{\tau}^{\mathrm{T}} M_0^{-\mathrm{T}} M_0^{-1}\Delta\dot{\tau} + \varepsilon_2.\end{aligned} \tag{3.80}$$

As the uncertainties of system satisfy (3.41), i.e., $\|\bar{\Delta}\| - (\vartheta_1 + \vartheta_2\|\dot{\xi}\| + \vartheta_3\|\dot{\xi}\|^2 + \vartheta_4\|\dot{\xi}\|^3 + \rho) < 0$, we can get

$$\begin{aligned}\dot{V}_3 \leq & -c_1 z_1^{\mathrm{T}} z_1 - z_1^{\mathrm{T}} K_2 z_1^{\beta} - c_2 Y_1^{\mathrm{T}} Y_1^{\beta} - c_3 Y_1^{\mathrm{T}} Y_1 - c_4 \chi_1^{\mathrm{T}} \chi_1^{\beta} - c_5 \chi_1^{\mathrm{T}} \chi_1 - c_6 z_2^{\mathrm{T}} z_2 \\ & - z_2^{\mathrm{T}} K_4 z_2^{\beta} - c_7 Y_2^{\mathrm{T}} Y_2^{\beta} - c_8 Y_2^{\mathrm{T}} Y_2 - c_9 \chi_2^{\mathrm{T}} \chi_2^{\beta} - c_{10} \chi_2^{\mathrm{T}} \chi_2 - z_3^{\mathrm{T}} K_5 z_3 - z_3^{\mathrm{T}} K_6 z_3^{\beta} \\ & - c_{11}\chi_3^{\mathrm{T}}\chi_3^{\beta} - c_{12}\chi_3^{\mathrm{T}}\chi_3 - \sum_{i=1}^{4}\frac{1-\beta}{4}\tilde{\theta}_i^2 - \sum_{i=1}^{4}\frac{1}{2}\tilde{\theta}_i^{\beta+1} + \sum_{i=1}^{4}\frac{1}{2}\theta_i^2 + \sum_{i=1}^{4}\frac{1-\beta}{4} \\ & + \frac{(1-\beta)(\lambda_{\min}(L_3)-1)}{2} + \frac{1}{2}\Delta\dot{\tau}^{\mathrm{T}} M_0^{-\mathrm{T}} M_0^{-1}\Delta\dot{\tau} + \epsilon_2 \\ \leq & -(c_1 z_1^{\mathrm{T}} z_1 + c_3 Y_1^{\mathrm{T}} Y_1 + c_5\chi_1^{\mathrm{T}}\chi_1 + c_6 z_2^{\mathrm{T}} z_2 + c_8 Y_2^{\mathrm{T}} Y_2 + c_{10}\chi_2^{\mathrm{T}}\chi_2 + z_3^{\mathrm{T}} K_5 z_3 \\ & + c_{12}\chi_3^{\mathrm{T}}\chi_3 + \sum_{i=1}^{4}\frac{1-\beta}{4}\tilde{\theta}_i^2) - (\lambda_{\min}(K_2)\|z_1\|^{2\mu} + c_2\|Y_1\|^{2\mu} + c_4\|\chi_1\|^{2\mu} \\ & + \lambda_{\min}(K_4)\|z_2\|^{2\mu} + c_7\|Y_2\|^{2\mu} + c_9\|\chi_2\|^{2\mu} + \lambda_{\min}(K_6)\|z_3\|^{2\mu} + c_{11}\|\chi_3\|^{2\mu} \\ & \sum_{i=1}^{4}\frac{1}{2}\|\tilde{\theta}_i\|^{2\mu}) + \epsilon_3 \\ \leq & -aV_3 - bV_3^{\mu} + \epsilon_3,\end{aligned} \tag{3.81}$$

where $c_{11} = \lambda_{\min}(L_3) - 1 > 0$, $c_{12} = \frac{(1-\beta)(\lambda_{\min}(L_3)-1)}{2}$, $\mu = \frac{1+\beta}{2}$, $a = \min(2c_1, 2c_3, 2c_5, 2c_6, 2c_8, 2c_{10}, 2\lambda_{\min}(K_5), 2c_{12}, \frac{1-\beta}{2})$, $b = \min(2^{\mu}\lambda_{\min}(\mathrm{K}_2), 2^{\mu}c_2, 2^{\mu}c_4, 2^{\mu}\lambda_{\min}(K_4), 2^{\mu}c_7, 2^{\mu}c_9, 2^{\mu}c_{11}, 2^{\mu}\lambda_{\min}(K_6), 2^{\mu-1})$, $\varepsilon_3 = \varepsilon_2 + \sum_{i=1}^{4}\frac{1-\beta}{4} +$

$\frac{(1-\beta)(\lambda_{\min}(L_3)-1)}{2} + \sum_{i=1}^{4} \frac{1}{2}\vartheta_i^2 + \frac{1}{2}\Delta\dot{\tau}^{\mathrm{T}} M_0(\xi)^{-\mathrm{T}} M_0(\xi)^{-1}\Delta\dot{\tau}$. From Lemma 2.1, all closed-loop signals are bounded, and the tracking errors are regulated to a small neighborhood around the origin within finite time. The analysis of the ultimate boundedness of the closed-loop signals is similar to Theorem 3.1, and we omit here for clarity. This completes the proof. □

3.3.4 Numerical Simulations

Consider the following double-link manipulator system, whose dynamic model is described as

$$M(\xi)\ddot{\xi} + H(\xi, \dot{\xi})\dot{\xi} + G_w + \Delta_w = \tau, \tag{3.82}$$

where $\xi = [\xi_1, \xi_2]^T$ and $\dot{\xi} = [\dot{\xi}_1, \dot{\xi}_2]^T$ represent the joint angle and angular velocity of the double-link manipulator, respectively. $M(\xi)$ is the inertia matrix, $H(\xi, \dot{\xi})\dot{\xi}$ denotes the Coriolis force and centrifugal force, G_w is the generalized potential force, $\Delta_w = [\Delta_{w1}, \Delta_{w2}]^T$ represents the external disturbance, and $\tau = [\tau_1, \tau_2]^T$ is the control input vector.

$$M(\xi) = \begin{bmatrix} p_1 + p_2 + 2p_3 \cos\xi_2, & p_2 + p_3 \cos\xi_2 \\ p_2 + p_3 \cos\xi_2, & p_2 \end{bmatrix},$$

$$H(\xi, \dot{\xi}) = \begin{bmatrix} -p_3\dot{\xi}_2 \sin\xi_2, & -p_3(\dot{\xi}_1 + \dot{\xi}_2)\sin\xi_2 \\ p_3\dot{\xi}_1 \sin\xi_2, & 0 \end{bmatrix}, \quad G_w = \begin{bmatrix} p_4 g \cos\xi_1 + p_5 g \cos(\xi_1 + \xi_2) \\ p_5 g \cos(\xi_1 + \xi_2) \end{bmatrix},$$

where $p_1 = 0.25m_1l_1^2 + m_2l_1^2 + J_1$, $p_2 = 0.25m_2l_2^2 + J_2$, $p_3 = 0.5m_2l_1l_2$, $p_4 = 0.5m_1l_2 + m_2l_1$, $p_5 = 0.5m_2l_2$, m_1 and m_2 are the masses of the two links, l_1 and l_2 are the lengths of the two links, J_1 and J_2 are the moments of inertia of the two links, g is the acceleration due to gravity. As there exist uncertainties, let $m_1 = m_{10} + \Delta m_{10}$, $m_2 = m_{20} + \Delta m_{20}$, $l_1 = l_{10} + \Delta l_{10}$, $l_2 = l_{20} + \Delta l_{20}$, $J_1 = J_{10} + \Delta J_{10}$, $J_2 = J_{20} + \Delta J_{20}$. The parameter settings are given in Table 3.1.

Table 3.1 Parameters of the double-link manipulator

Parameter	Value	Parameter	Value	Parameter	Value
m_{10}	0.95 kg	m_{20}	0.98 kg	l_{10}	1.0 m
l_{20}	1.0 m	J_{10}	$0.23\ \mathrm{kg \cdot m^2}$	J_{20}	$0.24\ \mathrm{kg \cdot m^2}$
Δm_{10}	0.1 kg	Δm_{20}	$0.2\sin(1.2t + 1)$ kg	Δl_{10}	0.0 m
Δl_{20}	$0.2\sin(0.8t - 3)$ m	ΔJ_{10}	$0.04\ \mathrm{kg \cdot m^2}$	ΔJ_{20}	$0.02\sin(1.5t)\ \mathrm{kg \cdot m^2}$
Δ_{w1}	$0.1\sin 0.5t$	Δ_{w2}	0.15		

Table 3.2 Parameters for simulations

Parameter	Value	Parameter	Value	Parameter	Value
K_1	diag(15.0, 15.0)	K_2	diag(1.0, 1.0)	K_3	diag(10.0, 10.0)
K_4	diag(1.0, 1.0)	K_5	diag(2.5, 2.5)	K_6	diag(1.0, 1.0)
ρ	0.01	β	0.5	$\tau_{1\max}$	[25.0;7.0]
$\tau_{1\min}$	[-25.0;-7.0]	$\gamma_1, \gamma_2, \gamma_3$	0.05	$\Gamma_1, \Gamma_2, \Gamma_3$	5.0
L_1	diag(2.0, 2.0)	L_2	diag(2.0, 2.0)		

(Example 1) Considering the double-link manipulator system in (3.82), we conduct the following backstepping-based constrained finite-time control based on Sect. 3.3.2:

$$\begin{aligned} v &= v_1 + v_2, \\ v_1 &= M_0(\xi)(-f_2(x_1, x_2) + \dot{\alpha}_1 - L_2\chi_2 - K_3 z_2 - K_4 z_2^{\beta} - z_1), \\ v_2 &= -M_0(\xi)(\hat{\vartheta}_1 + \hat{\vartheta}_2\|x_2\| + \hat{\vartheta}_3\|x_2\|^2)\text{sign}(z_2). \end{aligned} \tag{3.83}$$

For the details about the control design, please refer to Sect. 3.3.2, and we omit here for clarity. The control parameters are set as shown in Table 3.2.

A comparative simulation is conducted to verify the effectiveness of our proposed method. Upon the adaptive terminal sliding mode control in [14], its control input is designed as

$$\begin{aligned} \tau &= \tau_{eq} + \Delta\tau_1, \\ \tau_{eq} &= M_0(\ddot{\xi}_d - \beta K_5 \text{diag}(|z_1|^{\beta-1})\dot{z}_1 - K_8 S) + H_0(\xi, \dot{\xi}) + G_{w0}, \\ \Delta\tau_1 &= -(\hat{\vartheta}_1 + \hat{\vartheta}_2\|\dot{\xi}\| + \hat{\vartheta}_3\|\dot{\xi}\|^2 + \rho)\text{sgn}(M_0^{-1}S), \\ \dot{\hat{\vartheta}}_1 &= \gamma_1\|S^T M_0^{-1}\|, \ \dot{\hat{\vartheta}}_2 = \gamma_2\|S^T M_0^{-1}\|\|\dot{\xi}\|, \ \dot{\hat{\vartheta}}_3 = \gamma_3\|S^T M_0^{-1}\|\|\dot{\xi}\|^2, \end{aligned} \tag{3.84}$$

where $S = \dot{z}_1 + K_7 z_1$ and other control parameters are given in Table 3.2. Let the reference trajectory be $\xi_d = [0.8\cos 0.8t, 0.5\cos(0.6t - 0.5)]^T$, the initial state be $\xi(0) = [0, 0]^T$, and the initial state of the adaptive laws be $\hat{\vartheta}(0) = [0.15, 0.15, 0.15]^T$. The simulation results are shown in Figs. 3.1, 3.2, 3.3, 3.4, 3.5 and 3.6.

In the simulation results, FTC represents the finite-time control based on our backstepping-based constrained control in (3.83), and TSMC represents the terminal sliding mode control in (3.84). In Fig. 3.1, it shows the angular position tracking performance of the first and second joints under uncertainties and external disturbances. It can be observed that our FTC can track the reference trajectory within finite time, while TSMC guarantees the system to the sliding mode surface within finite time. The angular tracking errors of the first and second joints are depicted in Fig. 3.2, which clearly demonstrate that the tracking error under FTC converges to

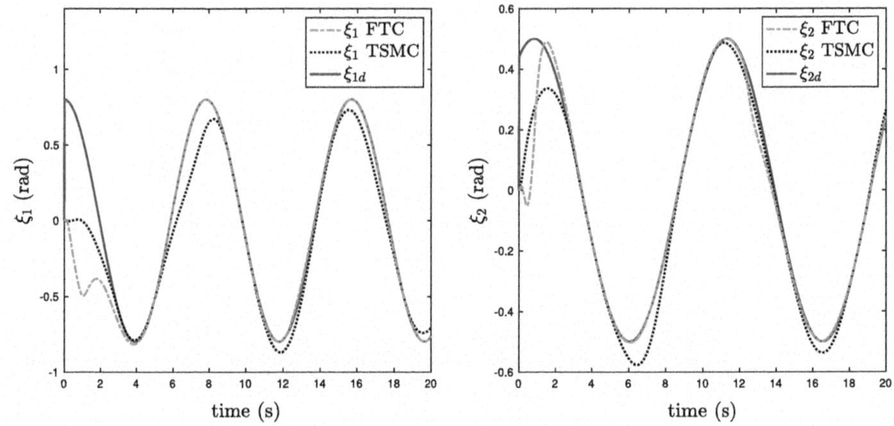

Fig. 3.1 The tracking performance of ξ_1; and the tracking performance of ξ_2

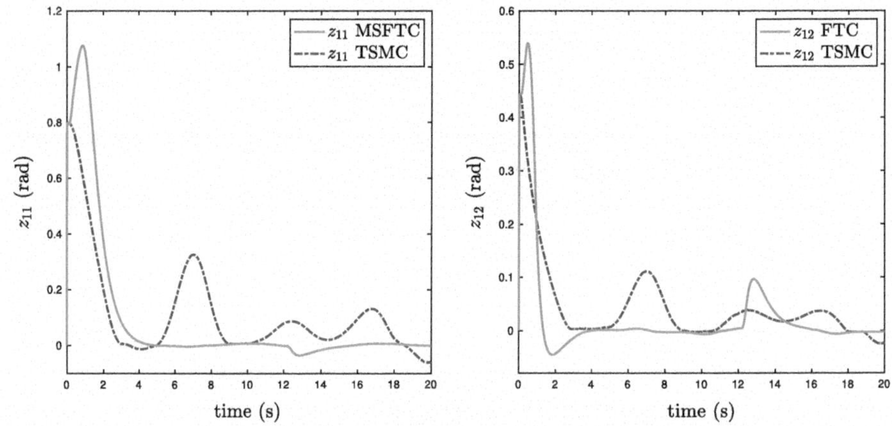

Fig. 3.2 The tracking performance of z_{11}; and the tracking performance of z_{12}

a neighborhood around the origin within finite time. For TSMC results, the tracking errors converge to the terminal sliding mode surface within finite time, then the tracking error converges to a small region around the origin. By comparing their simulation results, it can be concluded that FTC shows better robustness and tracking performance against unknown uncertainties and external disturbances compared to TSMC.

As system is subject to input saturation constraints, the control inputs τ_1 and τ_2 are shown in Fig. 3.3. According to the simulation results, the control inputs τ_1 and τ_2 under FTC are affected by saturation nonlinearity during the time intervals of $0-2s$ and $12-14s$. By introducing the virtual auxiliary systems in (3.13) and (3.14), the virtual auxiliary state variables $\chi_1 = [\chi_{11}, \chi_{12}]^T$ and $\chi_2 = [\chi_{21}, \chi_{22}]^T$ are rapidly

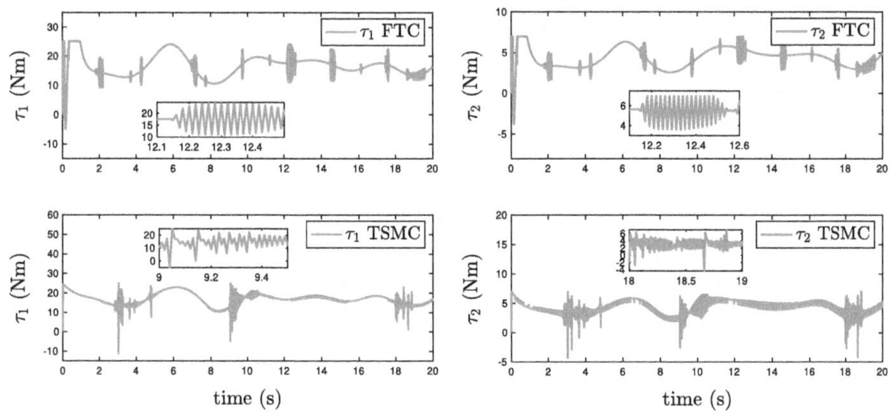

Fig. 3.3 The tracking performance of τ_1; and the tracking performance of τ_2

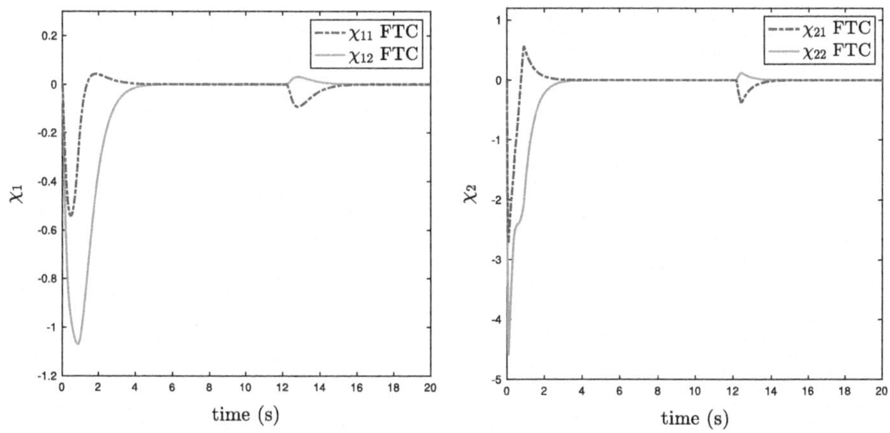

Fig. 3.4 The tracking performance of χ_1; and the tracking performance of χ_2

updated during this period based on the input saturation error signals to compensate for the saturation nonlinearity. When the saturation nonlinearity disappears, the virtual auxiliary state variables quickly converge to zero. Specifically, as shown in Fig. 3.2, when the tracking errors z_{11} and z_{12} are subjected to input saturation constraints, the tracking errors z_{11} and z_{12} under FTC exhibit significant tracking deviations, but they achieve rapid convergence with the assistance of the virtual auxiliary systems. The control inputs of FTC and TSMC remain within the predetermined reasonable range. However, since both FTC (3.83) and the terminal sliding mode controller (3.84) contain discontinuous sign functions, when the tracking error z_2 converges near the origin, this can easily cause chattering problem, which could affect the system's tracking performance. In Fig. 3.4, it plots the variations of the states of the virtual auxiliary system. According to the simulation results, when the

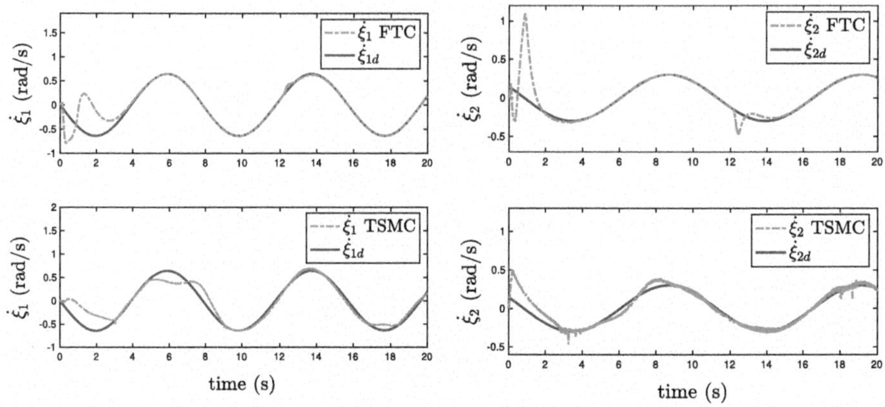

Fig. 3.5 The tracking performance of $\dot{\xi}_1$; and the tracking performance of $\dot{\xi}_2$

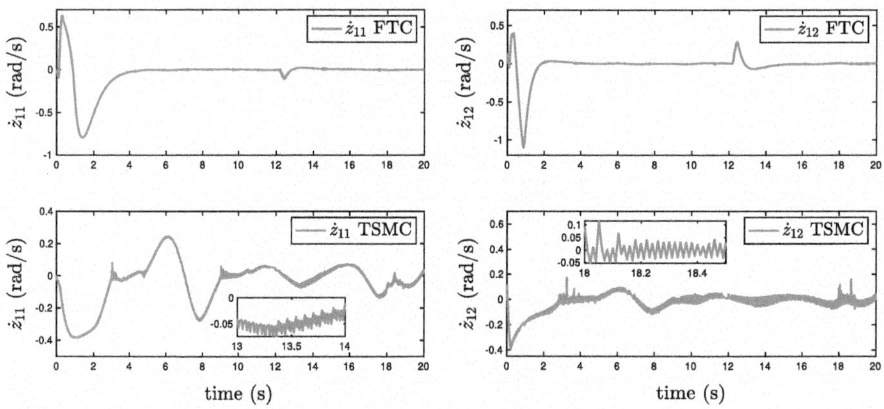

Fig. 3.6 The tracking performance of $\dot{z}_{11}$; and the tracking performance of $\dot{z}_{12}$

system is subjected to saturation nonlinearity, the virtual auxiliary systems rapidly update the state variables to compensate for the effects of the saturation nonlinearity, ensuring the tracking performance of the system. Once the saturation nonlinearity disappears, the virtual auxiliary system state variables quickly converge to zero.

Additionally, Fig. 3.5 shows the angular velocity tracking trajectories of the system. FTC ensures that the system converges to a neighborhood around the reference tracking trajectory within finite time, while TSMC ensures the convergence to the terminal sliding mode surface within finite time. The simulation results demonstrate that, compared to the TSMC method, FTC achieves faster convergence and exhibits robustness against unknown uncertainties and external disturbances. In Fig. 3.6, we show the angular velocity tracking error trajectories, further validating the excellent tracking performance of FTC. However, due to the chattering phenomenon, both FTC and TSMC exhibit noticeable chattering problem in the angular velocity tracking trajectories, which degrades the system's tracking performance.

(Example 2) We conduct comparative in this example to demonstrate the effectiveness of our proposed second-order dynamic surface control. For the Euler–Lagrange system in (3.82) under uncertainties and external disturbances, we design the following second-order constrained dynamic surface control developed in Sect. 3.3.3:

$$\begin{aligned}
\dot{\tau}_\xi =& \dot{\tau}_1 + \dot{\tau}_2,\\
\dot{\tau}_1 =& \frac{d}{dt}(H_0(\xi,\dot{\xi})\dot{\xi} + G_{w0}) - M_0\left(\frac{d}{dt}M_0(\xi)^{-1}(\tau - H_0(\xi,\dot{\xi})\dot{\xi} - G_{w0}\right) - \frac{\alpha_2 - \bar{y}_3}{h_2} + L_3\chi_3)\\
\dot{\tau}_2 =& - M_0(\xi)(z_2 + K_5 z_3 + K_6 z_3^{\beta} + (\hat{\vartheta}_1 + \hat{\vartheta}_2\|\dot{\xi}\| + \hat{\vartheta}_3\|\dot{\xi}\|^2 + \hat{\vartheta}_4\|\dot{\xi}\|^3 + \rho)\mathrm{sign}(z_3)), \quad (3.85)
\end{aligned}$$

where the control parameters are set in Table 3.3. In order to conduct a comparative simulation, we refer to the second-order sliding mode control in [7]:

$$\begin{aligned}
\dot{\tau} &= \dot{\tau}_{eq} + \dot{\tau}_{sw}\\
\dot{\tau}_{eq} &= M_0(-P(\xi,\dot{\xi}) - \bar{K}_1 P_0(\xi,\dot{\xi}) - \frac{1}{\bar{\beta}_1}\bar{K}_3^{-1}\dot{s}_1^{2-\bar{\beta}_1} - \bar{L}_2\dot{\chi}_2 - \bar{L}_3\chi_3\\
&\quad - \frac{\bar{\beta}_2}{\bar{\beta}_1}\bar{K}_3^{-1}\bar{K}_2\mathrm{diag}(|s_1|^{\bar{\beta}_2-1})\dot{s}_1^{2-\bar{\beta}_1} + \bar{K}_1(-\bar{L}_1\dot{\chi}_1 + \dot{\chi}_2)),\\
\dot{\tau}_{sw} &= -M_0[(\hat{\theta}_1 + \hat{\theta}_2\|\dot{\xi}\| + \hat{\theta}_3\|\dot{\xi}\|^2 + \hat{\theta}_4\|\dot{\xi}\|^3 + \rho)\mathrm{sgn}(s_2) + \frac{1}{\bar{\beta}_1}\bar{K}_3^{-1}\bar{K}_4 s_2], \quad (3.86)
\end{aligned}$$

where $\bar{K}_i \in \mathbb{R}^{p\times p}$, $i = 1, \ldots, 4$ are positive definite symmetric matrices, $1 < \bar{\beta}_1 < 2$, $\bar{\beta}_2 > \bar{\beta}_1$, $z_1 = \xi - \xi_d - \chi_1$, $z_2 = \dot{\xi} - \dot{\xi}_d - \chi_2$, $s_1 = z_2 + \bar{K}_1 z_1$, $s_2 = s_1 + \bar{K}_2 s_1^{\bar{\beta}_2} + \bar{K}_3\dot{s}_1^{\bar{\beta}_1}$, $P_0(\xi,\dot{\xi}) = M_0^{-1}(\tau - H_0\dot{\xi} - G_{w0}) - \ddot{\xi}_d$, $P(\xi,\dot{\xi}) = \frac{d}{dt}(M_0^{-1})(\tau_\xi - C_0\dot{\xi} - G_{w0}) - M_0^{-1}\frac{d}{dt}(H_0\dot{\xi} + G_{w0}) - \frac{d}{dt}\ddot{\xi}_d$, ρ is a positive real number, $\bar{L}_i \in$

Table 3.3 Parameters for simulations

Parameter	Value	Parameter	Value	Parameter	Value
K_1	diag(1.0, 1.0)	K_2	diag(0.2, 0.2)	K_3	diag(20.0, 20.0)
K_4	diag(1.3, 1.3)	K_5	diag(2.0, 2.0)	K_6	diag(1.0, 1.0)
$\bar{K}_1$	diag(1.5, 1.5)	$\bar{K}_2$	diag(1.5, 1.5)	$\bar{K}_3$	diag(1.2, 1.2)
$\bar{K}_4$	diag(100.0, 100.0)	L_1	diag(160.0, 160.0)	L_2	diag(40.0, 40.0)
L_3	diag(3.0, 3.0)	$\bar{L}_1$	diag(160.0, 160.0)	$\bar{L}_2$	diag(1.0, 1.0)
$\bar{L}_3$	diag(1.0, 1.0)	β	0.5	ρ	0.05
$\tau_{\max}$	[30; 8]	$\tau_{\min}$	[−30, −8]	$\Gamma_1, \Gamma_2, \Gamma_3$	0.1
$\bar{\Gamma}_1, \bar{\Gamma}_2, \bar{\Gamma}_3$	0.5	h_1	0.1	β_1	1.1
β_2	2.8	h_1	0.1		

$\mathbb{R}^{p\times p}$, $i = 1, \ldots, 3$ are positive definite symmetric matrices, $\hat{\vartheta}_i$, $i = 1, \ldots, 4$ are the estimates of ϑ_i, and we have

$$\dot{\hat{\vartheta}}_1 = \bar{\Gamma}_1(\|s_2^{\mathrm{T}}\bar{K}_3\mathrm{diag}(|\dot{s}_1|^{\beta_1-1})\| - \hat{\vartheta}_1),\ \dot{\hat{\vartheta}}_2 = \bar{\Gamma}_2(\|s_2^{\mathrm{T}}\bar{K}_3\mathrm{diag}(|\dot{s}_1|^{\beta_1-1})\|\|\dot{\xi}\| - \hat{\vartheta}_2),$$
$$\dot{\hat{\vartheta}}_3 = \bar{\Gamma}_3(\|s_2^{\mathrm{T}}\bar{K}_3\mathrm{diag}(|\dot{s}_1|^{\beta_1-1})\|\|\dot{\xi}\|^2 - \hat{\vartheta}_3),\ \dot{\hat{\vartheta}}_4 = \bar{\Gamma}_4(\|s_2^{\mathrm{T}}\bar{K}_3\mathrm{diag}(|\dot{s}_1|^{\beta_1-1})\|\|\dot{\xi}\|^3 - \hat{\vartheta}_4). \tag{3.87}$$

The control parameters of second-order sliding mode control are set in Table 3.3. The reference trajectory is $\xi_d = [0.8\cos 0.8t, 0.5\cos(0.6t - 0.5)]^T$, the initial state is $\xi(0) = [0, 0]^T$, and the initial state of the adaptive laws is $\hat{\vartheta}(0) = [0.15, 0.15, 0.15, 0.15]^T$. The simulation results are shown in Figs.

In the simulation results, SFTC represents our second-order finite-time control based on our proposed second-order constrained dynamic surface control in (3.85), and SOSMC denotes the second-order sliding mode control method in (3.86). The angular position tracking performance is shown in Figs. 3.7 and 3.8, where SFTC can ensure the convergence to the reference trajectory within finite time. The system states of SOSMC converge to the neighborhood of the sliding mode surfaces s_2 and s_1 within finite time, after which the tracking errors converge to a small neighbor around the equilibrium. The tracking errors of angle positions are depicted in Figs. 3.9 and 3.10. Compared to the convergence performance of SOSMC, the angular tracking error of SFTC can converge to the neighborhood of the origin more quickly, and it therefore demonstrates its robustness against unknown uncertainties and external disturbances.

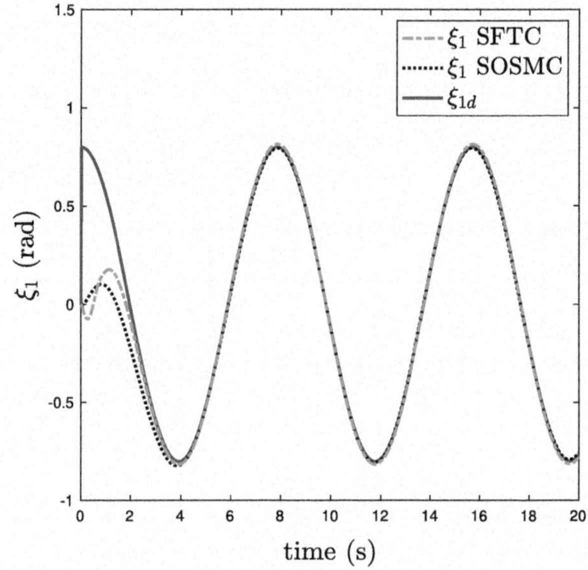

Fig. 3.7 The tracking performance of ξ_1

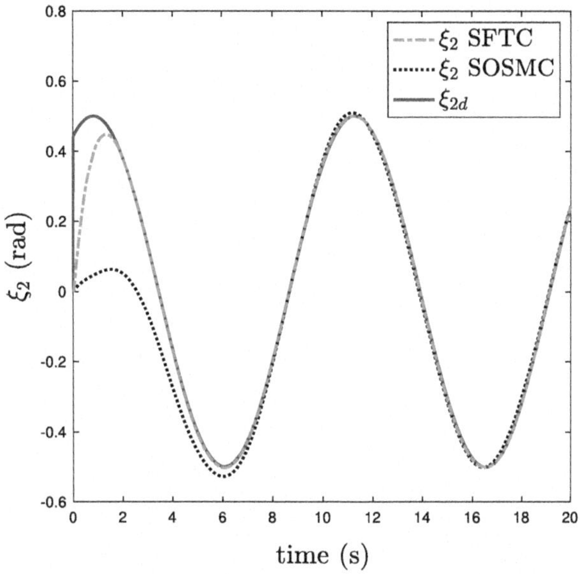

Fig. 3.8 The tracking performance of ξ_2

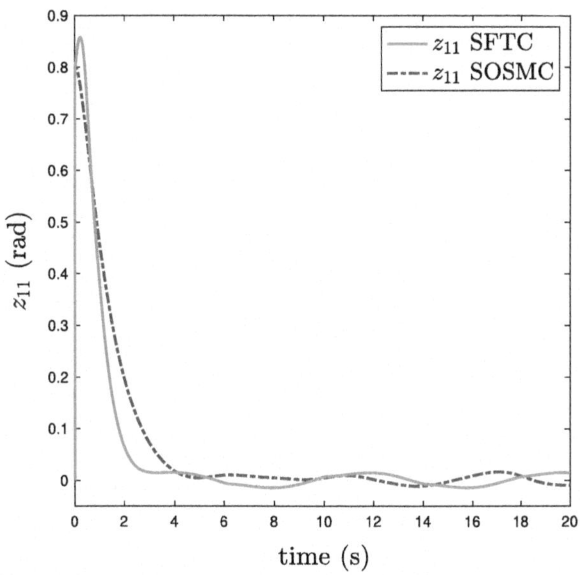

Fig. 3.9 The tracking performance of z_{11}

The angular velocity tracking performance is depicted in Figs. 3.11 and 3.12, where SFTC exhibits faster transient response performance and converges to the neighborhood around the reference trajectory within finite time. The tracking errors of angular velocity, $\dot{z}11$ and $\dot{z}12$, are shown in Figs. 3.13 and 3.14. SFTC not only converges to the neighbor of the origin more quickly but also demonstrates strong robustness against unknown uncertainties and external disturbances. Considering

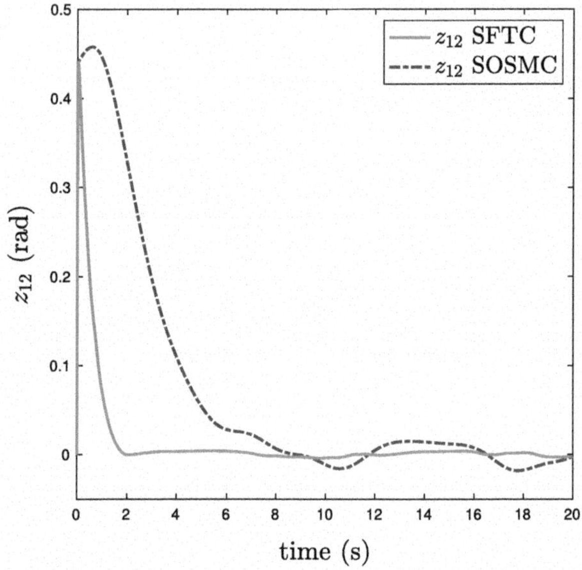

Fig. 3.10 The tracking performance of z_{12}

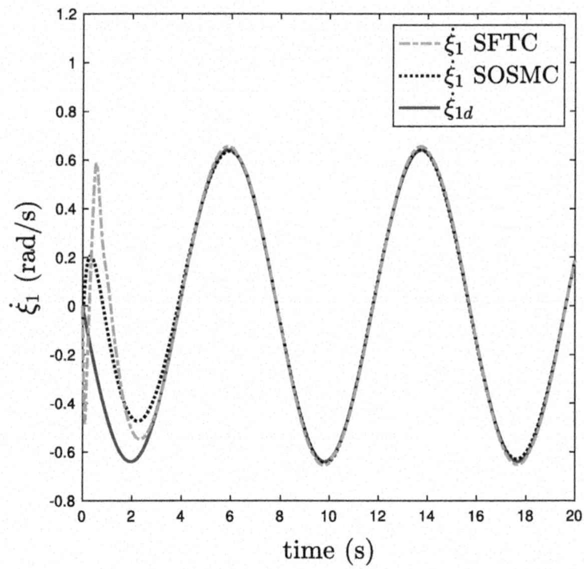

Fig. 3.11 The tracking performance of $\dot{\xi}_1$

the input saturation constraints, our proposed novel constraint mechanism in (3.45) effectively restricts control input signals τ_1 and τ_2 always within the predetermined range, as shown in Figs. 3.15 and 3.16. In addition, compared to (3.83), although (3.85) contain discontinuous sign functions, they act on the derivatives of the control inputs, and the actual control input is the integral of the discontinuous derivatives of

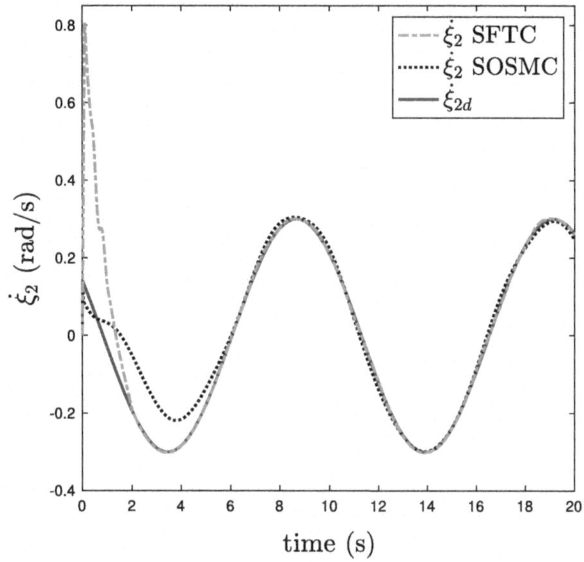

Fig. 3.12 The tracking performance of $\dot{\xi}_2$

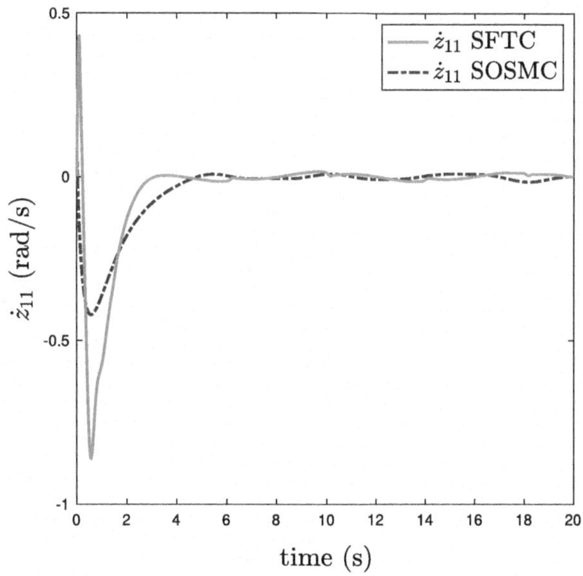

Fig. 3.13 The tracking performance of $\dot{z}_{11}$

control inputs. This significantly reduces the chattering phenomenon of the controller and enhances the system's tracking performance.

To ensure tracking performance under input saturation constraints, both SFTC and SOSMC adopt our proposed novel virtual auxiliary system (3.46) to (3.48) to compensate for the effects of saturation nonlinearities. According to the control inputs in Figs. 3.15 and 3.16, both SFTC and SOSMC are influenced by saturation

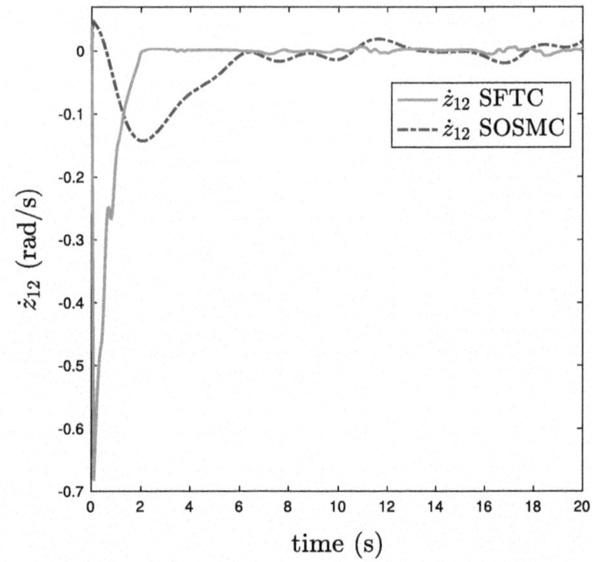

Fig. 3.14 The tracking performance of $\dot{z}_{12}$

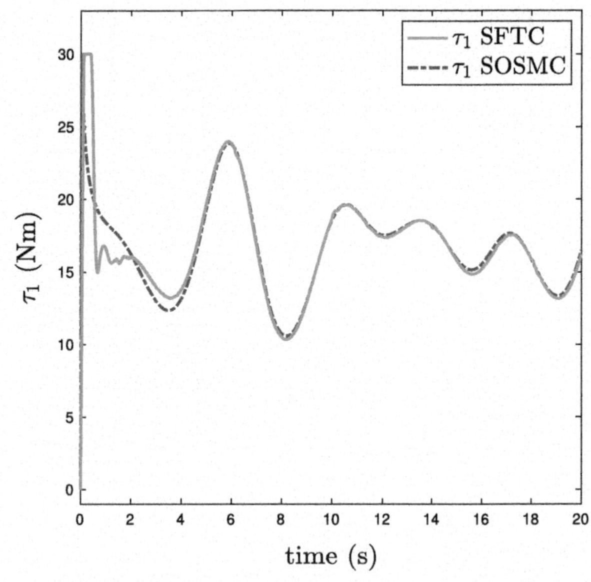

Fig. 3.15 The tracking performance of τ_1

nonlinearities during the initial tracking phase. With the help of our virtual auxiliary system, its virtual auxiliary system states χ_i, $i = 1, \ldots, 3$ are rapidly updated, as shown in Figs. 3.17 and 3.18, compensating for the effects of saturation nonlinearities and ensuring the system's tracking performance. Once saturation disappears, the virtual auxiliary system ensures that the virtual auxiliary system states rapidly converge to zero. In this way, under unknown uncertainties and external disturbances, our

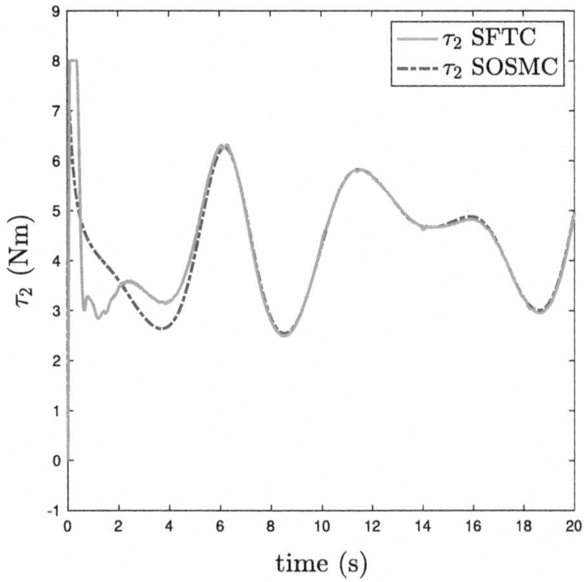

Fig. 3.16 The tracking performance of τ_2

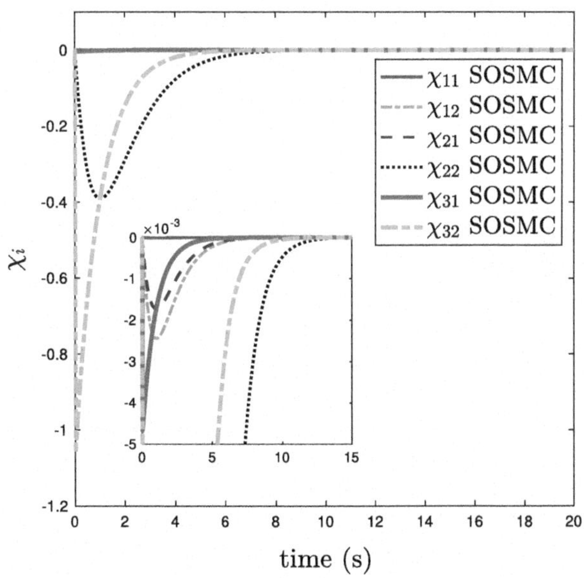

Fig. 3.17 The tracking performance of χ_1

proposed second-order dynamic surface control not only guarantees high-precision tracking but also avoids the chattering phenomenon, ensuring the system's tracking performance. By comparing with existing literature, it is evident that our second-order dynamic surface control exhibits faster convergence and superior robustness against uncertainties and external disturbances.

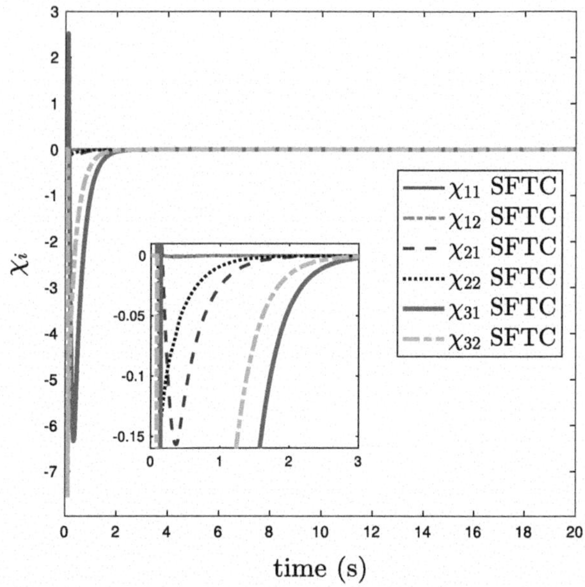

Fig. 3.18 The tracking performance of χ_2

3.4 Constrained Output-Feedback Control Under Input Saturation and Actuator Faults

In practical application, the controlled systems always exhibit significant nonlinearity and uncertainty, and the nonlinear functions within the system often do not satisfy the linear growth conditions described in [2, 4, 15]. This presents new challenges in the control design. Considering the excellent estimation capabilities of neural networks [5] and fuzzy logic systems [1] for nonlinearity, in recent years, numerous results have been investigated in combining neural networks and fuzzy logic systems with the backstepping technique, such as, [3, 5, 12, 17]. However, it should be pointed out that most control systems are suffering from saturation constraints and affected by unexpected factors, such as actuator faults and environment disturbances which could degrade tracking performance or even lead to catastrophic accidents. To address these issues, we, in this section, propose a novel control design for Euler–Lagrange systems with considering input saturation and actuator faults at the same time. Furthermore, in many applications, due to the limited sensor capabilities or cost considerations, especially in the formation or swarm scenarios, each agent can only obtain the position or attitude of the neighboring agents, indicating that only the output states are measurable. It renders the above state feedback control methods are not applicable anymore. Therefore, how to design a control scheme that guarantees finite-time stability of Euler–Lagrange systems under input saturation, actuator faults, and unmeasurable states makes the control design more challenging yet interesting.

3.4.1 Problem Formulation

As only the system output is measurable, we utilize fuzzy logic systems to approximate the unknown nonlinearities within the system, then establish a fuzzy state observer to estimate the unmeasurable states. To facilitate the design of fuzzy state observer, we rewrite Euler–Lagrange systems (3.1) in the following state-space representation:

$$\begin{cases} \dot{x}_{m,1} = x_{m,2} \\ \dot{x}_{m,2} = u_m + f_{m,2}(\bar{x}) \\ y_m = x_{m,1} \end{cases} \tag{3.88}$$

where $m = 1, \ldots, p$, $x_m = [x_{m,1}, x_{m,2}]^T$, $x_{m,1}$ denotes the position state variable, $x_{m,2}$ denotes the velocity state variable, $\bar{x} = [x_{1,1}, x_{1,2}, \cdots, x_{p,1}, x_{p,2}]^T$, $f_{m,2}(\cdot)$ represents the unknown nonlinearities. Based on dynamics in (3.1), it is easy to derive the expression of $f_{m,2}(\cdot)$, which is omitted here for brevity. Considering that the control input u_m is subject to saturation constraint and actuator faults, u_m is given as

$$u_m(t) = \rho_m(t)h_m(t) + \bar{v}_m(t), \tag{3.89}$$

where $\rho_m(t)$ and $\bar{v}_m(t)$ denotes the unknown effective coefficient of the actuator and the additional fault, respectively. $h_m(t)$ represents the control input with saturation constraint, which is given as

$$h_m(t) = \text{sat}(v_m) = \begin{cases} v_{\max}, & v_m > v_{\max} \\ v_m, & v_{\min} \le v_m \le v_{\max} \\ v_{\min}, & v_m < v_{\min} \end{cases} \tag{3.90}$$

where $v_{\max} > 0$ and $v_{\min} < 0$ denotes the upper and lower bounds of the control input, respectively, v_m is the actual control input to be designed later. To avoid the sharp turning points of the saturation function sat$(\cdot)$ and improve the control performance, we replace it with a smooth function $g(v_m(t))$ in the control design:

$$g(v_m(t)) = \begin{cases} v_{\max} \tanh\left(\frac{v_m}{v_{\max}}\right), & v_m \ge 0 \\ v_{\min} \tanh\left(\frac{v_m}{v_{\min}}\right), & v_m < 0 \end{cases} = \begin{cases} v_{\max} \frac{e^{\frac{v_m}{v_{\max}}} - e^{-\frac{v_m}{v_{\max}}}}{e^{\frac{v_m}{v_{\max}}} + e^{-\frac{v_m}{v_{\max}}}}, & v_m \ge 0 \\ v_{\min} \frac{e^{\frac{v_m}{v_{\min}}} - e^{-\frac{v_m}{v_{\min}}}}{e^{\frac{v_m}{v_{\min}}} + e^{-\frac{v_m}{v_{\min}}}}, & v_m < 0 \end{cases} \tag{3.91}$$

Hence, the equation (3.90) can be rewritten as

$$h_m(t) = g(v_m(t)) + d(v_m(t)), \tag{3.92}$$

where $|d(v_m(t))| = |\text{sat}(v_m(t)) - g(v_m(t))| \leq \max\{v_{\max}(1 - \tanh(1)), v_{\min}(\tanh(1) - 1)\} = D$. Moreover, based on the mean-value theorem, there exists a positive real number $0 < \mu < 1$ such that:

$$g(v_m(t)) = g(v_0) + g_{v_\mu}(v_m(t) - v_0), \tag{3.93}$$

where $g_{v_\mu} = \frac{\partial g(v_m)}{\partial v}|_{v_m = v_\mu}$ and $v_\mu = \mu v_m + (1 - \mu)v_0$, if we select $v_0 = 0$, we have $g(v_m) = g_{v_\mu} v_m$. Then, combining (3.89) and (3.93) yields

$$u_m(t) = \rho_m(t, t_p)(g_{v_\mu} v_m(t) + d(v_m(t))) + \bar{v}_m. \tag{3.94}$$

The control objective in this section is to develop constrained output-feedback control strategies for Euler–Lagrange systems such that all the closed-loop system states are bounded, and the tracking errors are regulated to a small neighborhood around the equilibrium after a finite period of time despite the coexistence of unmeasured system states, actuator saturations and faults.

To achieve the above control objective, the following assumptions and lemmas are presented.

Assumption 3.5 The actuator effectiveness parameters ρ_m and additive faults $\bar{v}_m$ are bounded, that is, $0 < \underline{\rho}_m \leq \rho_m \leq 1$ and $|\bar{v}_m| \leq \varpi$ where $\underline{\rho}_m$ and ϖ are unknown constants.

Assumption 3.6 The desired trajectories y_{md}, and their nth-order time derivatives $y_{md}^{(i)}$ are continuous and bounded.

3.4.2 *Fuzzy Logic Systems*

Fuzzy logic systems are composed of four parts: the knowledge base, the fuzzifier, the fuzzy inference, and the defuzzifier [16]. The following IF-THEN rules are adopted:

Rule$^{(l)}$: If x_1 is A_1^l, x_2 is A_2^l, $\cdots$, x_k is A_k^l,

THEN: y is B^l, $l = 1, \cdots, L$,

where L is the number of fuzzy rules, $x = [x_1, \cdots, x_k]^{\mathrm{T}}$ and y are input and output variables of the fuzzy logic systems, respectively. Let $\mu_{A_i^l}(x_i)$ and $\mu_{B^l}(y)$ be the membership functions of the designed sets A_i^l and B^l, and the fuzzy logic systems can, thus, be expressed as,

$$y(x) = \frac{\sum_{l=1}^{L} \bar{y}_l \prod_{j=1}^{n} \mu_{A_j^l}(x_j)}{\sum_{l=1}^{L}(\prod_{j=1}^{n} \mu_{A_j^l}(x_j))}, \tag{3.95}$$

where $\bar{y}_l = \max \mu_{B^l}(y_l)$. We define the following basic function,

$$\varphi_l(x) = \frac{\prod_{j=1}^{n} \mu_{A_j^l}(x_j)}{\sum_{l=1}^{L}(\prod_{j=1}^{n} \mu_{A_j^l}(x_j))}. \tag{3.96}$$

Denote $\varphi(x) = [\varphi_1, \cdots, \varphi_L]^{\mathrm{T}}$ and $\theta = [\theta_1, \cdots, \theta_L]^{\mathrm{T}}$. Therefore, the fuzzy logic systems (3.95) can be rewritten as $y(x) = \theta^{\mathrm{T}}\varphi(x)$.

Lemma 3.2 ([16]) *Suppose $f(x)$ is a continuous function over a compact set Ω. Let ϵ be a positive constant, and there exist fuzzy logic systems such that,*

$$\sup_{x\in\Omega} |f(x) - \theta^T\varphi(x)| \le \epsilon. \tag{3.97}$$

3.4.3 Fuzzy State Observer

From Euler–Lagrange systems in (3.88), we will design a fuzzy state observer to estimate the unmeasurable states $x_{m,2}$ with the only measurable system output y_m. Upon Lemma 3.2, we first estimate the unknown nonlinearities $f_{m,2}(\bar{x})$ by using fuzzy logic system:

$$\hat{f}_{m,2}(\hat{x}_m|\hat{\vartheta}_{m,2}) = \hat{\vartheta}_{m,2}^T\varphi_{m,2}(\hat{x}_m), \tag{3.98}$$

where $\hat{x}_m = [\hat{x}_{m,1}, \hat{x}_{m,2}]^T$, and $\hat{\vartheta}_{m,2}$ denotes the estimation of $\vartheta_{m,2}$. Based on fuzzy logic systems, it can be derived:

$$\vartheta_{m,2}^* = \arg \min_{\vartheta_{m,2}\in U_m} \left[\sup_{x_m,\hat{x}_m\in\Omega_m} |f_{m,2}(x_m) - \hat{f}_{m,2}(\hat{x}_m|\hat{\vartheta}_{m,2})| \right], \tag{3.99}$$

where U_m and Ω_m are both compact sets. According to Lemma 3.2, we have

$$\epsilon_{m,2} = f_{m,2}(x_m) - \hat{f}_{m,2}(\hat{x}_m|\theta_{m,2}^*), \tag{3.100}$$

where $|\epsilon_{m,2}| \le \bar{\epsilon}_{m,2}$ with $\bar{\epsilon}_{m,2}$ being a positive constant.

To facilitate the design of fuzzy state observer, the Euler–Lagrange systems in (3.88) are rewritten as

$$\begin{cases} \dot{x}_m = A_m x_m + K_m y_m + B_{m,2} f_{m,2}(x_m) + B_{m,2} u_m \\ y_m = C_m x_m \end{cases} \tag{3.101}$$

where $A_m = \begin{bmatrix} -k_{m,1} & 1 \\ -k_{m,2} & 0 \end{bmatrix}$, $K_m = \begin{bmatrix} k_{m,1} \\ k_{m,2} \end{bmatrix}$, and $C_m = [1, 0]$, $B_{m,2} = [0, 1]^T$. By appropriately selecting K_m, we can ensure that A_m is a Hurwitz matrix, that is, for any positive definite symmetric matrix $Q_m = Q_m^T$, there exists a positive definite symmetric matrix $P_m = P_m^T$ such that

$$A_m^{\mathrm{T}} P_m + P_m A_m = -Q_m. \tag{3.102}$$

Therefore, the fuzzy state observer is constructed as

$$\begin{cases} \dot{\hat{x}}_m = A_m \hat{x}_m + K_m y_m + B_{m,2} \hat{f}_{m,2}(\hat{x}_m|\hat{\theta}_{m,2}) + B_{m,2} g(v_m(t)) \\ y_m = C_m x_m \end{cases} \tag{3.103}$$

where $\hat{x}_m$ and $\hat{\vartheta}_{m,j}$ denotes the estimation of x_m and $\vartheta_{m,j}^*$, respectively.

Define the estimation error $e_m = x_m - \hat{x}_m$, and from (3.101) and (3.103), we have

$$\dot{e}_m = A_m e_m + B_{m,2} \tilde{\vartheta}_{m,2}^{\mathrm{T}} \varphi_{m,2}(\hat{x}_m) + B_{m,2}(\bar{v}_m + \rho_m \mathrm{sat}(v_m) - g(v_m(t))) + \epsilon_m, \tag{3.104}$$

where $\epsilon_m = [\epsilon_{m,1}, \epsilon_{m,2}]^T$ satisfying $\|\epsilon_m\| \leq \bar{\epsilon}_m$; $\tilde{\vartheta}_{m,2} = \vartheta_{m,2}^* - \hat{\vartheta}_{m,2}$ is the estimation error. Consider the following Lyapunov function:

$$V_{m,0} = e_m^{\mathrm{T}} P_m e_m, \tag{3.105}$$

where P_m denotes a positive definite symmetric matrix. Based on (3.102) and (3.104), there is

$$\begin{aligned} \dot{V}_{m,0} =& e_m^{\mathrm{T}}(A_m^{\mathrm{T}} P_m + P_m A_m)e_m + 2e_m^{\mathrm{T}} P_m(\epsilon_m + B_{m,2}(\bar{v}_m + \rho_m \mathrm{sat}(v_m) \\ & - g(v_m)) + B_{m,2} \tilde{\vartheta}_{m,2}^{\mathrm{T}} \varphi_{m,2}). \end{aligned} \tag{3.106}$$

According to Young's inequality, it gives

$$2e_m^{\mathrm{T}} P_m B_{m,2} \tilde{\vartheta}_{m,2}^{\mathrm{T}} \varphi_{m,2}(\hat{x}_m) \leq 2\|e_m\|^2 + \|P_m\|^2 \sum_{j=1}^{2} \tilde{\vartheta}_{m,j}^{\mathrm{T}} \tilde{\vartheta}_{m,j} \tag{3.107}$$

$$2e_m^{\mathrm{T}} P_m(\epsilon_m + B_{m,2}\bar{v}_m) \leq 2\|e_m\|^2 + \|P_m\|^2 \bar{\epsilon}_m^2 + \|P_m\|^2 \varpi^2, \tag{3.108}$$

and

$$\begin{aligned} & 2e_m^{\mathrm{T}} P_m B_{m,2}(\rho_m \mathrm{sat}(v_m) - g(v_m)) \\ & \leq \|e_m\|^2 + \|P_m\|^2 \|\rho_m \mathrm{sat}(v_m) - \rho_m g(v_m) + \rho_m g(v_m) - g(v_m)\|^2 \\ & \leq \|e_m\|^2 + 2\|P_m\|^2 D^2 + 2\|P_m\|^2 (\underline{\rho}_m - 1)^2 v_{\max}^2. \end{aligned} \tag{3.109}$$

Substituting (3.107)–(3.109) into (3.106) yields

$$\dot{V}_{m,0} \leq -\lambda_{m,0}\|e_m\|^2 + \|P_m\|^2 \tilde{\vartheta}_{m,2}^{\mathrm{T}} \tilde{\theta}_{m,2} + d_{m,0}, \tag{3.110}$$

where $\lambda_{m,0} = \lambda_{\min}(Q) - 5$ and $d_{m,0} = \|P_m\|^2\bar{\epsilon}_m^2 + \|P_m\|^2\varpi^2 + 2\|P_m\|^2 D^2 + 2\|P_m\|^2(\underline{\rho}_m - 1)^2 v_{\max}^2$.

3.4.4 Constrained Finite-Time Output-Feedback Control

Define the tracking errors:

$$z_{m,1} = x_{m,1} - y_{md}, \tag{3.111}$$

$$z_{m,2} = \hat{x}_{m,2} - \alpha_{m,2}^f, \tag{3.112}$$

$$\omega_{m,2} = \alpha_{m,2}^f - \alpha_{m,1}, \tag{3.113}$$

where $z_{m,1}$ and $z_{m,2}$ are tracking errors, y_{md} denotes the reference trajectory, $\alpha_{m,j}$ is the virtual control input, $\alpha_{m,2}^f$ is an output of a first-order system, and $\omega_{m,2}$ describes the filter error. Referring to $e_{m,2} = x_{m,2} - \hat{x}_{m,2}$ and (3.111), the time derivative of $z_{m,1}$ is given by

$$\dot{z}_{m,1} = x_{m,2} + f_{m,1} - \dot{y}_{md} = e_{m,2} + z_{m,2} + \omega_{m,2} + \alpha_{m,1} + f_{m,1} - \dot{y}_{md}. \tag{3.114}$$

where the virtual control input $\alpha_{m,1}$, the actual control input v_m, and the adaptive law $\vartheta_{m,1}$ are designed as

$$\alpha_{m,1} = -c_{m,1}z_{m,1} - h_{m,1}z_{m,1}^{\beta} + \dot{y}_{md}, \tag{3.115}$$

$$v_m = -k_{m,2}e_{m,1} - c_{m,2}z_{m,2} - h_{m,2}z_{m,2}^{\beta} - z_{m,1} + \dot{\alpha}_{m,2}^f, \tag{3.116}$$

$$\dot{\hat{\vartheta}}_{m,2} = \gamma_{m,2}(z_{m,2}\varphi_{m,2}(\hat{x}_{m,2}) - \sigma_{m,2}\hat{\vartheta}_{m,2}), \tag{3.117}$$

where $c_{m,1}$, $h_{m,1}$, $k_{m,2}$, $c_{m,2}$, $h_{m,2}$, $\gamma_{m,2}$, $\sigma_{m,2}$, $0 < \beta < 1$ are all positive constants. Based on the above control design, we can derive the following theorem.

Theorem 3.3 *Consider the Euler–Lagrange systems (3.1) with actuator faults (3.89) and input saturation (3.90). If the fuzzy state observer (3.103), the constrained finite-time output-feedback control design (3.116) associated with the virtual control input (3.115), and an adaptive law (3.117), all signals in the closed-loop system are bounded, and the tracking signals $z_{m,1}$, $z_{m,2}$, $\tilde{\vartheta}_{m,1}$, and em converge to a small neighborhood around the origin within finite time.*

Proof : First, consider the following Lyapunov function:

$$V_{m,1} = \dot{V}_{m,0} + \frac{1}{2}z_{m,1}^2. \tag{3.118}$$

Along with (3.114) and (3.115), the time derivative of $V_{m,1}$ is given by

$$\begin{aligned}\dot{V}_{m,1} =& \dot{V}_{m,0} + z_{m,1}\dot{z}_{m,1} \\ =& \dot{V}_{m,0} + z_{m,1}(e_{m,2} + z_{m,2} + \omega_{m,2} + \alpha_{m,1} + f_{m,1} - \dot{y}_{md}) \\ \leq& \dot{V}_{m,0} + z_{m,1}(e_{m,2} + z_{m,2} + \omega_{m,2} - c_{m,1}z_{m,1} - h_{m,1}z_{m,1}^{\beta} + \epsilon_{m,1}). \end{aligned} \tag{3.119}$$

Based on Young's inequality, we have

$$z_{m,1}(e_{m,2} + \epsilon_{m,1}) \leq z_{m,1}^2 + \frac{1}{2}(e_{m,2}^2 + \epsilon_{m,1}^2) \leq z_{m,1}^2 + \frac{1}{2}(\|e_m\|^2 + \bar{\epsilon}_m^2), \tag{3.120}$$

Taking the above inequality into (3.119) yields

$$\begin{aligned}\dot{V}_{m,1} \leq & -\lambda_{m,0}\|e_m\|^2 + d_{m,0} + z_{m,1}(z_{m,2} + \omega_{m,2}) - c_{m,1}z_{m,1}^2 - h_{m,1}z_{m,1}^{\beta+1} \\ & + 2z_{m,1}^2 + \frac{1}{2}(\|e_m\|^2 + \bar{\epsilon}_m^2) \\ \leq & -\lambda_{m,1}\|e_m\|^2 - \bar{c}_{m,1}z_{m,1}^2 - h_{m,1}z_{m,1}^{\beta+1} + d_{m,1} + z_{m,1}(z_{m,2} + \omega_{m,2}), \end{aligned} \tag{3.121}$$

where $\lambda_{m,1} = \lambda_{m,0} - \frac{1}{2}$, $\bar{c}_{m,1} = c_{m,1} - 2$, and $d_{m,1} = d_{m,0} + \frac{1}{2}\bar{\epsilon}_m^2$. To avoid tedious calculations in backstepping design, let the virtual control input $\alpha_{m,1}$ pass through a first-order filter, then we have $\alpha_{m,2}^f$:

$$b_{m,2}\dot{\alpha}_{m,2}^f + \alpha_{m,2}^f = \alpha_{m,1}, \tag{3.122}$$

where $b_{m,2}$ is a positive constant. The initial condition is set as $\alpha_{m,2}^f(0) = \alpha_{m,1}(0)$, based on (3.113) and (3.122), we have

$$\dot{\omega}_{m,2} = \dot{\alpha}_{m,2}^f - \dot{\alpha}_{m,1} \leq -\frac{\omega_{m,2}}{b_{m,2}} + H_{m,2}, \tag{3.123}$$

where $H_{m,2} = \frac{\partial \alpha_{m,1}}{\partial x_{m,1}}\dot{x}_{m,1} + \frac{\partial \alpha_{m,1}}{\partial \hat{\theta}_{m,1}}\dot{\hat{\theta}}_{m,1} + \frac{\partial \alpha_{m,1}}{\partial y_{md}}\dot{y}_{md}$ is bounded. From the tracking error definition (3.112) and the first-order filter (3.122), we have

$$\dot{z}_{m,2} = \dot{\hat{x}}_{m,2} - \dot{\alpha}_{m,2}^f = k_{m,2}e_{m,1} + \hat{\theta}_{m,2}^{\mathrm{T}}\varphi_{m,2}(\hat{x}_m) + g(v_m) - \dot{\alpha}_{m,2}^f. \tag{3.124}$$

Consider the following Lyapunov function:

$$V_{m,2} = V_{m,1} + \frac{1}{2}z_{m,2}^2 + \frac{1}{2}\omega_{m,2}^2 + \frac{1}{2\gamma_{m,2}}\tilde{\vartheta}_{m,2}^{\mathrm{T}}\tilde{\vartheta}_{m,2}. \tag{3.125}$$

Its time derivative is derived as

$$\begin{aligned}\dot{V}_{m,2} =&\dot{V}_{m,1} + z_{m,2}\dot{z}_{m,2} + \omega_{m,2}\dot{\omega}_{m,2} - \frac{1}{\gamma_{m,2}}\tilde{\vartheta}_{m,2}^{\mathrm{T}}\dot{\hat{\vartheta}}_{m,2}\\ =&\dot{V}_{m,1} + z_{m,2}(\hat{\vartheta}_{m,2}^{\mathrm{T}}\varphi_{m,2}(\hat{x}_m) + g(v_m) + k_{m,2}e_{m,1} - \dot{\alpha}_{m,2}^{f})\\ &+ \omega_{m,2}\dot{\omega}_{m,2} - \frac{1}{\gamma_{m,2}}\tilde{\vartheta}_{m,2}^{\mathrm{T}}\dot{\hat{\vartheta}}_{m,2}.\end{aligned} \tag{3.126}$$

Based on the definition of the estimation error $\tilde{\vartheta}_{m,2} = \vartheta_{m,2}^{*} - \hat{\vartheta}_{m,2}$, the above $z_{m,2}\hat{\theta}_{m,2}^{\mathrm{T}}\varphi_{m,2}(\hat{x}_m)$ can be rewritten as

$$\begin{aligned}z_{m,2}\hat{\vartheta}_{m,2}^{\mathrm{T}}\varphi_{m,2}(\hat{x}_m) &= z_{m,2}\vartheta_{m,2}^{*\mathrm{T}}\varphi_{m,2}(\hat{x}_m) - z_{m,2}\tilde{\vartheta}_{m,2}^{\mathrm{T}}\varphi_{m,2}(\hat{x}_m)\\ &\le \frac{1}{2}z_{m,2}^2 + \frac{1}{2}\vartheta_{m,2}^{*\mathrm{T}}\vartheta_{m,2}^{*} - z_{m,2}\tilde{\vartheta}_{m,2}^{\mathrm{T}}\varphi_{m,2}(\hat{x}_m),\end{aligned} \tag{3.127}$$

Substituting the control input (3.116), the adaptive law (3.117), and (3.127) into $\dot{V}_{m,2}$ yields

$$\begin{aligned}\dot{V}_{m,2} \le&\dot{V}_{m,1} + \omega_{m,2}\dot{\omega}_{m,2} + \frac{1}{2}z_{m,2}^2 + z_{m,2}(-\dot{\alpha}_{m,2}^{f} + g_{v_\mu}(\dot{\alpha}_{m,2}^{f} - z_{m,1}\\ &- k_{m,2}e_{m,1} - c_{m,2}z_{m,2} - h_{m,2}z_{m,2}^{\beta}) + k_{m,2}e_{m,1}) - z_{m,2}\tilde{\vartheta}_{m,2}^{\mathrm{T}}\varphi_{m,2}(\hat{x}_m)\\ &+ \tilde{\vartheta}_{m,2}^{\mathrm{T}}(\sigma_{m,2}\hat{\vartheta}_{m,2} + z_{m,2}\varphi_{m,2}(\hat{x}_m)) + \frac{1}{2}\theta_{m,2}^{*\mathrm{T}}\theta_{m,2}^{*}\\ \le&\dot{V}_{m,1} - g_{v_\mu}(c_{m,2}z_{m,2}^2 + h_{m,2}z_{m,2}^{\beta+1} + z_{m,2}z_{m,1}) + z_{m,2}(1 - g_{v_\mu})(k_{m,2}e_{m,1}\\ &- \dot{\alpha}_{m,2}^{f}) + \frac{1}{2}z_{m,2}^2 + \frac{1}{2}\vartheta_{m,2}^{*\mathrm{T}}\vartheta_{m,2}^{*} + \omega_{m,2}\dot{\omega}_{m,2} + \sigma_{m,2}\tilde{\vartheta}_{m,2}^{\mathrm{T}}\hat{\vartheta}_{m,2}.\end{aligned} \tag{3.128}$$

From Young's inequality, we have

$$z_{m,2}e_{m,1} \le \frac{1}{2}(z_{m,2}^2 + \|e_m\|^2) \tag{3.129}$$

$$z_{m,2}z_{m,1} \le \frac{1}{2}(z_{m,2}^2 + z_{m,1}^2) \tag{3.130}$$

$$z_{m,2}\dot{\alpha}_{m,2}^{f} = -\frac{z_{m,2}\omega_{m,2}}{b_{m,2}} \ge -\frac{(z_{m,2}^2 + \omega_{m,2}^2)}{2b_{m,2}} \tag{3.131}$$

$$\sigma_{m,2}\tilde{\vartheta}_{m,2}^{\mathrm{T}}\hat{\vartheta}_{m,2} = \sigma_{m,2}\tilde{\vartheta}_{m,2}^{\mathrm{T}}(\vartheta_{m,2}^{*} - \tilde{\vartheta}_{m,2}) \le -\frac{\sigma_{m,2}}{2}\tilde{\vartheta}_{m,2}^{\mathrm{T}}\tilde{\vartheta}_{m,2} + \frac{\sigma_{m,2}}{2}\vartheta_{m,2}^{*\mathrm{T}}\vartheta_{m,2}^{*}. \tag{3.132}$$

Substituting (3.129)–(3.132) into (3.128), one gets

$$\begin{aligned}\dot{V}_{m,2} \le & -\lambda_{m,2}\|e_m\|^2 - \bar{c}_{m,1}z_{m,1}^2 + d_{m,2} - h_{m,1}z_{m,1}^{\beta+1} - \frac{\sigma_{m,2}-1}{2}\tilde{\vartheta}_{m,2}^{\mathrm{T}}\tilde{\vartheta}_{m,2} \\ & + \|P_m\|^2\tilde{\vartheta}_{m,2}^{\mathrm{T}}\tilde{\vartheta}_{m,2} + \omega_{m,2}(-\frac{\omega_{m,2}}{b_{m,2}} + H_{m,2}) + z_{m,1}\omega_{m,2} - \bar{c}_{m,2}z_{m,2}^2 \\ & - g_{v_\mu}h_{m,2}z_{m,2}^{\beta+1} + \frac{1-g_{v_\mu}}{2b_{m,2}}\omega_{m,2}^2 + \frac{1-g_{v_\mu}}{2}z_{m,1}^2, \end{aligned} \tag{3.133}$$

where $\bar{c}_{m,2} = g_{v_\mu}c_{m,2} - \frac{(1-g_{v_\mu})k_{m,2}}{2} - \frac{1-g_{v_\mu}}{2b_{m,2}} - \frac{1}{2} - \frac{1-g_{v_\mu}}{2}$, $\lambda_{m,2} = \lambda_{m,1} - \frac{(1-g_{v_\mu})k_{m,2}}{2}$, and $d_{m,2} = d_{m,1} + \frac{1+\sigma_{m,2}}{2}\vartheta_{m,2}^{*\mathrm{T}}\vartheta_{m,2}^{*}$. For $\omega_{m,2}H_{m,2}$ and $z_{m,1}\omega_{m,2}$, we get

$$\omega_{m,2}H_{m,2} \le \frac{1}{2}\omega_{m,2}^2 + \frac{1}{2}\bar{H}_{m,2}^2, \tag{3.134}$$

$$z_{m,1}\omega_{m,2} \le \frac{1}{2}z_{m,1}^2 + \frac{1}{2}\omega_{m,2}^2, . \tag{3.135}$$

Then, it gives

$$\begin{aligned}\dot{V}_{m,2} \le & -\lambda_{m,2}\|e_m\|^2 - \sum_{k=1}^{2} r_{m,k}z_{m,k}^2 - \sum_{k=1}^{2} l_{m,k}z_{m,k}^{\beta+1} \\ & - \bar{\sigma}_{m,2}\tilde{\vartheta}_{m,2}^{\mathrm{T}}\tilde{\vartheta}_{m,2} - \bar{b}_{m,2}\omega_{m,2}^2 + \bar{d}_{m,2}, \end{aligned} \tag{3.136}$$

where $r_{m,1} = \bar{c}_{m,1} - \frac{1}{2}$, $r_{m,2} = \bar{c}_{n,2}$, $l_{m,1} = h_{m,1}$, $l_{m,2} = g_{v_\mu}h_{m,2}$, $\bar{\sigma}_{m,2} = \frac{\sigma_{m,2}-1}{2} - \|P_m\|^2$, $\bar{b}_{m,2} = \frac{1}{b_{m,2}} - \frac{1-g_{v_\mu}}{2b_{m,2}} - 1$, and $\bar{d}_{m,2} = d_{m,2} + \frac{1}{2}\bar{H}_{m,2}$. Upon Lemma 2.4, one has

$$\begin{aligned}-\|e_m\|^2 &= -\sum_{k=1}^{2} e_{m,k}^2 - \sum_{k=1}^{2}(e_{m,k}^2)^{\frac{\beta+1}{2}} + \sum_{k=1}^{2}(e_{m,k}^2)^{\frac{\beta+1}{2}} \\ &\le -\frac{1-\beta}{2}\sum_{k=1}^{2} e_{m,k}^2 - \sum_{k=1}^{2}(e_{m,k}^2)^{\frac{\beta+1}{2}} + \frac{2(1-\beta)}{2}. \end{aligned} \tag{3.137}$$

For $\bar{\sigma}_{m,2}\tilde{\vartheta}_{m,2}^{\mathrm{T}}\tilde{\vartheta}_{m,2}$ and $\bar{b}_{m,2}\omega_{m,2}^2$, they follow the similar process in the above derivation. Therefore, we can obtain

$$\begin{aligned}\dot{V}_{m,2} \le & -\frac{\lambda_{m,2}(1-\beta)}{2}\sum_{k=1}^{2} e_{m,k}^2 - \sum_{k=1}^{2} r_{m,k}z_{m,k}^2 - \frac{1-\beta}{2}(\bar{\sigma}_{m,2}\tilde{\vartheta}_{m,2}^{\mathrm{T}}\tilde{\vartheta}_{m,2} - \bar{b}_{m,2}\omega_{m,2}^2) \\ & - \lambda_{m,2}\sum_{k=1}^{2}(e_{m,k}^2)^{\frac{1+\beta}{2}} - \sum_{k=1}^{2} l_{m,k}(z_{m,k}^2)^{\frac{1+\beta}{2}} - (\bar{\sigma}_{m,2}\tilde{\vartheta}_{m,k}^{\mathrm{T}}\tilde{\vartheta}_{m,2})^{\frac{1+\beta}{2}} \\ & - (\bar{b}_{m,2}\omega_{m,2}^2)^{\frac{1+\beta}{2}} + \bar{d}_{m,2} + 2(1-\beta) + \frac{2\lambda_{m,2}(1-\beta)}{2}. \end{aligned} \tag{3.138}$$

With Lemma 2.4, the above inequality can be rewritten as

$$
\begin{aligned}
\dot{V}_{m,2} \leq & - c_1\left(\frac{1}{2}\|e_m\|^2 + \frac{1}{2}\sum_{k=1}^{2} z_{m,k}^2 + \frac{1}{2}\omega_{m,2}^2 + \frac{1}{2\gamma_{m,2}}\tilde{\vartheta}_{m,2}^{\mathrm{T}}\tilde{\vartheta}_{m,2}\right) - c_2\left(\frac{1}{2}\|e_m\|^2\right. \\
& \left. + \frac{1}{2\gamma_{m,2}}\tilde{\vartheta}_{m,2}^{\mathrm{T}}\tilde{\vartheta}_{m,2} + \frac{1}{2}\sum_{k=1}^{2} z_{m,k}^2 + \frac{1}{2}\omega_{m,2}^2\right)^{\frac{1+\beta}{2}} + \varepsilon_1 \\
\leq & - c_1 V_{m,2} - c_2 V_{m,2}^{\varrho} + \varepsilon_1,
\end{aligned}
\tag{3.139}
$$

where $c_1 = \min(\lambda_{m,2}(1-\beta), 2r_{m,k}, \gamma_{m,2}\bar{\sigma}_{m,k}(1-\beta), \bar{b}_{m,2}(1-\beta))$, $c_2 = \min(\lambda_{m,2}2^{\varrho}, l_{m,k}2^{\varrho}, 2^{\varrho}\bar{\sigma}_{m,k}^{\varrho}\gamma_{m,2}^{\varrho}, 2^{\varrho}\bar{b}_{m,2}^{\varrho})$, and $\varrho = \frac{1+\beta}{2}$, $\varepsilon_1 = \bar{d}_{m,2} + 2(1-\beta) + \frac{2\lambda_{m,2}(1-\beta)}{2}$. Based on Lemma 2.1, we can conclude that all signals in the closed-loop system are bounded, and the tracking signals $z_{m,1}$, $z_{m,2}$, $\tilde{\vartheta}_{m,1}$, and e_m converge to a small neighborhood around the origin within finite time. This completes the proof. □

3.4.5 Numerical Simulations

In this section, we conduct simulation on the Euler–Lagrange systems developed in (3.82) with parameter settings in Table 3.1. As discussed in this section, with considering input saturation, measurement constraint and actuator faults, the proposed constrained finite-time output feedback control for the Euler–Lagrange systems is given by

$$
\begin{aligned}
\alpha_{m,1} &= -c_{m,1}z_{m,1} - h_{m,1}z_{m,1}^{\beta} + \dot{y}_{md}, \\
v_m &= -k_{m,2}e_{m,1} - c_{m,2}z_{m,2} - h_{m,2}z_{m,2}^{\beta} - z_{m,1} + \dot{\alpha}_{m,2}^{f}, \\
\dot{\hat{\vartheta}}_{m,2} &= \gamma_{m,2}(z_{m,2}\varphi_{m,2}(\hat{x}_{m,2}) - \sigma_{m,2}\hat{\vartheta}_{m,2}),
\end{aligned}
$$

where the control parameters are set in Table 3.4. Based on (3.103), we establish the fuzzy state observer:

Table 3.4 Parameters for simulations

Parameter	Value	Parameter	Value	Parameter	Value
$c_{1,1}, c_{2,1}$	2.0	$c_{1,2}, c_{2,2}$	20.0	$h_{1,1}, h_{2,1}$	0.1
$\gamma_{1,2}, \gamma_{2,2}$	1.0	$\sigma_{1,2}, \sigma_{2,2}$	1.0	$b_{1,2}, b_{2,2}$	0.1
β	0.8	$k_{1,1}, k_{2,1}$	1.0	$k_{1,2}, k_{2,2}$	250.0
$\tau_{\max}$	[24.0; 27.0]	$\tau_{\min}$	[−24.0; −27.0]	ρ_1	$1 - 0.4\lvert\sin 0.3t\rvert$
ρ_2	0.5	ρ_3	$1 - 0.2\lvert\cos 0.45t\rvert$	$\bar{v}_1$	$0.3\sin t$
$\bar{v}_2$	$0.2\sin 0.3t$	$\bar{v}_3$	0.2		

$$\begin{cases} \dot{\hat{x}}_m = A_m \hat{x}_m + K_m y_m + B_{m,2} \hat{f}_{m,2}(\hat{x}_m | \hat{\theta}_{m,2}) + B_{m,2} g(v_m(t)) \\ y_m = C_m x_m \end{cases}$$

where its parameters are given in the following table. To better demonstrate the effectiveness of our proposed method, comparative simulations are conducted with the finite-time output feedback control (FTOFC) and the dynamic surface control (DSC) method in [11]. Upon the control design in [11], DSC follows:

$$\alpha_{m,1} = -c_{m,1} z_{m,1} - h_{m,1} z_{m,1}^{\beta} + \dot{y}_{md}, \tag{3.140}$$

$$v_m = -k_{m,2} e_{m,1} - h_{m,2} z_{m,2}^{\beta} - z_{m,1} + \dot{\alpha}_{m,2}^{f}, \tag{3.141}$$

$$\dot{\hat{\vartheta}}_{m,2} = \gamma_{m,2}(z_{m,2} \varphi_{m,2}(\hat{x}_{m,2}) - \sigma_{m,2} \hat{\vartheta}_{m,2}), \tag{3.142}$$

where $c_{m,1}$, $h_{m,1}$, $k_{m,2}$, $h_{m,2}$, $\gamma_{m,2}$, $\sigma_{m,2}$ are given in Table 3.4. v_m is subject to input saturation constraint with $u_m = \text{sat}(v_m)$. To estimate the unmeasurable states, the fuzzy state observer used in DSC is:

$$\begin{cases} \dot{\hat{x}}_m = A_m \hat{x}_m + K_m y_m + B_{m,2} \hat{f}_{m,2}(\hat{x}_m | \hat{\theta}_{m,2}) + B_{m,2} v_m(t) \\ y_m = C_m x_m \end{cases} \tag{3.143}$$

where the positive definite symmetric matrix K_m is consistent with the one in the FTOFC. The reference trajectory is set as $\xi_d = [0.8 \cos 0.8t, 0.5 \cos(0.6t - 0.5)]^T$, the initial position is $\xi(0) = [0, 0]^T$, and the initial condition of the adaptive law is $\hat{\vartheta}(0) = [0.2, 0.2, 0.2]^T$. The simulation results are shown in Figs. 3.19, 3.20, 3.21, 3.22, 3.23, 3.24, 3.25 and 3.30.

The angular position tracking results of the Euler–Lagrange system are shown in Figs. 3.19 and 3.20. Under FTOFC and DSC, both methods are able to track the desired trajectory within finite time. The angular tracking errors z_{11} and z_{12} are depicted in Figs. 3.21 and 3.22. From simulation results, both FTOFC and DSC can achieve the convergence of the system tracking error to the neighborhood of the origin within finite time, but the tracking performance of FTOFC is significantly better than that of DSC. Due to the input saturation constraint in the initial stage of tracking, the system control input is limited to a given reasonable range as shown in Figs. 3.23 and 3.24. Therefore, in the initial process of tracking, there exists a large estimation error of the system state by the fuzzy state observer, and the system angular motion trajectories exhibit significant overshoot. Compared to DSC, FTOFC fully considers the influence of input saturation constraints in the control design, and its tracking trajectory can achieve faster convergence

The joint angular velocity tracking trajectories are shown in Figs. 3.25 and 3.26. Both control methods can converge to the desired angular velocity within finite time. However, due to the impact of uncertainties and unknown actuator faults, the angular velocity tracking trajectory ξ_2 of DSC exhibits a noticeable deviation. In comparison to DSC, our FTOFC demonstrates fast transient response, quick convergence speed, and high tracking accuracy.

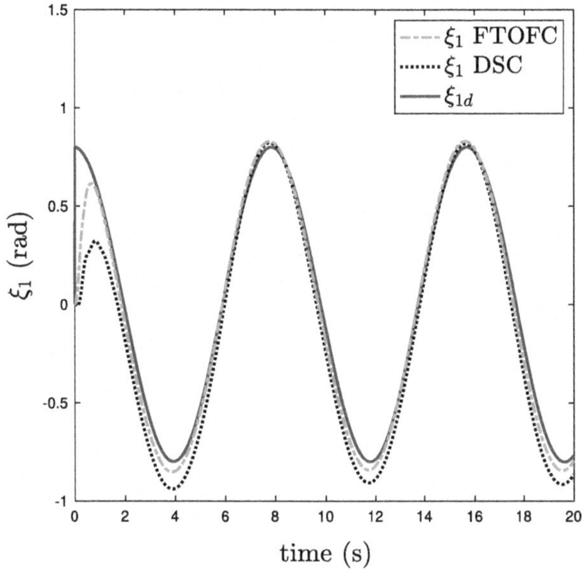

Fig. 3.19 The tracking performance of ξ_1

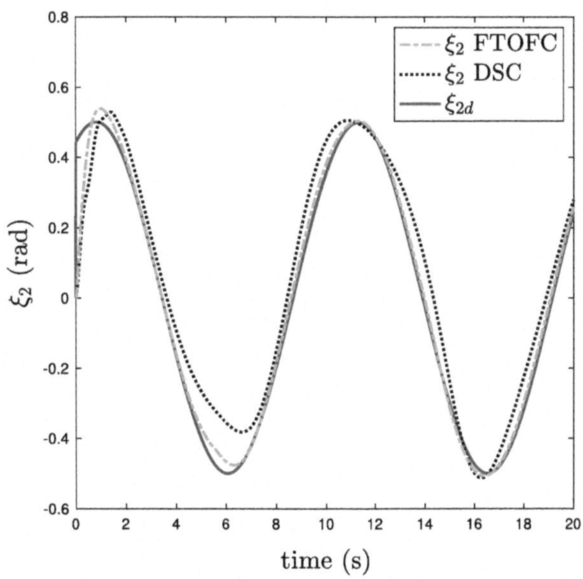

Fig. 3.20 The tracking performance of ξ_2

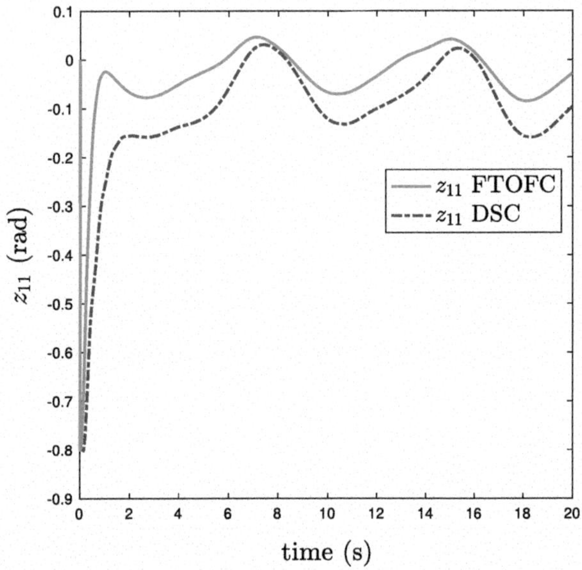

Fig. 3.21 The tracking performance of z_{11}

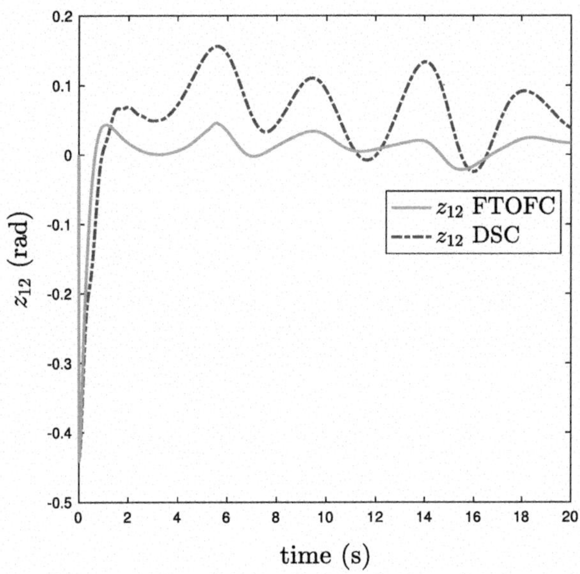

Fig. 3.22 The tracking performance of z_{12}

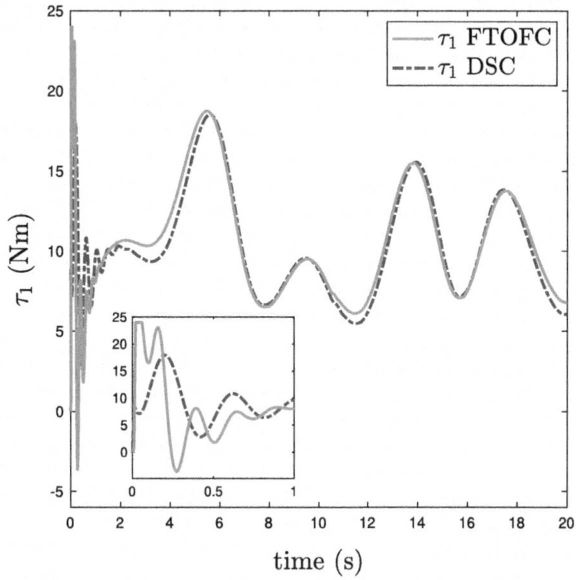

Fig. 3.23 The tracking performance of τ_1

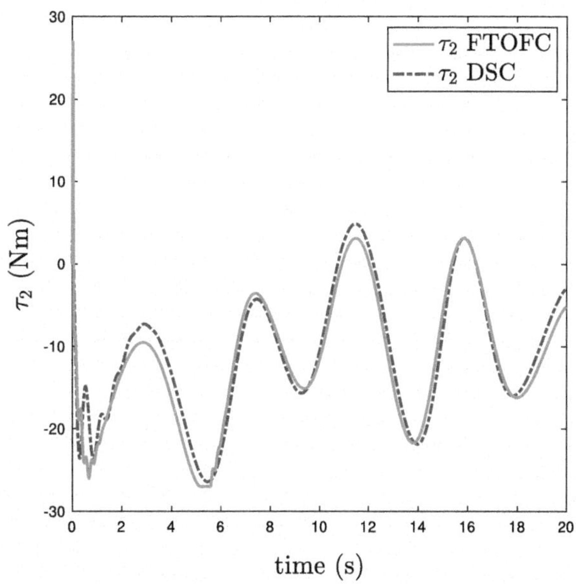

Fig. 3.24 The tracking performance of τ_2

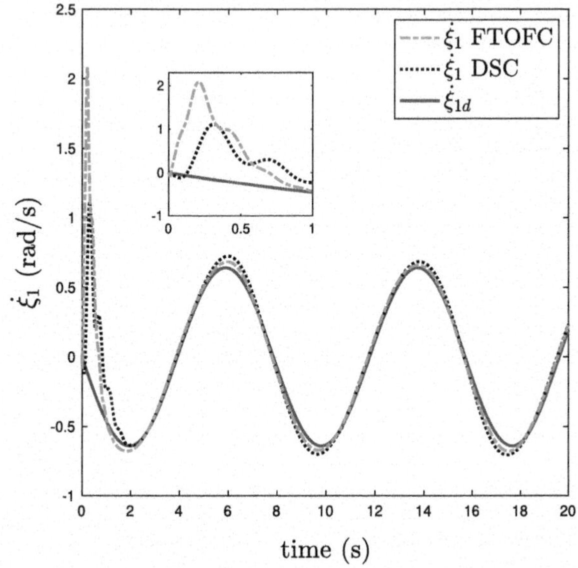

Fig. 3.25 The tracking performance of $\dot{\xi}_1$

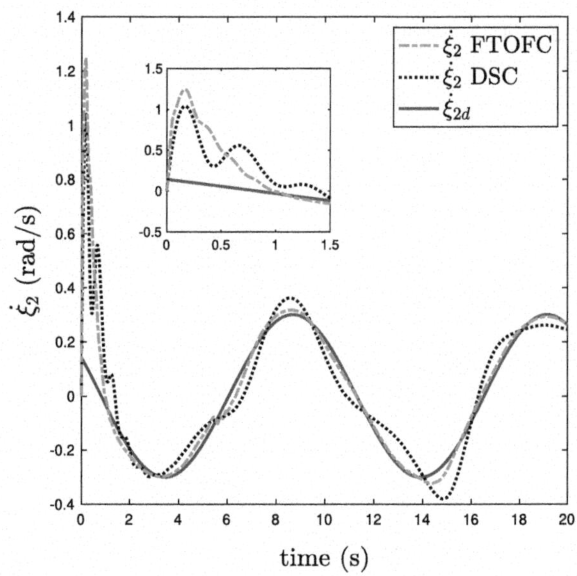

Fig. 3.26 The tracking performance of $\dot{\xi}_2$

As shown in Figs. 3.23 and 3.24, FTOFC are suffering from input saturation constraints during the $5-6$s. Although the angular velocity tracking signal $\dot{\xi}_2$ exhibits tracking deviation, with the utilization of fuzzy state observer, FTOFC quickly adjusts and converges to the desired angular velocity signal, indicating strong robustness of our proposed method. The state variables of the fuzzy state observer are shown in Figs. 3.27, 3.28, 3.29 and 3.30. Both DSC and FTOFC are capable to approximate the unmeasurable system states, with the estimation errors regulated to a neighborhood around the origin within finite time. However, as FTOFC takes the effects of input saturation constraints into account in its design, it exhibits higher estimation accuracy and faster convergence even under input saturation constraints as shown in Figs. 3.27, 3.28, 3.29 and 3.30. Specifically, during the $5-6s$ period, when the system's control input is constrained, the system state ξ_2 exhibits noticeable chattering

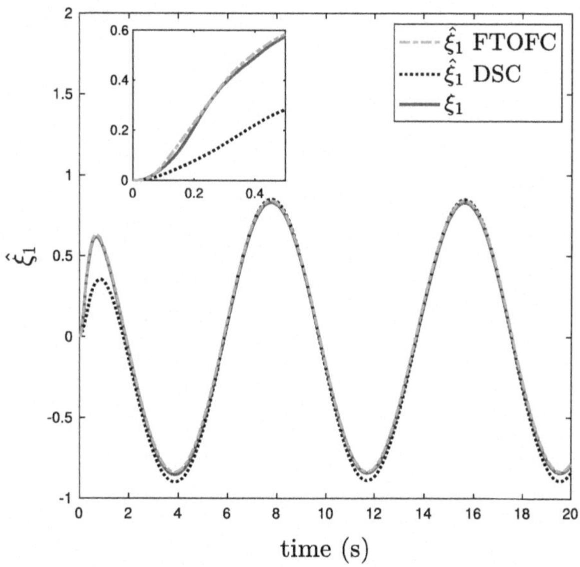

Fig. 3.27 The tracking performance of $\hat{\xi}_1$

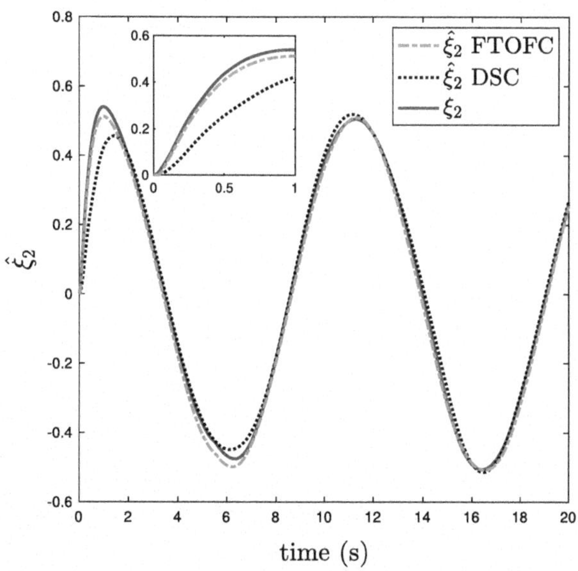

Fig. 3.28 The tracking performance of $\hat{\xi}_1$

problem. While the fuzzy state observer of FTOFC also shows slight oscillations, it enables $\dot{\hat{\xi}}_2$ to rapidly converge to the real value, indicating superior estimation performance over DSC.

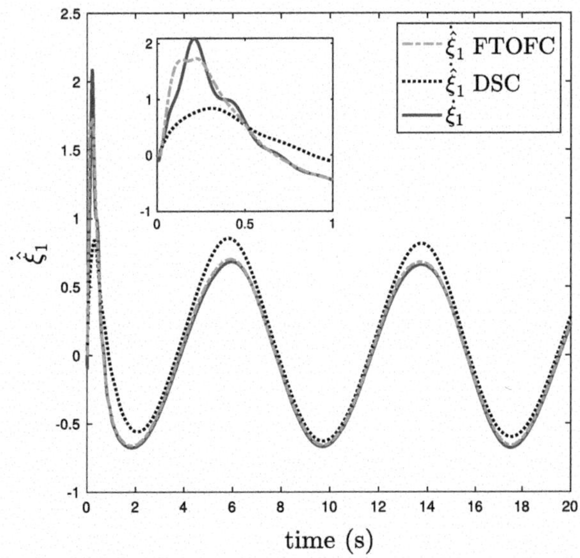

Fig. 3.29 The tracking performance of $\dot{\hat{\xi}}_1$

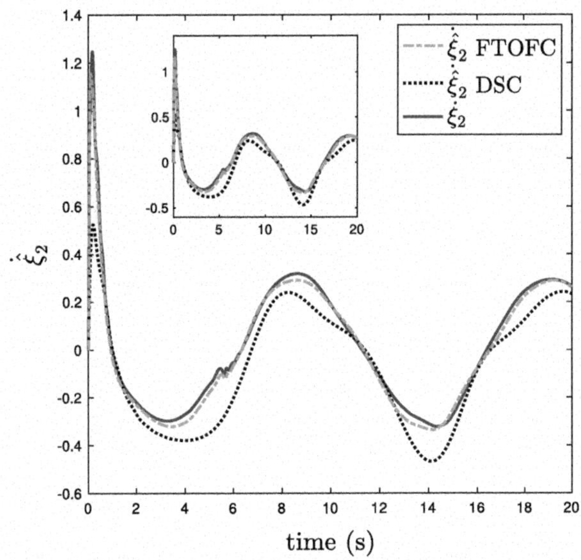

Fig. 3.30 The tracking performance of $\dot{\hat{\xi}}_2$

3.5 Conclusions

In this chapter, we have investigated constrained control designs for Euler–Lagrange systems under input saturation, actuator faults, and measurement constraints. First, based on the input saturation error signal, a virtual auxiliary system is constructed to compensate for saturation nonlinearity and a backstepping-based constrained finite-time control is designed. Second, a novel saturation constraint mechanism is designed for the derivative of the control input, and a new virtual auxiliary system is therefore established based on the input derivative error signal. Finally, fuzzy logic systems are used to approximate the unknown nonlinear functions in systems, and a fuzzy state observer is established to estimate the unmeasurable system states. With the aid of the fuzzy state observer, a constrained finite-time output feedback control is proposed, along with its stability analysis. The effectiveness of all constrained nonlinear control designs has been demonstrated by simulation results.

References

1. Chen, B., Liu, X., Liu, K., Shi, P., Lin, C.: Direct adaptive fuzzy control for nonlinear systems with time-varying delays. Information Sciences **180**(5), 776–792 (2010)
2. Chen, B., Liu, X.P., Ge, S.S., Lin, C.: Adaptive fuzzy control of a class of nonlinear systems by fuzzy approximation approach. IEEE Transactions on Fuzzy Systems **20**(6), 1012–1021 (2012)
3. Chen, Z., Li, Z., Chen, C.P.: Adaptive neural control of uncertain mimo nonlinear systems with state and input constraints. IEEE Trans. Neural Netw. Learning Syst. **28**(6), 1318–1330 (2016)
4. Ding, S., Li, S., Zheng, W.X.: Nonsmooth stabilization of a class of nonlinear cascaded systems. Automatica **48**(10), 2597–2606 (2012)
5. He, W., Huang, H., Ge, S.S.: Adaptive neural network control of a robotic manipulator with time-varying output constraints. IEEE Trans. Cybern. **47**(10), 3136–3147 (2017)
6. Ji, R., Li, D., Ma, J.: Adaptive second-order fast nonsingular terminal sliding mode control for a tilting quadcopter. In: 2022 13th Asian Control Conference (ASCC), pp. 403–408. IEEE (2022)
7. Ji, R., Ma, J., Ge, S.S., Ji, R.: Adaptive second-order sliding mode control for a tilting quadcopter with input saturations. IFAC-PapersOnLine **53**(2), 3910–3915 (2020)
8. Li, D., Ma, G., Li, C., He, W., Mei, J., Ge, S.S.: Distributed attitude coordinated control of multiple spacecraft with attitude constraints. IEEE Transactions on Aerospace and Electronic Systems **54**(5), 2233–2245 (2018)
9. Li, D., Ma, G., Xu, Y., He, W., Ge, S.S.: Layered affine formation control of networked uncertain systems: A fully distributed approach over directed graphs. IEEE transactions on cybernetics **51**(12), 6119–6130 (2020)
10. Li, D., Zhang, W., He, W., Li, C., Ge, S.S.: Two-layer distributed formation-containment control of multiple euler–lagrange systems by output feedback. IEEE Transactions on Cybernetics **49**(2), 675–687 (2018)
11. Li, Y., Li, K., Tong, S.: Finite-time adaptive fuzzy output feedback dynamic surface control for mimo nonstrict feedback systems. IEEE Transactions on Fuzzy Systems **27**(1), 96–110 (2018)
12. Li, Y., Tong, S.: Adaptive fuzzy output-feedback stabilization control for a class of switched nonstrict-feedback nonlinear systems. IEEE Trans. Cybern. **47**(4), 1007–1016 (2016)
13. Longuski, J.M., Guzmán, J.J., Prussing, J.E.: Optimal control with aerospace applications. Springer (2014)

14. Neila, M.B.R., Tarak, D.: Adaptive terminal sliding mode control for rigid robotic manipulators. International J. Automation Computing **8**(2), 215–220 (2011)
15. Tong, S., Li, Y., Shi, P.: Observer-based adaptive fuzzy backstepping output feedback control of uncertain mimo pure-feedback nonlinear systems. IEEE Trans. Fuzzy Syst. **20**(4), 771–785 (2012)
16. Tong, S., Li, Y., Sui, S.: Adaptive fuzzy tracking control design for siso uncertain nonstrict feedback nonlinear systems. IEEE Transactions on Fuzzy System **24**(6), 1441–1454 (2016)
17. Tong, S., Wang, T., Li, Y., Chen, B.: A combined backstepping and stochastic small-gain approach to robust adaptive fuzzy output feedback control. IEEE Trans. Fuzzy Syst. **21**(2), 314–327 (2012)

Chapter 4
Constrained Control of Nonlinear System with Input Saturation and Tunnel Prescribed Performance Constraints

4.1 Introduction

This chapter addresses the tracking control problem for a class of nonlinear systems with simultaneously considering input saturation and prescribed performance constraints. We formally propose a novel Tunnel Prescribed Performance with brief expressions, which not only simplifies the design process, but also achieves smaller overshoot performance owing to the tight feasible region established by the parallel performance. Based on the proposed Tunnel Prescribed Performance, we further develop a unified yet concise Finite-time Tunnel Prescribed Performance, which can adaptive adjusted its orientation based on different initial conditions. Compared with the classical funnel-shape prescribed performance functions, our tunnel-shape Finite-time Tunnel Prescribed Performance can not only provide tight allowable set leading to better transient performance, but also user-specified settling time in advance instead of relying on the initial conditions as depicted in finite-time control. Therefore, two control strategies, namely, Tunnel Prescribed Control and Finite-time Tunnel Prescribed Control, are proposed for nonlinear systems such that the tracking errors can evolve within the tunnel prescribed performance even in the presence of uncertainties, such as actuator faults, time delay, and input saturation. Both control strategies can achieve robust tracking performance against uncertainties and less conservatism control design. The effectiveness of the proposed control strategies is verified by numerical simulations.

R. Ji et al., *Constrained Nonlinear Control: Performance and Robustness*,
https://doi.org/10.1007/978-981-96-2971-8_4

4.2 Tunnel Prescribed Control of Nonstrict Feedback Nonlinear Systems with Actuator Faults

4.2.1 Problem Formulation

Consider the following nonstrict feedback nonlinear system:

$$\begin{cases} \dot{x}_i = f_i(x) + x_{i+1}, \quad i = 1, \ldots, n-1 \\ \dot{x}_n = f_n(x) + u(t) \\ y = x_1 \end{cases} \tag{4.1}$$

where x_i denotes the system state, y is the only measurable state, and $x = [x_1, \ldots, x_n]^T$ are system state vector, $f_i(\cdot)$ represents an unknown nonlinear smooth function, and $u(t)$ denotes the control input with considering input saturation and actuator faults, which is given as

$$u(t) = g(t, t_0)h(v(t)) + h_s(t, t_1), \tag{4.2}$$

where $g(t, t_0)$ and $h_s(t, t_1)$ represent the actuator effectiveness and additive actuation faults where t_0 and t_1 being the instants of their occurrences. For $h(v)$, it follows

$$h(v) = \begin{cases} v_{\max}, \ v > v_{\max} \\ v(t), \ v_{\min} \le v(t) \le v_{\max} \\ v_{\min}, \ v(t) < v_{\min} \end{cases} \tag{4.3}$$

where $v_{\max}$ and $v_{\min}$ are known constants, and $v(t)$ denotes the actual control input to be designed later. For the actuator faults, we have the following assumption.

Assumption 4.1 The actuator effectiveness $g(t, t_0)$ and the additive actuation faults $h_s(t, t_1)$ are bounded: $\underline{\rho} \le g(t, t_0) \le 1$ and $|h_s(t, t_1)| \le \bar{h}_s$ with $\underline{\rho}$ and $\bar{h}_s$ being positive constants.

The control objective of this section is to design a control scheme for the nonlinear systems described in (4.1) such that (a) Tunnel prescribed performance is first proposed in a concise form to simplify the control design, by which smaller overshoot performance can be achieved. (b) Tunnel prescribed control is then proposed to not only guarantee finite-time stability of the closed-loop system, but also ensure the tracking errors evolving within the tunnel prescribed performance even under actuator faults and input saturation.

4.2.2 Tunnel Prescribed Performance

Before we present our tunnel prescribed performance, we review the conventional prescribed performance functions (PPFs). Define the tracking error $e(t) = y - y_d$ with y_d being the desired trajectory. Therefore, the tracking error $e(t)$ is constrained by PPFs as

$$\begin{cases} -\delta\varrho(t) < e(t) < \varrho(t), \ e(0) \geq 0 \\ -\varrho(t) < e(t) < \delta\varrho(t), \ e(0) < 0 \end{cases} \tag{4.4}$$

where $0 \leq \delta \leq 1$ and $\varrho(t) = (\varrho_0 - \varrho_\infty)e^{-\kappa t} + \varrho_\infty$ with $\varrho_0 > \varrho_\infty > 0$ and $\kappa > 0$. Based on the approaches in [1, 11, 19, 23, 26, 29, 31], the traditional prescribed performance control can indeed guarantee the tracking errors to satisfy the user-specified performance constraints (4.4). However, it is worth noting that PPFs are separately defined according to initial tracking errors. Namely, the control design and stability analysis should be re-developed due to different initial conditions. This mechanism imposes undesirable complexities in the control design. Supposing double integrators system $\ddot{x} = u$, the boundaries constraints of PPFs are distributed on both sides of the origin and gradually converge to the residual sets as shown in Fig. 4.1a and c. It is easy to find that the initial shape of PPFs is similar to a "horn", which cannot have a positive constraint on the transient performance and obvious overshoot performance can be seen.

Motivated by the above constraints of the traditional PPFs, we, in this chapter, propose tunnel prescribed performance (TPP) to handle these problems:

$$-e_l(t) < e(t) < e_u(t), \tag{4.5}$$

with

$$\begin{cases} e_l(t) = (\delta_l - \text{sign}(e(0)))\varrho(t) + \varrho_\infty \text{sign}(e(0)) \\ e_u(t) = (\delta_u + \text{sign}(e(0)))\varrho(t) - \varrho_\infty \text{sign}(e(0)) \end{cases} \tag{4.6}$$

where $0 \leq \delta_l \leq 1$ and $0 \leq \delta_u \leq 1$. According to (4.5) and (4.6), TPP is defined with a concise form, where performance boundaries are with only one group expression. Thus, TPP can dramatically reduce complexities of the control design and stability analysis. And the performance boundaries of TPP are described in Fig. 4.1b and d. Compared with the traditional PPFs, there is a significant shape change, where TPP is exactly similar to a"tunnel" with more tight space. The boundaries of TPP are distributed on the same side and converge to the final residual sets, which can make TPP have an ability of achieving the transient performance with smaller overshoot by appropriate selecting TPP.

To better show TPP design mechanism and extend it to a more general concept, the performance functions of e_u and e_l in (4.6) are exhibited in Fig. 4.2, where the corresponding areas of functions are shaded. Here, we give detail analysis of $e(0) > 0$

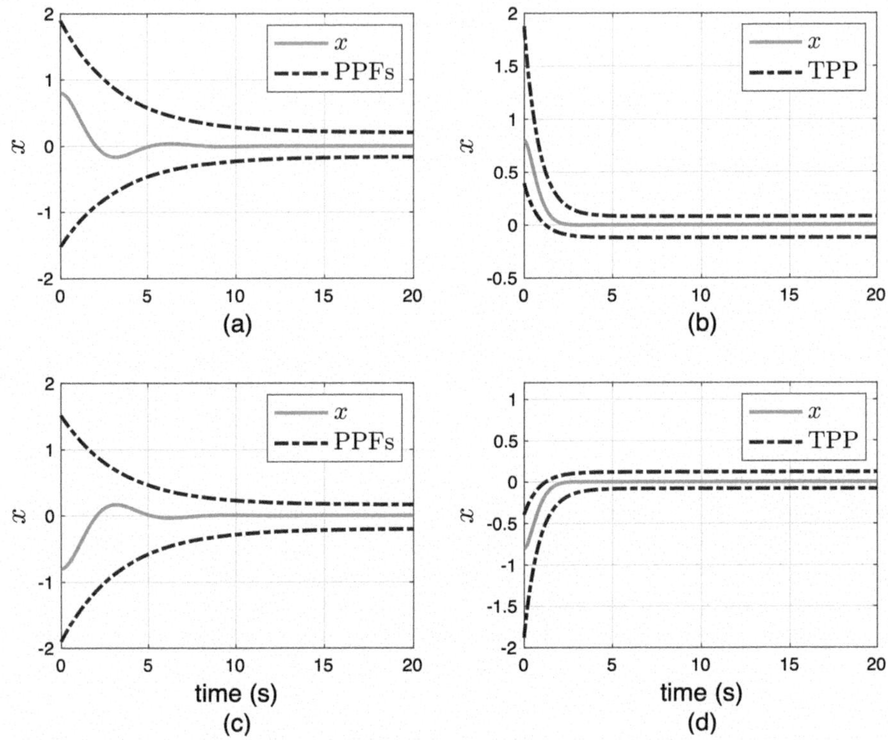

Fig. 4.1 The performance of PPFs and TPP

conditions and $e(0) < 0$ follows the same process. First, according to (4.6) under $e(0) > 0$, the trajectories of $\varrho(t)$, e_l and e_u are described in Fig. 4.2a. And the areas less than e_l and e_u are shaded by black and gray respectively. Then according to (4.5), the area of $-e_l(t) < e(t) < e_u(t)$ is exhibited in Fig. 4.2b. The boundary constraints are distributed at the same side of the origin with more tight space, and gradually converge to the prescribed set. The corresponding designed area is shaped like a "Tunnel". Form a more general perspective, the objective of TPP is to design prescribed performance similar to "Tunnel", but not limit to (4.6). We can develop more modified TPP, such as finite-time convergence, small overshoot or even no overshoot, fast convergence rate, etc. Thus, we extend our TPP design to a more general concept for our future research. Compared with the traditional PPFs designed in [1, 7, 9–11, 19, 23, 24, 26, 29, 31, 36, 37], we propose TPP with concise form, which can not only simplify the control design, but also small overshoot performance can be achieved in a positive manner. We further extend TPP design mechanism to a more general concept, which can pave the way for more expressions of TPP.

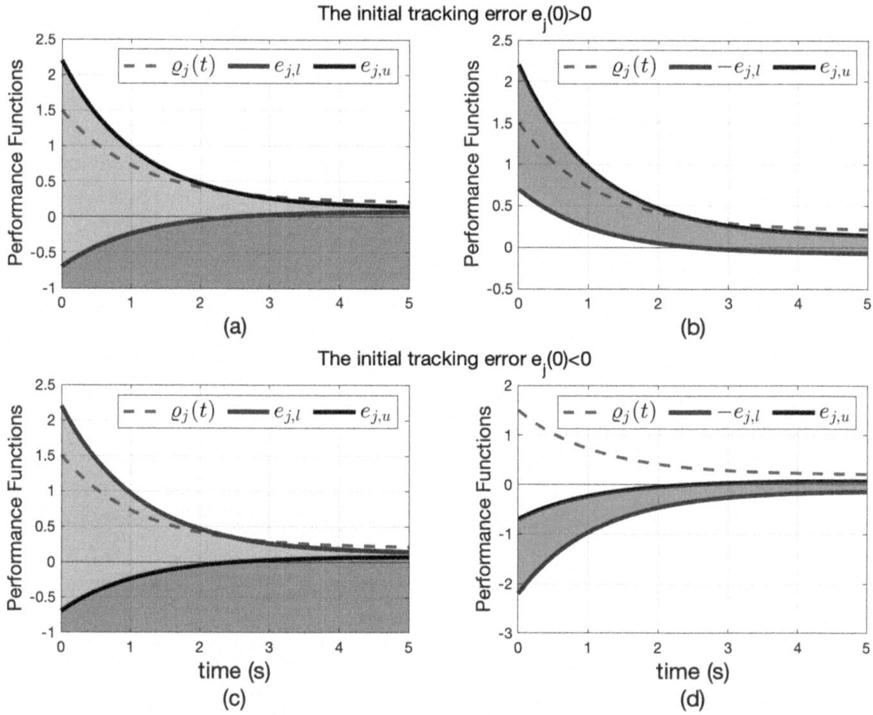

Fig. 4.2 The performance functions of e_u and e_l

4.2.3 Fuzzy State Observer

The only measurable state of the nonlinear system (4.1) is the output y. Utilizing fuzzy logic systems, a fuzzy state observer is expected to be constructed to estimate the system state $x_i,\ i = 2, \ldots, n$ for the control design. For the unknown function $f_i(\cdot)$, we have

$$\hat{f}_i(\hat{x}|\hat{\theta}_i) = \hat{\theta}_i^{\mathrm{T}} \varphi_i(\hat{x}), \tag{4.7}$$

where $\hat{x}$ and $\hat{\theta}_i$ are the estimated vectors of x and θ_i. Then the nonlinear system (4.1) can be rewritten as

$$\begin{cases} \dot{x} = Ax + Ky + F(x) + B_n u \\ y = Cx \end{cases} \tag{4.8}$$

where $F(x) = \sum_{k=1}^{n} B_k f_k(x)$, $B_k = [0, \ldots, 1, \ldots 0]^{\mathrm{T}}$, $B_n = [0, \ldots, 1]^{\mathrm{T}}$, and $C = [1, \ldots, 0]$. A and K are with

$$A = \begin{bmatrix} -k_1 & & \\ \vdots & I_{n-1} & \\ -k_n & 0 & \dots & 0 \end{bmatrix}, \quad K = \begin{bmatrix} k_1 \\ \vdots \\ k_n \end{bmatrix}, \tag{4.9}$$

where $k_i > 0, i = 1, \dots, n$. And the state observer, hereby, can be constructed as

$$\begin{cases} \dot{\hat{x}} = A\hat{x} + Ky + \hat{F}(\hat{x}) + B_n h(v) \\ y = Cx \end{cases} \tag{4.10}$$

where $\hat{F}(\hat{x}) = \sum_{k=1}^{n} B_k \hat{f}_k(\hat{x})$. For the sake of simplicity, the variables are left out in the brackets in the following design process. Combining (4.8) and (4.10), we can obtain

$$\dot{e}_0 = Ae_0 + F - \hat{F} + B_n(u - h(v)), \tag{4.11}$$

where $e_0 = x - \hat{x}$ and $\hat{x} = [\hat{x}_1, \dots, \hat{x}_n]^T$ with $e_0 = [e_0^1, \dots, e_0^n]^{\mathrm{T}}$. We adopt the candidate Lyapunov function as $V_0 = e_0^{\mathrm{T}} P e_0$ with P being the positive definite matrix. Upon Lemma 3.2 and substituting (4.2), (4.7) and (4.11) into the time derivative of V_0 yields

$$\dot{V}_0 = 2e_0^{\mathrm{T}} P[\sum_{k=1}^{n} B_k(\varepsilon_k + \tilde{\theta}_k \varphi_k(\hat{x})) + B_n(h_s - h + gh)] + e_0^{\mathrm{T}}(A^{\mathrm{T}} P + PA)e_0, \tag{4.12}$$

where $\tilde{\theta}_k = \theta_k^* - \hat{\theta}_k$ with θ_k^* being the optimal parameter. From the Young's inequality, we have

$$2e_0^{\mathrm{T}} P B_n(h_s - (1-g)h) \le 2\|e_0\|^2 + \|P\|^2(\bar{h}_s^2 + (1-g)^2 h^2) \tag{4.13}$$

$$2e_0^{\mathrm{T}} P \sum_{k=1}^{n} B_k(\tilde{\theta}_k \varphi_k(\hat{x}) + \varepsilon_k) \le 2n\|e_0\|^2 + \|P\|^2 \sum_{k=1}^{n} (\tilde{\theta}_k^{\mathrm{T}} \tilde{\theta}_k + \varepsilon_k^2). \tag{4.14}$$

And we can design the parameter in A to make it be a strict Hurwitz matrix. Namely, there exists a positive definite matrix Q such that $A^{\mathrm{T}} P + PA = -Q$. Taking (4.13) and (4.14) into (4.12), we have

$$\begin{aligned} \dot{V}_0 \le & -e_0^{\mathrm{T}} Q e_0 + (2+n)\|e_0\|^2 + \|P\|^2(\bar{h}_s^2 + \sum_{k=1}^{n} \varepsilon_k^2 + (1-g)^2 h^2) + \|P\|^2 \sum_{k=1}^{n} \tilde{\theta}_k^{\mathrm{T}} \tilde{\theta}_k \\ \le & -\lambda_0 \|e_0\|^2 + \|P\|^2 \sum_{k=1}^{n} \tilde{\theta}_k^{\mathrm{T}} \tilde{\theta}_k + \epsilon_0, \end{aligned} \tag{4.15}$$

where $\lambda_0 = \lambda_{\min}(Q) - 2 - 2n$ and $\epsilon_0 = \|P\|^2(\bar{h}_s^2 + \sum_{k=1}^{n} \varepsilon_k^2 + (1-g)^2 h^2)$.

4.2.4 Tunnel Prescribed Control

In this section, we propose our Tunnel Prescribed Control (TPC) for the nonlinear system with input saturation, actuator faults and unmeasurable states.

Following error transform function $T(\cdot)$, we can transform the constraint error e into an equivalent unconstrained variable z_1

$$e(t) = \frac{e_u + e_l}{2} T(z_1) + \frac{e_u - e_l}{2}, \tag{4.16}$$

where $T(\cdot)$ follows the properties in [1] that (i) $-1 < T(z_1) <$ for $-\infty < z_1 < \infty$, and (ii) $\lim_{z_1 \to +\infty} T(z_1) = 1$, $\lim_{z_1 \to -\infty} T(z_1) = -1$. Thus, the bounded equivalent variable z_1 can guarantee that the tracking errors $e(t)$ can satisfy the inequality (4.5) for all time. Define the following auxiliary variables as

$$z_i = \hat{x}_i - \alpha_i^f, \; i = 2, \ldots, n, \tag{4.17}$$

where α_i^f is the first-order filter output by directly passing through α_i to the filter. The virtual and actual control inputs can be developed as

$$\alpha_1 = -\hat{\theta}_1^{\mathrm{T}} \varphi(\hat{x}_1) + \dot{y}_d - \Lambda^{\mathrm{inv}}(\Lambda_u \dot{e}_u + \Lambda_l \dot{e}_l) - \Lambda^{\mathrm{inv}}(c_1 z_1 + b_1 z_1^{\beta}), \tag{4.18}$$

$$\alpha_2 = -k_2 e_0^1 - \frac{\omega_2}{\delta_2} - \Lambda z_1 - \hat{\theta}_2^{\mathrm{T}} \varphi_2(\hat{\bar{x}}_2) - c_2 z_2 - b_2 z_2^{\beta}, \tag{4.19}$$

$$\alpha_i = -k_i e_0^1 - \frac{\omega_i}{\delta_i} - z_{i-1} - \hat{\theta}_i^{\mathrm{T}} \varphi_i(\hat{\bar{x}}_i) - c_i z_i - b_i z_i^{\beta}, \; i = 3, \ldots, n-1, \tag{4.20}$$

$$v = -z_{n-1} - \frac{\omega_n}{\delta_n} - c_n z_n - b_n z_n^{\beta} - k_n e_0^1, \tag{4.21}$$

with

$$\dot{\hat{\theta}}_1 = \gamma_1(z_1 \Lambda \varphi_1(\hat{x}_1) - \sigma_1 \hat{\theta}_1), \tag{4.22}$$

$$\dot{\hat{\theta}}_2 = \gamma_2(z_2 \varphi_2(\hat{\bar{x}}_2) - \sigma_2 \hat{\theta}_2), \tag{4.23}$$

$$\dot{\hat{\theta}}_i = \gamma_i(z_i \varphi_i(\hat{\bar{x}}_i) - \sigma_i \hat{\theta}_i), \; i = 3, \ldots, n-1, \tag{4.24}$$

$$\dot{\hat{\theta}}_n = -\gamma_n(z_n \varphi_n(\hat{x}_n) + \sigma_n \hat{\theta}_n), \tag{4.25}$$

where $\Lambda = \partial z_1/\partial e$, $\Lambda_u = \partial z_1/\partial e_u$, $\Lambda_l = \partial z_1/\partial e_l$, c_i, b_i, δ_i, γ_i and σ_i are positive constants, $\omega_{i+1} = \alpha_{i+1}^f - \alpha_i$, $\hat{\bar{x}}_i = [\hat{x}_1, \ldots, \hat{x}_i]^{\mathrm{T}}$, and $0 < \beta < 1$. Now, we are at the position to present the main result of this section.

Theorem 4.1 *Consider the nonlinear system (4.1) suffering from actuator faults (4.2), the state observer (4.10), virtual control inputs (4.18)–(4.20), actuate control input (4.21), adaptive laws (4.22)–(4.25), and tunnel prescribed performance (4.5). Then, all the system states can be regulated to a small neighborhood around the* *desired trajectories within finite time, and the tracking errors are guaranteed within the tunnel prescribed performance even under actuator faults.*

Proof *Step 1:* From (4.16), the time derivative of z_1 is,

$$\dot{z}_1 = \Lambda(x_2 + f_1(x) - \dot{y}_d) + \Lambda_u \dot{e}_u + \Lambda_l \dot{e}_l. \tag{4.26}$$

Let $z_2 = \hat{x}_2 + \eta_{2,l} - \eta_{2,u} - \alpha_2^f$ and $\omega_2 = \alpha_2^f - \alpha_1$, where α_2^f is the first-order filter output with the designed virtual control α_1 as input. Due to $e_0^2 = x_2 - \hat{x}_2$, we have,

$$\dot{z}_1 = \Lambda(e_0^2 + z_2 + \omega_2 + \alpha_1 + f_1(x)) + \Lambda_u \dot{e}_u + \Lambda_l \dot{e}_l - \Lambda \dot{y}_d, \tag{4.27}$$

Consider the candidate Lyapunov function as,

$$V_1 = V_0 + \frac{1}{2} z_1^2 + \frac{1}{2\gamma_1} \tilde{\theta}_1^{\mathrm{T}} \tilde{\theta}_1. \tag{4.28}$$

Substituting (4.18), (4.22) and (4.27) into the time derivative of V_1 yields,

$$\begin{aligned}
\dot{V}_1 =& \dot{V}_0 + z_1 \dot{z}_1 - \frac{1}{\gamma_1} \tilde{\theta}_1^{\mathrm{T}} \dot{\hat{\theta}}_1 \\
=& \dot{V}_0 + z_1 \Lambda(\theta_1^{\mathrm{T}} \varphi_1(\hat{x}_1) - \hat{\theta}_1^{\mathrm{T}} \varphi_1(\hat{x}_1) + \omega_2 + \theta_1^{\mathrm{T}} \varphi_1(\hat{x}) - \theta_1^{\mathrm{T}} \varphi_1(\hat{x}_1)) \\
& - \tilde{\theta}_1^{\mathrm{T}} (z_1 \Lambda \varphi_1(\hat{x}_1) - \sigma_1 \hat{\theta}_1) - c_1 z_1^2 + z_1 \Lambda(e_0^2 + z_2 + \varepsilon_1) - b_1 |z_1|^{\beta+1}.
\end{aligned} \tag{4.29}$$

According to the Young's inequality, we can obtain,

$$z_1 \Lambda(\theta_1^{*\mathrm{T}} \varphi_1(\hat{x}) - \theta_1^{*\mathrm{T}} \varphi_1(\hat{x}_1)) \le z_1^2 + \Lambda^2 \|\theta_1^*\|^2 \tag{4.30}$$

$$z_1 \Lambda(e_0^2 + \varepsilon_1) \le z_1^2 + \frac{1}{2} \Lambda^2 \|e_0\|^2 + \frac{1}{2} \Lambda^2 \varepsilon_1^2. \tag{4.31}$$

Due to $\tilde{\theta}_1 = \theta_1^* - \hat{\theta}_1$, we further have,

$$\sigma_1 \tilde{\theta}_1^{\mathrm{T}} \hat{\theta}_1 = \sigma_1 \tilde{\theta}_1^{\mathrm{T}} (\theta_1^* - \tilde{\theta}_1) \le -\frac{\sigma_1}{2} \|\tilde{\theta}_1\|^2 + \frac{\sigma_1}{2} \|\theta_1^*\|^2. \tag{4.32}$$

Substituting (4.30)–(4.32) into (4.29) yields, we can obtain,

$$\begin{aligned}
\dot{V}_1 \le& -\lambda_1 \|e_0\|^2 + \|P\|^2 \sum_{k=1}^{n} \|\tilde{\theta}_k\|^2 - \bar{c}_1 z_1^2 \\
& - b_1 |z_1|^{\beta+1} + z_1 \Lambda z_2 + z_1 \Lambda \omega_2 - \frac{\sigma_1}{2} \|\tilde{\theta}_1\|^2 + \epsilon_1,
\end{aligned} \tag{4.33}$$

where $\lambda_1 = \lambda_0 - \frac{1}{2}\Lambda^2$, $\bar{c}_1 = c_1 - 2$ and $\epsilon_1 = \frac{\sigma_1}{2}\|\theta_1^*\|^2 + \epsilon_0 + \Lambda^2\|\theta_1^*\|^2 + \frac{1}{2}\Lambda^2\varepsilon_1^2$. A first-order filter is introduced to avoid the prohibitively complicated computation in the backstepping process,

$$\delta_2\dot{\alpha}_2^f + \alpha_2^f = \alpha_1,\ \alpha_2^f(0) = \alpha_1(0), \tag{4.34}$$

Therefore, we have,

$$\dot{\omega}_2 = -\frac{\omega_2}{\delta_2} - \dot{\alpha}_1. \tag{4.35}$$

Step 2: According to (4.17), the time derivative of z_2 is,

$$\dot{z}_2 = \dot{\hat{x}}_2 + \dot{\eta}_{2,l} - \dot{\eta}_{2,u} - \dot{\alpha}_2^f = k_2 e_0^1 + \hat{x}_3 + \hat{\theta}_2^{\mathrm{T}}\varphi_2(\hat{x}) + \frac{\omega_2}{\delta_2}. \tag{4.36}$$

Let $z_3 = \hat{x}_3 - \alpha_3^f$ be the tracking error and $\omega_3 = \alpha_3^f - \alpha_2$, where α_3^f is obtained by through the virtual control α_2 into the filter. We then have,

$$\dot{z}_2 = z_3 + \omega_3 + \alpha_2 + k_2 e_0^1 + \frac{\omega_2}{\delta_2} + \hat{\theta}_2^{\mathrm{T}}\varphi_2(\hat{x}_j). \tag{4.37}$$

The virtual control α_2 and the adaptive law for $\hat{\theta}_2$ are designed as (4.19) and (4.23). Construct the Lyapunov function,

$$V_2 = V_1 + \frac{1}{2}z_2^2 + \frac{1}{2\gamma_2}\tilde{\theta}_2^{\mathrm{T}}\tilde{\theta}_2 + \frac{1}{2}\omega_2^2. \tag{4.38}$$

Substituting (4.19), (4.23), and (4.35)–(4.37) into the time derivative of V_2, we have,

$$\begin{aligned}\dot{V}_2 =& \dot{V}_1 + z_2(z_3 + \omega_3 - \Lambda z_1 - c_2 z_2 + \theta_2^{*\mathrm{T}}\varphi_2(\hat{\bar{x}}_2) - \hat{\theta}_2^{\mathrm{T}}\varphi_2(\hat{\bar{x}}_2) + \theta_2^{*\mathrm{T}}\varphi_2(\hat{x})\\ &- \theta_2^{*\mathrm{T}}\varphi_2(\hat{\bar{x}}_2) - \tilde{\theta}_2^{\mathrm{T}}\varphi_2(\hat{x}) - b_2 z_2^{\beta}) - \omega_2(\frac{\omega_2}{\delta_2} + \dot{\alpha}_1) - \tilde{\theta}_2^{\mathrm{T}}(z_2\varphi_2(\hat{\bar{x}}_2) - \sigma_2\hat{\theta}_2).\end{aligned} \tag{4.39}$$

According to the Young's inequality and $\tilde{\theta}_2 = \theta_2^* - \hat{\theta}_2$, we have,

$$z_2(\theta_2^{*\mathrm{T}}\varphi_2(\hat{x}) - \theta_2^{*\mathrm{T}}\varphi_2(\hat{\bar{x}}_2)) \le z_2^2 + \|\theta_2^*\|^2, \tag{4.40}$$

$$-z_2\tilde{\theta}_2^{\mathrm{T}}\varphi_2(\hat{x}) \le \frac{1}{2}z_2^2 + \frac{1}{2}\|\tilde{\theta}_2\|^2 \tag{4.41}$$

$$\sigma_2\tilde{\theta}_2^{\mathrm{T}}\hat{\theta}_2 = \sigma_2\tilde{\theta}_2^{\mathrm{T}}(\theta_2^* - \tilde{\theta}_2) \le -\frac{\sigma_2}{2}\|\tilde{\theta}_2\|^2 + \frac{\sigma_2}{2}\|\theta_2^*\|^2. \tag{4.42}$$

Taking (4.33) and (4.40)–(4.42) into (4.39), we can obtain,

$$\begin{aligned}\dot{V}_2 \leq & -\lambda_1\|e_0\|^2-\bar{c}_1z_1^2-b_1|z_1|^{\beta+1}-(c_2-\frac{3}{2})z_2^2-b_2|z_2|^{\beta+1}+z_1\Lambda\omega_2\\ & -\frac{\sigma_1}{2}\|\tilde{\theta}_1\|^2-\frac{\sigma_2-1}{2}\|\tilde{\theta}_2\|^2+\epsilon_1+\|P\|^2\sum_{k=1}^{n}\|\tilde{\theta}_k\|^2+z_2(z_3+\omega_3)\\ & -\omega_2(\frac{\omega_2}{\delta_2}+\dot{\alpha}_1)+\frac{\sigma_2+2}{2}\|\theta_2^*\|^2\\ \leq & -\lambda_1\|e_0\|^2-\sum_{k=1}^{2}\bar{c}_kz_k^2-\sum_{k=1}^{2}b_k|z_k|^{\beta+1}-\sum_{k=1}^{2}\bar{\sigma}_k\|\tilde{\theta}_k\|^2+\|P\|^2\sum_{k=1}^{n}\|\tilde{\theta}_k\|^2\\ & +z_2(z_3+\omega_3)+z_1\Lambda\omega_2+\epsilon_2-\omega_2(\frac{\omega_2}{\delta_2}+\dot{\alpha}_1).\end{aligned} \tag{4.43}$$

where $\bar{c}_2=c_2-\frac{3}{2}$, $\bar{\sigma}_1=\frac{\sigma_1}{2}$, $\bar{\sigma}_2=\frac{\sigma_2-1}{2}$, and $\epsilon_2=\epsilon_1+\frac{\sigma_2+2}{2}\|\theta_2^*\|^2$. Let α_2 pass through the following first-order filter,

$$\delta_3\dot{\alpha}_3^f+\alpha_3^f=\alpha_2,\ \alpha_3^f(0)=\alpha_3(0), \tag{4.44}$$

The time derivative of ω_3 can be described as,

$$\dot{\omega}_3=-\frac{\omega_3}{\delta_3}-\dot{\alpha}_2. \tag{4.45}$$

Step i ($3\leq i\leq n-1$)*:* We define the tracking error z_i as $z_i=\hat{x}_i-\alpha_i^f$. Then the time derivative of z_i can be formulated as,

$$\dot{z}_i=\dot{\hat{x}}_i-\dot{\alpha}_i^f=k_ie_0^1+\hat{x}_{i+1}+\hat{\theta}_i^{\mathrm{T}}\varphi_i(\hat{x})+\frac{\omega_i}{\delta_i}. \tag{4.46}$$

Let $z_{i+1}=\hat{x}_{i+1}-\alpha_{i+1}^f$ and $\omega_{i+1}=\alpha_{i+1}^f-\alpha_i$. The tracking error z_i can be rewritten as,

$$\dot{z}_i=z_{i+1}+\omega_{i+1}+\alpha_i+k_ie_0^1+\frac{\omega_i}{\delta_i}+\hat{\theta}_i^{\mathrm{T}}\varphi_i(\hat{x}). \tag{4.47}$$

The virtual control input α_i and the adaptive law $\hat{\theta}_i$ can be designed as (4.20) and (4.24). Then, the Lyapunov function is constructed as,

$$V_i=V_{i-1}+\frac{1}{2}z_i^2+\frac{1}{2\gamma_i}\tilde{\theta}_i^{\mathrm{T}}\tilde{\theta}_i+\frac{1}{2}\omega_i^2. \tag{4.48}$$

Taking (4.20), (4.24), (4.45), and (4.47) into the time derivative of V_i yields,

$$\begin{aligned}\dot{V}_i =&\dot{V}_{i-1} + z_i(z_{i+1} + \omega_{i+1} - z_{i-1} + \theta_i^{*\mathrm{T}}\varphi_i(\hat{\bar{x}}_i) - \hat{\theta}_i^{\mathrm{T}}\varphi_i(\hat{\bar{x}}_i) - b_i z_i^{\beta} + \theta_i^{*\mathrm{T}}\varphi_i(\hat{x})\\&- \theta_i^{*\mathrm{T}}\varphi_i(\hat{\bar{x}}_i) - c_i z_i - \tilde{\theta}_i^{\mathrm{T}}\varphi_i(\hat{x})) - \tilde{\theta}_i^{\mathrm{T}}(z_i\varphi_i(\hat{\bar{x}}_i) - \sigma_i\hat{\theta}_i) - \omega_i(\frac{\omega_i}{\delta_i} + \dot{\alpha}_{i-1}).\end{aligned} \tag{4.49}$$

According to the Young's inequality and the definition $\tilde{\theta}_i = \theta_i^* - \hat{\theta}_i$, we can have,

$$z_i(\theta_i^{*\mathrm{T}}\varphi_i(\hat{x}) - \theta_i^{*\mathrm{T}}\varphi_i(\hat{\bar{x}}_i)) \le z_i^2 + \|\theta_i^*\|^2, \tag{4.50}$$

$$- z_i\tilde{\theta}_i^{\mathrm{T}}\varphi_i(\hat{x}) \le \frac{1}{2}z_i^2 + \frac{1}{2}\|\tilde{\theta}_i\|^2 \tag{4.51}$$

$$\sigma_i\tilde{\theta}_i^{\mathrm{T}}\hat{\theta}_i = \sigma_i\tilde{\theta}_i^{\mathrm{T}}(\theta_i^* - \tilde{\theta}_i) \le -\frac{\sigma_i}{2}\|\tilde{\theta}_i\|^2 + \frac{\sigma_i}{2}\|\theta_i^*\|^2. \tag{4.52}$$

Taking (4.50)–(4.52) into (4.49), we have,

$$\begin{aligned}\dot{V}_i \le&\dot{V}_{i-1} + z_i(z_{i+1} + \omega_{i+1} - z_{i-1} + \tilde{\theta}_i^{\mathrm{T}}\varphi_i(\hat{\bar{x}}_i) - b_i z_i^{\beta} - c_i z_i)\\&+ \frac{1}{2}z_i^2 + \frac{1}{2}\|\tilde{\theta}_i\|^2 + z_i^2 + \|\theta_i^*\|^2 - \tilde{\theta}_i^{\mathrm{T}}z_i\varphi_i(\hat{\bar{x}}_i) - \frac{\sigma_i}{2}\|\tilde{\theta}_i\|^2\\&+ \frac{\sigma_i}{2}\|\theta_i^*\|^2 - \omega_i(\frac{\omega_i}{\delta_i} + \dot{\alpha}_{i-1})\\\le&- \lambda_1\|e_0\|^2 - \sum_{k=1}^{i}\bar{c}_k z_k^2 - \sum_{k=1}^{i} b_k|z_k|^{\beta+1} - \sum_{k=1}^{i}\bar{\sigma}_k\|\tilde{\theta}_k\|^2 + \|P\|^2\sum_{k=1}^{n}\|\tilde{\theta}_k\|^2\\&+ \sum_{k=2}^{i} z_k\omega_{k+1} + z_1\Lambda\omega_2 + z_i z_{i+1} - \sum_{k=2}^{i}\omega_k(\frac{\omega_k}{\delta_k} + \dot{\alpha}_{k-1}) + \epsilon_i,\end{aligned} \tag{4.53}$$

where $\bar{c}_i = c_i - \frac{3}{2}$, $\epsilon_i = \epsilon_{i-1} + \frac{\sigma_i+2}{2}\|\theta_i^*\|^2$, and $\bar{\sigma}_i = \frac{\sigma_i-1}{2}$. Let a_i through the first-order filter,

$$\delta_{i+1}\dot{\alpha}_{i+1}^f + \alpha_{i+1}^f = \alpha_i,\ \alpha_{i+1}^f(0) = \alpha_i(0), \tag{4.54}$$

where δ_{i+1} is a positive constant. Therefore, we have,

$$\dot{\omega}_{i+1} = -\frac{\omega_{i+1}}{\delta_{i+1}} - \dot{\alpha}_i. \tag{4.55}$$

Step n: Let $z_n = \hat{x}_n - \alpha_n^f$. And according to the definition $\omega_n = \alpha_n^f - \alpha_{n-1}$, we obtain the time derivative of z_n as,

$$\dot{z}_n = \dot{\hat{x}}_n - \dot{\alpha}_n^f = h(v) + \hat{\theta}_n^{\mathrm{T}}\varphi_n(\hat{x}) + \frac{\omega_n}{\delta_n} + k_n e_0^1. \tag{4.56}$$

The Lyapunov function is designed as,

$$V_n = V_{n-1} + \frac{1}{2}z_n^2 + \frac{\tilde{\theta}_n^{\mathrm{T}}\tilde{\theta}_n}{2\gamma_n} + \frac{1}{2}\omega_n^2. \tag{4.57}$$

According to (4.21), (4.25), and (4.56), we have the time derivative of V_n as,

$$\begin{aligned}\dot{V}_n =& \dot{V}_n + z_n(-z_{n-1} - c_n z_n + \theta_n^{*\mathrm{T}}\varphi_n(\hat{x}) - \tilde{\theta}_n^{\mathrm{T}}\varphi_n(\hat{x}) - b_n z_n^{\beta} + h(v) - v)\\ &+ \tilde{\theta}_n^{\mathrm{T}}(\sigma_n\hat{\theta}_n + z_n\varphi_n(\hat{x}_n)) + \omega_n\dot{\omega}_n.\end{aligned} \tag{4.58}$$

Due to the Young's inequality and $\tilde{\theta}_n = \theta_n^* - \hat{\theta}_n$, we have,

$$z_n\theta_n^{*\mathrm{T}}\varphi_n(\hat{x}) \le \frac{1}{2}z_n^2 + \frac{1}{2}\|\theta_n^*\|^2 \tag{4.59}$$

$$\sigma_n\tilde{\theta}_n^{\mathrm{T}}\hat{\theta}_n = \sigma_n\tilde{\theta}_n^{\mathrm{T}}(\theta_n^* - \tilde{\theta}_n) \le -\frac{\sigma_n}{2}\|\tilde{\theta}_n\|^2 + \frac{\sigma_n}{2}\|\theta_n^*\|^2. \tag{4.60}$$

Substituting (4.53), (4.55), (4.59), and (4.60) into (4.58), we have,

$$\begin{aligned}\dot{V}_n \le& -\lambda_1\|e_0\|^2 - \sum_{k=1}^{n}\bar{c}_k z_k^2 + z_1\Lambda\omega_2 - \sum_{k=1}^{n} b_k|z_k|^{\beta+1} - \sum_{k=1}^{n}\bar{\sigma}_k\|\tilde{\theta}_k\|^2\\ &+ \|P_j\|^2\sum_{k=1}^{n}\|\tilde{\theta}_k\|^2 + \sum_{k=2}^{n-1} z_k\omega_{k+1} - \sum_{k=2}^{n}\omega_k(\frac{\omega_k}{\delta_k} + \dot{\alpha}_{k-1}) + \epsilon_n,\end{aligned} \tag{4.61}$$

where $\bar{c}_n = c_n - 1$, $\bar{\sigma}_n = \frac{\sigma_n}{2}$, and $\epsilon_n = \epsilon_{n-1} + \frac{\sigma_n+1}{2}\|\theta_n^*\|^2 + |h(v) - v|^2$. From Young's inequality, we have,

$$z_1\Lambda\omega_2 \le \frac{1}{2}z_1^2 + \frac{1}{2}\Lambda^2\omega_2^2 \tag{4.62}$$

$$\sum_{k=2}^{n-1} z_k\omega_{k+1} \le \sum_{k=2}^{n-1}\frac{1}{2}z_k^2 + \sum_{k=3}^{n}\frac{1}{2}\omega_k^2 \tag{4.63}$$

$$-\sum_{k=2}^{n}\omega_k\dot{\alpha}_{k-1} \le \sum_{k=2}^{n}\frac{1}{2}\omega_k^2 + \sum_{k=2}^{n}\frac{1}{2}\dot{\alpha}_{k-1}^2. \tag{4.64}$$

Taking (4.62)–(4.64) into (4.61) yields,

$$\dot{V}_n \le -\lambda_1\|e_0\|^2 - \sum_{k=1}^{n} l_k z_k^2 - \sum_{k=1}^{n} b_k|z_k|^{\beta+1} - \sum_{k=2}^{n}\bar{\delta}_k\omega_k^2 - \sum_{k=1}^{n} p_k\|\tilde{\theta}_k\|^2 + \bar{\epsilon}_n, \tag{4.65}$$

where $\bar{\epsilon}_n = \sum_{k=2}^{n} \frac{1}{2}\dot{\alpha}_{k-1}^2 + \epsilon_n$, $p_k = \bar{\sigma}_k - \|P_j\|^2$, $l_k = \bar{c}_k - \frac{1}{2}$ for $k = 1, \ldots, n$, $\bar{\delta}_2 = \frac{1}{\delta_2} - \frac{1+\Lambda^2}{2}$, and $\bar{\delta}_k = \frac{1}{\delta_k} - 1$ for $k = 3, \ldots, n$. From Lemma 2.2, we can obtain,

$$
\begin{aligned}
-\|\tilde{\theta}_k\|^2 &= -\|\tilde{\theta}_k\|^2 + \|\tilde{\theta}_k\|^{\beta+1} - \|\tilde{\theta}_k\|^{\beta+1} \\
&\leq \frac{1-\beta}{2} - \frac{1-\beta}{2}\|\tilde{\theta}_k\|^2 - \|\tilde{\theta}_k\|^{\beta+1}
\end{aligned} \tag{4.66}
$$

The terms $-\lambda_1\|e_0\|^2$ and $-\sum_{k=2}^{n}\bar{\delta}_k\omega_k^2$ follow the same process. The inequality (4.65) can be rewritten as

$$
\begin{aligned}
\dot{V}_n \leq &-\frac{1-\beta}{2}\lambda_1\|e_0\|^2 - \lambda_1\|e_0\|^{\beta+1} + \bar{\epsilon} \\
&-\frac{1-\beta}{2}\sum_{k=1}^{n} p_k\|\tilde{\theta}_k\|^2 - \sum_{k=1}^{n} p_k\|\tilde{\theta}_k\|^{\beta+1} \\
&-\frac{1-\beta}{2}\sum_{k=1}^{n}\bar{\delta}_k\omega_k^2 - \sum_{k=1}^{n}\bar{\delta}_k\omega_k^{\beta+1} \\
&-\sum_{k=1}^{n} l_k z_k^2 - \sum_{k=1}^{n} b_k|z_k|^{\beta+1}
\end{aligned} \tag{4.67}
$$

where $\bar{\epsilon} = \bar{\epsilon}_n + \frac{\lambda_1(1-\beta)}{2} + \sum_{k=1}^{n}\frac{\bar{\delta}_k(1-\beta)}{2} + \sum_{k=1}^{n}\frac{p_k(1-\beta)}{2}$. According to Lemma 2.4, the inequality (4.67) can be formulated as,

$$
\begin{aligned}
\dot{V}_n \leq &- c_1(\frac{1}{2}\|e_0\|^2 + \frac{1}{2}\sum_{k=1}^{n} z_k^2 + \sum_{k=2}^{n}\frac{1}{2}\omega_k^2 + \sum_{k=1}^{n}\frac{\tilde{\theta}_k^{\mathrm{T}}\tilde{\theta}_k}{2\gamma_k}) - c_2(\frac{1}{2}\|e_0\|^2 + \frac{1}{2}\sum_{k=1}^{n} z_k^2 \\
&+ \sum_{k=2}^{n}\frac{1}{2}\omega_k^2 + \sum_{k=1}^{n}\frac{\tilde{\theta}_k^{\mathrm{T}}\tilde{\theta}_k}{2\gamma_k})^{\bar{\beta}} + \bar{\epsilon} \\
\leq &-c_1 V_n - c_2(V_n)^{\bar{\beta}} + \bar{\epsilon},
\end{aligned} \tag{4.68}
$$

where $c_1 = \min((1-\beta)\lambda_1, 2l_k, (1-\beta)\bar{\delta}_k, (1-\beta)p_k\gamma_k)$, and $c_2 = \min(2^{\bar{\beta}}\lambda_1, 2^{\bar{\beta}}b_k, 2^{\bar{\beta}}\bar{\delta}_k, 2^{\bar{\beta}}p_k)$ with $\bar{\beta} = \frac{1+\beta}{2}$.

From (4.68) and Lemma 2.1, the following inequality holds

$$
\dot{V}_n \leq -c_1 V_n - c_2 c_3 V_n^{\bar{\beta}} + \bar{\epsilon} - c_2(1-c_3)V_n^{\bar{\beta}}. \tag{4.69}
$$

where c_3 is a positive constant. We define $\Omega_V = \{\xi \,|\, V_n^{\bar{\beta}} \leq \frac{\bar{\epsilon}}{c_2(1-c_3)}\}$ and $\bar{\Omega}_V = \{\xi \,|\, V_n^{\bar{\beta}} > \frac{\bar{\epsilon}}{c_2(1-c_3)}\}$.

Case 1: If $\xi \in \bar{\Omega}_V$, the following inequality holds

$$\dot{V}_n \leq -c_1 V_n - c_2 c_3 V_n^{\bar{\beta}}. \tag{4.70}$$

Integrating the above inequality over $[t_0, T]$, we have

$$\int_{t_0}^{T} \frac{V_n^{-\bar{\beta}}}{c_2 c_3 + c_1 V_n^{1-\bar{\beta}}} \mathrm{dV} \leq -\int_{t_0}^{T} \mathrm{dt}. \tag{4.71}$$

Then the settling time can be described as

$$T = t_0 + \frac{1}{c_1(1-\bar{\beta})} \ln\left(\frac{c_2 c_3 + c_1 V_n^{1-\bar{\beta}}(t_0)}{c_2 c_3}\right). \tag{4.72}$$

Namely, $\xi \in \Omega_V$ for $\forall t \geq T$.

Case 2: Considering the system state ξ in the region Ω_V, from the case 1 above, ξ cannot escape the set Ω_V. This completes the proof.

4.2.5 Numerical Simulations

In this section, we conduct numerical simulations to verify the effectiveness of our proposed tunnel prescribed control for nonlinear systems with input saturation, tunnel prescribed performance, and actuator faults. The following nonlinear system is considered:

$$\begin{cases} \dot{x}_1 = f_1(x) + x_2 \\ \dot{x}_2 = f_2(x) + u(t) \\ y = x_1 \end{cases} \tag{4.73}$$

where $x = [x_1, x_2]^T$, $f_1(x) = 0$, and $f_2(x) = 0.7227 + 0.49\cos(x_1)$. According to the Tunnel Prescribed Control introduced in the above section, the control scheme follows:

$$\alpha_1 = \dot{y}_d - \Lambda^{\text{inv}}(\Lambda_u \dot{e}_u + \Lambda_l \dot{e}_l) - \Lambda^{\text{inv}}(c_1 z_1 + b_1 z_1^{\beta}), \tag{4.74}$$

$$v_j = -\Lambda z_1 - \frac{\omega_2}{\delta_2} - c_2 z_2 - b_2 z_2^{\beta} - k_2 e_0^1, \tag{4.75}$$

$$\dot{\theta}_2 = -\gamma_2(z_2 \varphi_2(\hat{x}_2) + \sigma_2 \hat{\theta}_2), \tag{4.76}$$

with the control parameters are set in Table 4.1. The simulation results are shown in Figs. 4.3, 4.4, 4.5, 4.6 and 4.7. From Fig. 4.3, it depicts the tracking error $e(t)$ which is always evolving within the constraints defined by tunnel prescribed performance,

Table 4.1 Parameters for simulations

Parameter	Value	Parameter	Value	Parameter	Value
c_1, c_2	3.0	b_1, b_2	1.0	b_2	10.0
b_2	20.0	$v_{\max}$	25	g	0.8
$v_{\min}$	−25	k_1	5.0	k_2	60.0
σ_2	2.0	$\delta_2,$	0.1	γ_2	1.2
β	0.75	$\varrho(t)$	$0.7e^{-t} + 0.1$	h_s	$0.2\sin(0.3t)$
y_d	$0.5\cos(0.6t - 0.5)$	$\hat{\theta}_2(0)$	$[0.2, \ldots, 0.2]^{\mathrm{T}}$		
$e_l(t)$	$(0.2 - \mathrm{sign}(e_j(0)))\varrho(t) + 0.1\mathrm{sign}(e_j(0))$				
$e_u(t)$	$(0.7 + \mathrm{sign}(e_j(0)))\varrho(t) - 0.1\mathrm{sign}(e_j(0))$				

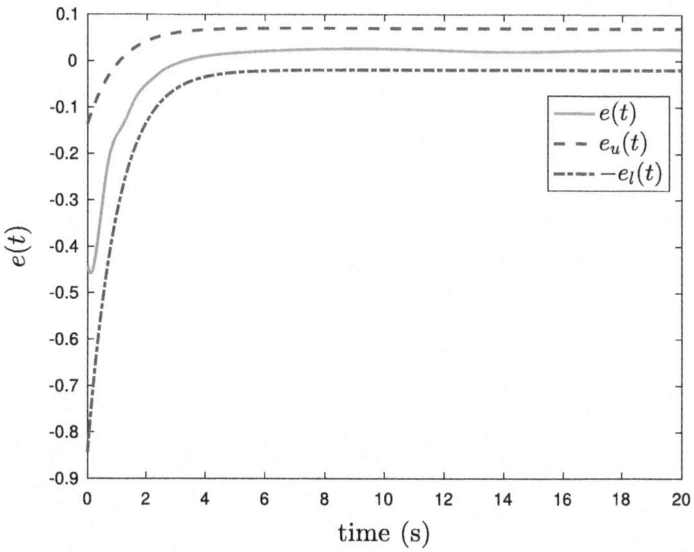

Fig. 4.3 The tracking error $e(t)$ under tunnel prescribed control

even in the presence of actuator faults and input saturation. This shows the robustness of our proposed control method in maintaining the tracking error within the performance constraints despite system uncertainties. Moreover, $e(t)$ is regulated to a small neighborhood around the equilibrium within finite time, showcasing fast convergence ability and high precision tracking of our control method. The control input signal $h(v(t))$, suffering from input saturation, is shown in Fig. 4.4, where $h(v(t))$ is restricted to the predefined region. The estimated system states $\hat{x}_1$ and $\hat{x}_2$ are shown in Figs. 4.5 and 4.6, respectively. The adaptive parameter $\|\hat{\theta}_2\|$ is depicted in Fig. 4.7. All the estimation errors, e_0^1, e_0^2, and $\tilde{\theta}_2$, are driven to a region around zero in finite time. In summary, the simulation results illustrate our proposed tunnel prescribed control for nonlinear systems subject to input saturation, tunnel prescribed

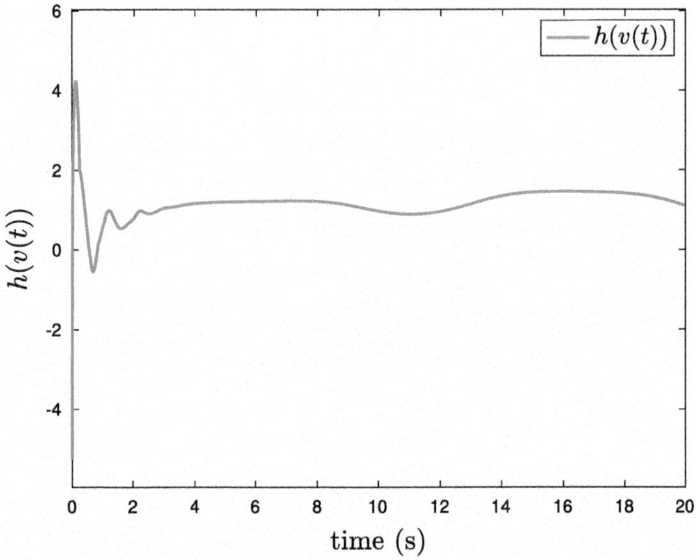

Fig. 4.4 The control input $h(v(t))$ with input saturation

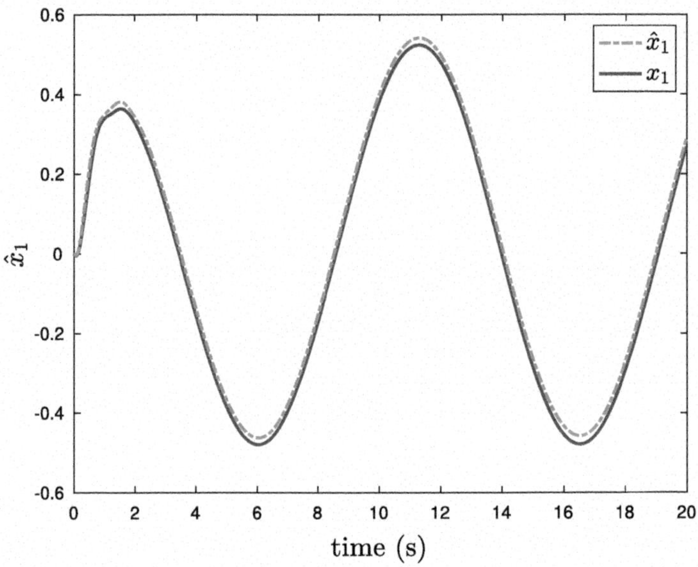

Fig. 4.5 The estimated system state $\hat{x}_1$ of the fuzzy state observer

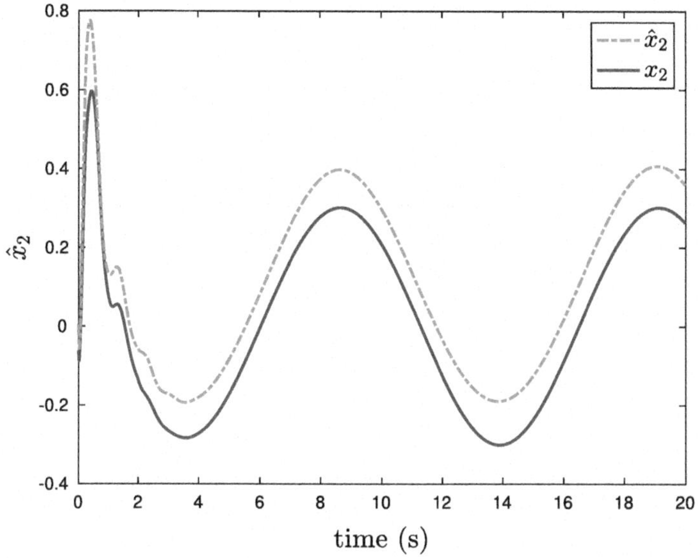

Fig. 4.6 The estimated system state $\hat{x}_2$ of the fuzzy state observer

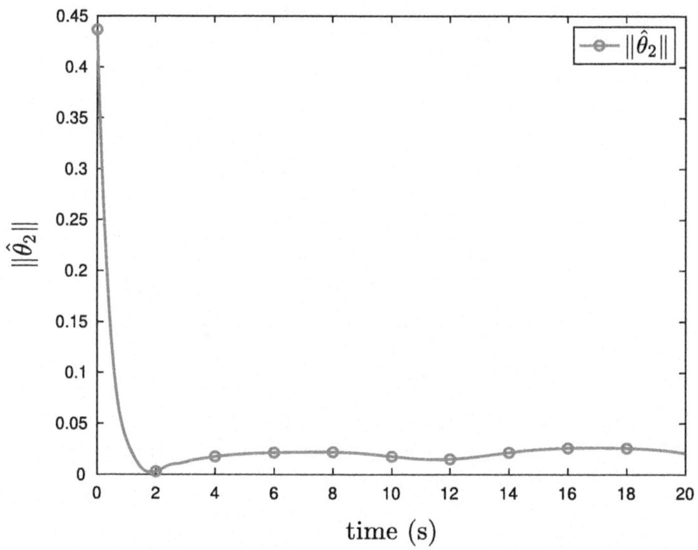

Fig. 4.7 The adaptive parameter $\|\hat{\theta}_2\|$

performance constraints, and actuator faults. It not only ensures that the tracking error remains within the prescribed performance boundaries but also guarantees the stability and reliability of the system under unknown conditions.

4.3 Finite-Time Tunnel Prescribed Control of Time-Delay Nonlinear Systems with Unknown Control Directions

4.3.1 Problem Formulation

Consider a class of nonlinear time-delay systems in the presence of unknown control directions as:

$$\begin{cases} \dot{x}_i = f_i(\bar{x}_i) + g_i(\bar{x}_i)x_{i+1} + q_i(\bar{x}_i(t-\tau_i)) \\ \dot{x}_n = f_n(\bar{x}_n) + g_n(\bar{x}_n)u + q_n(\bar{x}_n(t-\tau_n)) \\ y = x_1 \end{cases} \tag{4.77}$$

where $i \in \mathbb{I}_{1:n-1}$, $\bar{x}_i = [x_1, \ldots, x_i]^T \in \mathbb{R}^i$ and $\bar{x}_n = [x_1, \ldots, x_n]^T \in \mathbb{R}^n$ stand for the state vectors. $y \in \mathbb{R}$ is the system output which is the only measurable system state. $f_i(\cdot) : \mathbb{R}^i \to \mathbb{R}$ and $q_i(\cdot) : \mathbb{R}^i \to \mathbb{R}$ denote unknown and smooth functions. τ_i denotes unknown continuously differentiable time-varying delay satisfying $0 < \tau_i \leq \tau_{\max}$ and $\dot{\tau}_i \leq \tau \leq 1$ with $\tau_{\max}$ and τ being positive constants. $g_i(\cdot) : \mathbb{R}^i \to \mathbb{R}$ is unknown smooth function. $u \in \mathbb{R}$ denotes control input given by

$$u = \begin{cases} u_{\max}, & v_i > u_{\max} \\ v, & u_{\min} \leq v \leq u_{\max} \\ u_{\min}, & v < u_{\min} \end{cases} \tag{4.78}$$

where $u_{\max}$ and $u_{\min}$ are the known boundaries on u, and v denotes the actual control input to be designed.

Control objective: For a class of nonlinear time-delay systems (4.77) with considering input saturation, prescribed performance and unknown control direction, we design Finite-time Tunnel Prescribed Control such that: (i) The output y tracks the reference trajectory y_r, and the tracking error $e_1 = y - y_r$ are uniformly bounded when $v \neq u$; (ii) The tracking error e_1 always evolves within the region established by the Finite-time Tunnel Prescribed Performance even under the existence of unknown time delay and input saturation, i.e., $\{e_1 : -e_l(t) < e_1(t) < e_u(t)\}$; (iii) The developed Finite-time Tunnel Prescribed Control can guarantee that the tracking error e_1 is regulated to the predefined region within finite time.

Before processing the control design, we introduce some useful assumptions as follows.

Assumption 4.2 There exist positive constants $\underline{g}_i$ and $\bar{g}_i$ such that $0 < \underline{g}_i \leq |g_i(\cdot)| \leq \bar{g}_i$.

Assumption 4.3 Consider the nonlinear system (4.77) subject to input saturation (4.78) and unknown time-delay. There should exist an appropriate control strategy v such that the tracking control objective is established.

Remark 4.1 From Assumption 4.2, it indicates that both the virtual and actual control coefficients are unknown but bounded, which is widely used in the existing literature [5, 30, 32, 39, 40] to address the unknown control directions. Assumption 4.3 is also reasonable since there often exists a feasible control strategy in practice to achieve the control objectives even under input saturation, which is extensively used in the existing studies [5, 35]. Moreover, the user may verify the feasibility of Assumption 4.3 according to the reachability analysis in [20].

Remark 4.2 In the related studies, $f_i(\bar{x}_i), i \in \mathbb{I}_{1:n}$ or their bounding functions are needed to be partial known in [6, 33, 34]. The knowledge of the bound or the sign of $g_i(\bar{x}_i)$ is necessary in [4, 12, 21] for the control design. However, except Assumption 4.2, no additional knowledge about the system nonlinearities or their bounding functions is required. Therefore, such control design for (4.77) can be applied to many practical systems, such as robot systems, aerospace systems and missile systems.

4.3.2 Finite-Time Tunnel Prescribed Performance

We first recall the classical Prescribed Performance Functions (PPFs), which are given by:

$$\begin{cases} -\delta\bar{\rho}(t) < e_1(t) < \bar{\rho}(t), \text{ if } e_1(0) \geq 0 \\ -\bar{\rho}(t) < e_1(t) < \delta\bar{\rho}(t), \text{ if } e_1(0) < 0 \end{cases} \tag{4.79}$$

where $\bar{\rho}(t) = (\rho_0 - \rho_\infty)e^{-lt} + \rho_\infty$, $0 < \rho_\infty < \rho_0$, $0 < \delta \leq 1$ and $l > 0$. Its performance constraints on the tracking trajectories are shown in Fig. 4.8. According to the simulation results, it should be noted that: (i) The performance boundaries of PPFs are distributed on different sides, leading to a funnel-shape and loose allowable set as shown in Fig. 4.8, which is not applicable to situations with more strict requirements for tracking process (i.e., flight systems are expected within a tight allowable set with smaller overshoot to avoid losing tracking target due to the limited perception region). (ii) The convergence time of the classical PPFs to the steady state approaches to infinity. However, in most practical systems, it is desirable that the system states can converge to the predefined set within finite time.

Therefore, we propose the following Finite-time Tunnel Prescribed Performance (FTPP):

$$-e_l(t) < e_1(t) < e_u(t), \tag{4.80}$$

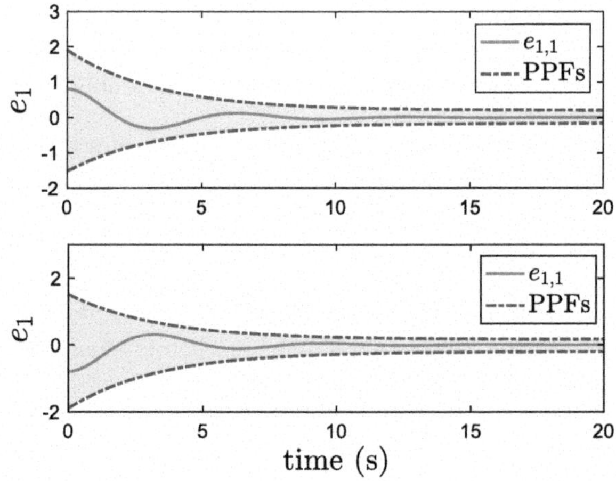

Fig. 4.8 The tracking error e_1 under PPFs for $e_1(0) \geq 0$ and $e_1(0) < 0$

where $e_l(t) = (\delta_l - \text{sign}(e_1(0)))\rho(t) + \rho_{t_f}\text{sign}(e_1(0))$ and $e_u(t) = (\delta_u + \text{sign}(e_1(0)))\rho(t) - \rho_{t_f}\text{sign}(e_1(0))$ with $0 < \delta_l, \delta_u \leq 1$ and $\rho(t)$ is specified as

$$\rho(t) = \begin{cases} (\rho_0 - \rho_{t_f})e^{\varpi(t)} + \rho_{t_f}, & 0 \leq t \leq t_f \\ \rho_{t_f}, & t > t_f \end{cases} \tag{4.81}$$

where $0 < \rho_{t_f} < \rho_0$, $0 < t_f < \infty$ and $\varpi(t) = -\frac{l t_f t}{t_f - t}$ with $l > 0$. Note that the constant t_f denotes the preset time for $\rho(t)$ from the initial value ρ_0 to the steady-state ρ_{t_f}.

To better show the salient features of FTPP, we take $e_1(0) > 0$ and $e_1(0) < 0$ in (4.80) as examples. According to the FTPP and the tracking trajectory $e_1(t)$ in Fig. 4.9, FTPP provides the following distinct advantages: (i) Compared with the classical PPFs in (4.79), FTPP unifies different initial conditions into one concise expression where, by using sign function and scaling factor, FTPP is capable of automatically adjusting its orientation according to different initial conditions as shown in Fig. 4.9. In this way, it facilitates the design procedure and control implementation. (ii) The initial performance boundaries of FTPP are distributed on the same side. Therefore, different from the funnel-shape PPFs, FTPP establishes a tight tunnel-shape allowable set, which can provide better transient performance. Moreover, its convergence time to the steady state is preassigned by users instead of relying on the system initial condition or the control parameters.

We further present a more general definition of FTPP to stimulate more studies.

Definition 4.1 Given a smooth function $\rho(t)$, we can call it Finite-time Prescribed Performance if it has he following properties: (i) $\rho(t) > 0$, (ii) $\dot{\rho}(t) \leq 0$, (iii) $\lim_{t \to t_f} \rho(t) = \rho_{t_f}$ with $\rho_{tf}, t_f > 0$, (iv) $\rho_i(t) = \rho_{i,t_f}$, $\forall t > t_f$. Moreover, if the performance constraints on the tracking error satisfy $\dot{e}_u(t)\dot{e}_l(t) \leq 0$, we can call it Finite-

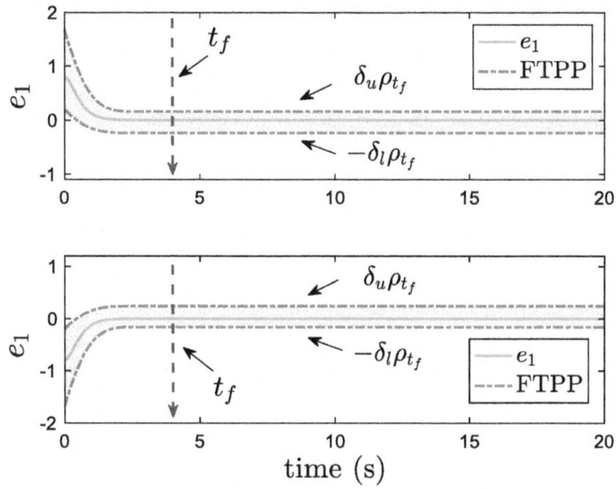

Fig. 4.9 The tracking error e_1 under FTPP for $e_1(0) > 0$ and $e_1(0) < 0$

time Tunnel Prescribed Performance, that is, the performance functions $-e_l(t)$ and $e_u(t)$ are parallel and establish a tunnel allowable set.

Remark 4.3 To clearly illustrate the distinct advantage provided by the FTPP, some comparative simulations are conducted in comparison with the classical PPFs in [1, 2]. For (4.79) and (4.81), let $\rho_0 = 2$, $\rho_\infty = 0.4$, $l = 0.5$ and $t_f = 2, 5, 30$, where we select the same parameters to obtain the reasonable results. As the simulation results in Fig. 4.10, it can be derived that FTPP $\rho(t)$ can provide faster convergence rate than the classical PPFs $\bar{\rho}(t)$ under the same initial conditions. Moreover, if we increase the value of the settling time (i.e., $t_f \to \infty$), our FTPP will be gradually close to the classical PPFs $\bar{\rho}(t)$. Therefore, we can user-specify the desirable tracking performance by appropriately selecting parameters of FTPP.

Remark 4.4 Compared with the funnel-shape PPFs in the existing studies [3, 17, 22], FTPP can impose more strict performance constraints on the tracking error since its tight tunnel-shape allowable set. On the other hand, different from previous methods in [8, 15, 26, 27] where fractional power items are needed to achieve finite-time convergence and their settling times rely on the initial tracking condition, FTPP provides a simple way to guarantee finite-time convergence without introducing additional items, and its settling time is user-specified in advance. In this way, FTPP provides a more flexible yet simple way to preassign the tracking performance by appropriately selecting parameters in FTPP.

4.3.3 Input-Driven Observer

We adopt the coordinate transformation $\xi_i = x_i \prod_{j=i}^{n} g_j^{-1}(\bar{x}_j)$ and then the nonlinear time-delay system (4.77) is rewritten as

$$\begin{cases} \dot{\xi}_i = \xi_{i+1} + \bar{f}_i(\bar{\xi}_i) + \bar{q}_i(\bar{\xi}_i(t - \tau_i)) \\ \dot{\xi}_n = u + \bar{f}_n(\bar{\xi}_n) + \bar{q}_n(\bar{\xi}_n(t - \tau_n)) \end{cases} \tag{4.82}$$

Then, to estimate the unmeasurable state ξ_i, an input-driven filter [14, 16, 38] is employed

$$\begin{cases} \dot{\hat{\xi}}_i = \hat{\xi}_{i+1} - k_i \hat{\xi}_1 \\ \dot{\hat{\xi}}_n = u - k_n \hat{\xi}_1 \end{cases} \tag{4.83}$$

where $\hat{\xi}_i$ denotes the estimated value and k_i is the positive parameter such that $p(s) = s^n + k_1 s^{n-1} + \cdots + k_{n-1}s + k_n$ to be Hurwitz. Define $\tilde{\xi}_i = \xi_i - \hat{\xi}_i$ as the estimate error and $\tilde{\xi} = [\tilde{\xi}_1, \ldots, \tilde{\xi}_n]^T$. The dynamics of $\tilde{\xi}$ is

$$\dot{\tilde{\xi}} = A\tilde{\xi} + K\xi_1 + F(\bar{\xi}_n) + Q(\bar{\xi}_n(t - \tau_n)) \tag{4.84}$$

where $F(\bar{\xi}_n) = [\bar{f}_1(\bar{\xi}_1), \ldots, \bar{f}_n(\bar{\xi}_n)]^T$ and $Q(\bar{\xi}_n(t - \tau_n)) = [\bar{q}_1(\bar{\xi}_1(t - \tau_1)), \ldots, \bar{q}_n(\bar{\xi}_n(t - \tau_n))]^T$. The matrices A and K are

$$A = \begin{bmatrix} -k_1 & & & \\ \vdots & & I_{n-1} & \\ -k_n & 0 & \cdots & 0 \end{bmatrix}, \quad K = \begin{bmatrix} k_1 \\ \vdots \\ k_n \end{bmatrix}. \tag{4.85}$$

According to the definition of k_i, for the given positive matrix $P = P^T$, we have $A^T P + PA = -Q$ with $Q = Q^T$ being a positive definite matrix.

Remark 4.5 Compared with most existing fuzzy state observer in [13, 18, 25] where unknown nonlinearities are usually estimated by fuzzy logic systems or feedback

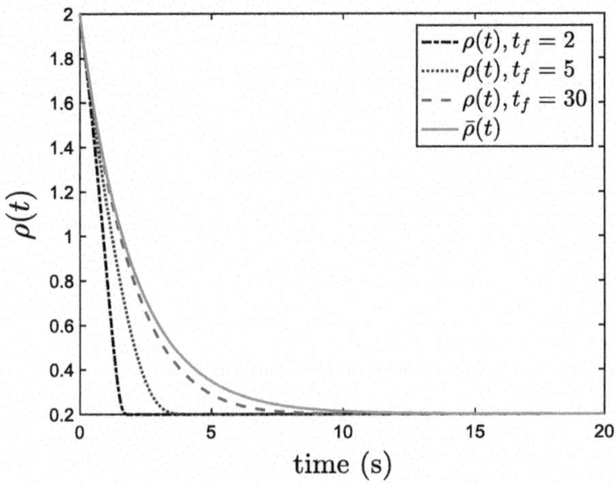

Fig. 4.10 Illustration of $\rho(t)$ and $\bar{\rho}(t)$

terms are needed, we only use the control input signal in our filter design as [14, 16, 38] to facilitate the control design. Although such filter is not robust, the purpose is to together handle all uncertainties $F(\bar{\xi}_n)$ in the control design process by using fuzzy logic systems. In this way, it can be found later that only one adaptive law is needed instead of estimating each $\bar{f}_i(\bar{\xi}_n)$ as in the previous studies, which can dramatically reduce the computation burden. Therefore, the so called estimation $\hat{\xi}_i$ acts as an auxiliary variable for stability analysis instead of reconstructing the original system state ξ_i. The stability analysis will also show that $\|\tilde{\xi}\| \in L^\infty$ is guaranteed, rather than $\lim_{t\to\infty} \|\tilde{\xi}\| = 0$.

4.3.4 Nussbaum-Based Tunnel Prescribed Control

In this section, we develop a Low-complexity Nussbaum-based Tunnel Prescribed Control for nonlinear time-delay systems (4.77). The following coordinate transformation is employed:

$$z_1 = \ln\left(\frac{e_l + e_1}{e_u - e_1}\right), \tag{4.86}$$

$$z_i = \hat{\xi}_i - \alpha_i^f, \ i \in \mathbb{I}_{2:n} \tag{4.87}$$

$$\omega_i = \alpha_i^f - \alpha_{i-1}, \ i \in \mathbb{I}_{2:n}, \tag{4.88}$$

where α_i and α_i^f denote the input and output states of the filter to be defined later. And we introduce the following assumption about the time-delay component $q_i(\bar{x}_i(t - \tau_i))$ in (4.77).

Assumption 4.4 For $i \in \mathbb{I}_{1:n}$, an unknown positive smooth function $h_{i,j}(Z_j)$ exists such that

$$|q_i(\bar{x}_i)| \le \sum_{j=1}^{i} |z_j| h_{i,j}(Z_j) \tag{4.89}$$

where $Z_j = [\bar{z}_j, \bar{\alpha}_j^f, \bar{\bar{\xi}}_j]^T$, $\alpha_1^f = y_r$, $\bar{z}_j = [z_1, \ldots, z_j]^T$, $\bar{\alpha}_j^f = [\bar{\alpha}_1^f, \ldots, \bar{\alpha}_j^f]^T$, and $\bar{\bar{\xi}}_j = [\bar{\xi}_1, \ldots, \bar{\xi}_j]^T$,

The above restrictions are assumed on the time-delay components, which is essential and reasonable in the control design process as assumed in [28, 41]. We then present the control design as follows.

Step 1: Before the backstepping design process, let

$$\theta = \max\{\|\theta_i^*\|^2 : i \in \mathbb{I}_{0:n}\} \tag{4.90}$$

where θ denotes the maximum value of $\|\theta_i^*\|$ defined in (4.96), (4.109), (4.101), (4.116), and (4.124). Let $\hat{\theta}$ as the estimate of θ updated by (4.121). From (4.86)–(4.88), we have the time derivative of z_1

$$\dot{z}_1 = \Lambda\bar{g}(\alpha_1 + \tilde{\xi}_2 + z_2 + \omega_2) + \Lambda f_1(\bar{x}_1) + \Lambda q_1(\bar{x}_1(t - \tau_1)) + M_1, \tag{4.91}$$

where $\Lambda = \partial z_1/\partial e_1$, $\Lambda_l = \partial z_1/\partial e_l$, $\Lambda_u = \partial z_1/\partial e_u$, $\bar{g} = \prod_{j=1}^{n} g_j(\bar{x}_j)$, $M_1 = \Lambda_u \dot{e}_u(t) + \Lambda_l \dot{e}_l(t) - \Lambda\dot{y}_r$, and $\Lambda = \partial z_1/\partial e_1$. Take the following Lyapunov candidate function:

$$\begin{aligned}
V_e &= \tilde{\xi}^T P\tilde{\xi} + \frac{1}{2}z_1^2 + \frac{1}{2r}\tilde{\theta}^2, \\
V_{q_1} &= \frac{1}{2}e^{-(t-\tau_1)}\int_{t-\tau_1}^{t} e^{\tau} z_1^2(\tau)h_{1,1}^2(Z_1(\tau))d\tau \\
&\quad + \sum_{i=1}^{n}\sum_{j=1}^{i} e^{-(t-\tau_i)}\int_{t-\tau_i}^{t} e^{\tau} z_j^2(\tau)h_{i,j}^2(Z_j(\tau))d\tau, \\
V_1 &= V_e + V_{q_1},
\end{aligned} \tag{4.92}$$

where P denotes a positive matrix and $r > 0$. Then, the time derivative of V_{q_1} is derived

$$\begin{aligned}
\dot{V}_{q_1} =& -\bar{\lambda}V_{q_1} + \frac{1}{2}e^{\tau_1}z_1^2h_{1,1}^2(Z_1) + \sum_{i=1}^{n}\sum_{j=1}^{i} e^{\tau_i}z_j^2h_{i,j}^2(Z_j) \\
&- \sum_{i=1}^{n}\sum_{j=1}^{i}\bar{\lambda}z_j^2(t-\tau_i)h_{i,j}^2(Z_j(t-\tau_i)) \\
&- \bar{\lambda}\frac{1}{2}z_1^2(t-\tau_1)h_{1,1}^2(Z_1(t-\tau_1))
\end{aligned} \tag{4.93}$$

where $\bar{\lambda} = 1 - \tau$ and based on $\dot{\tau}_i \le \tau \le 1$, we have $-(1-\dot{\tau}_i) \le -(1-\tau) \le -\bar{\lambda}$. According to (4.84) and (4.91), it yields

$$\begin{aligned}
\dot{V}_1 =& -\tilde{\xi}^T Q\xi + 2\tilde{\xi}^T P(U_0(\bar{\xi}_n) + Q(\bar{\xi}_n(t-\tau_n))) + \dot{V}_{q_1} \\
&+ z_1[\Lambda(\bar{g}(\alpha_1 + \tilde{\xi}_2 + z_2 + \omega_2) + f_1(\bar{x}_1) + q_1(\bar{x}_1(t-\tau_1))) + M_1] - \frac{1}{r}\tilde{\theta}\dot{\hat{\theta}}.
\end{aligned} \tag{4.94}$$

where $U_0(\bar{\xi}_n) = K\xi_1 + F(\bar{\xi}_n)$. By Lemma 3.2 and (4.91), for any given $\varepsilon_{i0} > 0$ with $i = 1, \ldots, n$, there exist fuzzy logic systems such that each element of $U_0(\bar{\xi}_n)$ satisfies

$$U_{i0}(\bar{\xi}_i) = \theta_{i0}^{*T}\varphi_{i0} + \varepsilon_{i0}, \ |\varepsilon_{0i}| \le \bar{\varepsilon}_{i0}, \tag{4.95}$$

where $U_{i0}(\bar{\xi}_i) = \bar{f}_i(\bar{\xi}_i) + k_i \xi_1$, $\bar{\varepsilon}_{0i} > 0$ and therefore we have

$$U_0(\bar{\xi}_n) = \theta_0^{*T} \varphi_0 + \varepsilon_0, \ \|\varepsilon_0\| \le \bar{\varepsilon}_0, \tag{4.96}$$

where $\theta_0^* = \text{diag}(\theta_{10}^*, \dots, \theta_{n0}^*)$, $\varphi_0 = [\varphi_{10}, \dots, \varphi_{n0}]^T$, and $\bar{\varepsilon}_0 > 0$. Due to $\varphi_0^T \varphi_0 \le 1$ and (4.91), we can get

$$2\tilde{\xi}^T P U_0(\bar{\xi}_n) = 2\tilde{\xi}^T P(\theta_0^{*T} \varphi_0 + \varepsilon_0) \le 2\kappa_0 \|\tilde{\xi}\|^2 + \frac{\|P\|^2}{\kappa_0}(\theta + \bar{\varepsilon}_0^2). \tag{4.97}$$

where $\kappa_0 > 0$. From (4.84) and Assumption 4.4 yields,

$$2\tilde{\xi}^T P Q \le \sum_{i=1}^{n} \sum_{j=1}^{i} \bar{\lambda} z_j^2 (t - \tau_i) h_{i,j}^2 (Z_j(t - \tau_i)) + \frac{(1+n)n}{2\bar{\lambda}} \|P\|^2 \|\tilde{\xi}\|^2, \tag{4.98}$$

Identically, we can also have

$$z_1 \Lambda q_1(\bar{x}_1(t - \tau_1)) \le \frac{\bar{\lambda}}{2} z_1^2 (t - \tau_1) h_{1,1}^2 (Z_1(t - \tau_1)) + \frac{\Lambda^2}{2\bar{\lambda}} z_1^2. \tag{4.99}$$

Substituting the above inequalities into (4.94), the derivative of V_1 is rewritten as

$$\begin{aligned} \dot{V}_1 \le & -\lambda_0 \|\tilde{\xi}\|^2 + z_1[\Lambda \bar{g} \alpha_1 + F_1 + M_1] - \frac{1}{r} \tilde{\theta} \dot{\hat{\theta}} - \bar{\lambda} V_{q_1} + \sum_{i=2}^{n} \sum_{j=2}^{i} e^{\tau_i} z_j^2 h_{i,j}^2 (Z_j) \\ & + \frac{3}{2} \Lambda^2 \bar{g}^2 z_1^2 + \frac{1}{2\bar{\lambda}} \Lambda^2 z_1^2 + \frac{1}{2} z_2^2 + \frac{1}{2} \omega_2^2 + \frac{\|P\|^2}{\kappa_0} (\theta + \bar{\varepsilon}_0^2) \end{aligned} \tag{4.100}$$

where $\lambda_0 = \lambda_{\max}(Q) - \frac{(1+n)n}{2\bar{\lambda}} \|P\|^2 - \frac{1}{2} - 2\kappa_0$. Let $F_1 = \Lambda f_1(\bar{x}_1) + \frac{1}{2} e^{\tau_1} z_1 h_{1,1}^2 (Z_1) + \sum_{i=1}^{n} e^{\tau_i} z_1 h_{i,1}^2 (Z_1)$ and according to Lemma 3.2, there exist fuzzy logic systems such that

$$F_1 = \theta_1^{*T} \varphi_1 + \varepsilon_1, \ \|\varepsilon_1\| \le \bar{\varepsilon}_1, \tag{4.101}$$

where $\varepsilon_1 > 0$. The virtual control input is developed by

$$\begin{aligned} \alpha_1 &= N(\zeta) \bar{\alpha}_1, \ \dot{\zeta} = \Lambda z_1 \bar{\alpha}_1, \\ \bar{\alpha}_1 &= \Lambda^{\text{inv}} (c_1 z_1 + M_1 + \frac{1}{2b_1^2} z_1 \hat{\theta}), \end{aligned} \tag{4.102}$$

where $N(\cdot)$ denotes the Nussbaum type function, and $c_1, b_1 > 0$. Then, taking (4.101) and (4.102) into (4.100), we can get

$$\dot{V}_1 \leq -\lambda_0\|\tilde{\xi}\|^2 - \bar{\lambda}V_{q_1} - \bar{c}_1 z_1^2 - \frac{1}{r}\tilde{\theta}\dot{\hat{\theta}} + \frac{1}{2b_1^2}z_1^2\tilde{\theta} + \frac{1}{2}z_2^2 + \epsilon_1$$
$$+ \frac{1}{2}\omega_2^2 + (\bar{g}N(\zeta)+1)\dot{\zeta} + \sum_{i=2}^{n}\sum_{j=2}^{i} e^{\tau_i} z_j^2 h_{i,j}^2(Z_j) \tag{4.103}$$

where $\bar{c}_1 = c_1 - \frac{\kappa_{\varepsilon_1}}{2} - \frac{3}{2}\Lambda^2\bar{g}^2 - \frac{1}{2\bar{\lambda}}\Lambda^2, \epsilon_1 = \frac{1}{2}b_1^2 + \frac{1}{2\kappa_{\varepsilon_1}}\bar{\varepsilon}_1^2 + \frac{\|P\|^2}{\kappa_0}(\theta + \bar{\varepsilon}_0^2)$ and $\kappa_{\varepsilon_1} > 0$. To alleviate the computation burden, we adopt the following first-order filter

$$\delta_2\dot{\alpha}_2^f + \alpha_2^f = \alpha_1,\ \alpha_2^f(0) = \alpha_1(0), \tag{4.104}$$

where δ_2 denotes a positive constant.

Step 2: From the definition of z_2, z_3 and ω_2, the time derivative of z_2 is

$$\dot{z}_2 = z_3 + \omega_3 + \alpha_2 + M_2, \tag{4.105}$$

where $M_2 = -k_2\hat{\xi}_1 + \frac{\omega_2}{\delta_2}$. Consider the Lyapunov V_2 as

$$V_2 = V_1 + \frac{1}{2}z_2^2 + \frac{1}{2}\omega_2^2. \tag{4.106}$$

Define the virtual control input α_2 as

$$\alpha_2 = -c_2 z_2 - M_2 - \frac{1}{2b_2^2}z_2\hat{\theta}, \tag{4.107}$$

where b_2 denotes a positive constant. From (4.103)–(4.107), the time derivative of V_2 is

$$\dot{V}_2 \leq -\lambda_0\|\tilde{\xi}\|^2 - \bar{\lambda}V_{q_1} - \bar{c}_1 z_1^2 - (c_2 - \frac{3}{2})z_2^2 - \bar{\delta}_2\omega_2^2 + z_2F_2 + (\bar{g}N(\zeta)+1)\dot{\zeta} + \frac{1}{2}z_3^2 + \frac{1}{2}\omega_3^2$$
$$-\frac{1}{2b_2^2}z_2^2\hat{\theta} - \frac{1}{r}\tilde{\theta}\dot{\hat{\theta}} + \frac{1}{2b_1^2}z_1^2\tilde{\theta} + \frac{1}{2\kappa_{\alpha_2}}\dot{\alpha}_1^2 + \sum_{i=3}^{n}\sum_{j=3}^{i} e^{\tau_i} z_j^2 h_{i,j}^2(Z_j) + \epsilon_1 \tag{4.108}$$

where $\bar{\delta}_2 = \frac{1}{\delta_2} - \frac{1}{2} - \frac{\kappa_{\alpha_2}}{2}$, $F_2 = \sum_{i=2}^{n} e^{\tau_i} z_2 h_{i,2}^2(Z_2)$ and $\kappa_{\alpha_2} > 0$. Based on Lemma 3.2, we have

$$F_2 = \theta_2^{*T}\varphi_2 + \varepsilon_2,\ \|\varepsilon_2\| \leq \bar{\varepsilon}_2, \tag{4.109}$$

where $\bar{\varepsilon}_2 > 0$. Substituting (4.109) into (4.108) yields

$$\dot{V}_2 \leq -\lambda_0\|\tilde{\xi}\|^2 - \bar{\lambda}V_{q_1} - \sum_{j=1}^{2}\bar{c}_j z_j^2 - \bar{\delta}_2\omega_2^2 + \sum_{j=1}^{2}\frac{1}{2b_j^2}z_j^2\tilde{\theta}$$
$$+\frac{1}{2}z_3^2 + \frac{1}{2}\omega_3^2 - \frac{1}{r}\tilde{\theta}\dot{\hat{\theta}} + \sum_{i=3}^{n}\sum_{j=3}^{i}e^{\tau_i}z_j^2 h_{i,j}^2(Z_j) + (\bar{g}N(\zeta)+1)\dot{\zeta} + \epsilon_2, \tag{4.110}$$

where $\bar{c}_2 = c_2 - \frac{3}{2} - \frac{\kappa_{\varepsilon 2}}{2}$, $\epsilon_2 = \epsilon_1 + \frac{1}{2\kappa_{\varepsilon_2}}\bar{\varepsilon}_2^2 + \frac{1}{2}b_2^2 + \frac{1}{2\kappa_{\alpha_2}}\dot{\alpha}_1^2$, and $\kappa_{\varepsilon_2} > 0$. Let α_2 pass the first-order filter

$$\delta_3\dot{\alpha}_3^f + \alpha_3^f = \alpha_2,\ \alpha_3^f(0) = \alpha_2(0), \tag{4.111}$$

where δ_3 is a positive constant.

Step m ($3 \leq m \leq n-1$): From (4.87) and (4.88), the time derivative of z_m can be described as

$$\dot{z}_m = z_{m+1} + \omega_{m+1} + \alpha_m + M_m, \tag{4.112}$$

where $M_m = -k_m\hat{\xi}_1 + \frac{1}{\delta_m}\omega_m$. Let the virtual control input α_m as

$$\alpha_m = -c_m z_m - M_m - \frac{1}{2b_m^2}z_m\hat{\theta}, \tag{4.113}$$

where c_m and b_m are positive constants. Next, we take the following Lyapunov candidate function

$$V_m = V_{m-1} + \frac{1}{2}z_m^2 + \frac{1}{2}\omega_m^2. \tag{4.114}$$

According to the definition of ω_m, (4.111)–(4.113), we have

$$\dot{V}_m \leq -\lambda_0\|\tilde{\xi}\|^2 - \bar{\lambda}V_{q_1} - \sum_{j=1}^{m-1}\bar{c}_j z_j^2 - \sum_{j=2}^{m}\bar{\delta}_j\omega_j^2 - \frac{1}{r}\tilde{\theta}\dot{\hat{\theta}} + \sum_{j=1}^{m-1}\frac{1}{2b_j^2}z_j^2\tilde{\theta}$$
$$+\sum_{i=m+1}^{n}\sum_{j=m+1}^{i}e^{\tau_i}z_j^2 h_{i,j}^2(Z_j) - c_m z_m^2 + \frac{3}{2}z_m^2 + \frac{1}{2}z_{m+1}^2$$
$$+\frac{1}{2}\omega_{m+1}^2 - \frac{1}{2b_m^2}z_m^2\hat{\theta} + z_m F_m + (\bar{g}N(\zeta)+1)\dot{\zeta} + \epsilon_{m-1} + \frac{1}{2\kappa_{\alpha_m}}\dot{\alpha}_m^2 \tag{4.115}$$

where $\bar{\delta}_m = \frac{1}{\delta_m} - \frac{1}{2} - \frac{\kappa_{\alpha m}}{2}$, $F_m = \sum_{i=m}^{n} e^{\tau_i} z_m h_{i,m}^2(Z_m)$ and $\kappa_{\alpha_m > 0}$. Based on the Lemma 3.2, there exist fuzzy logic systems such that

$$F_m = \theta_m^{*T} \varphi_m + \varepsilon_m, \ \|\varepsilon_m\| \leq \bar{\varepsilon}_m, \tag{4.116}$$

where $\bar{\varepsilon}_m > 0$ and then we have

$$\begin{aligned} \dot{V}_m \leq & - \lambda_0 \|\tilde{\xi}\|^2 - \bar{\lambda} V_{q_1} - \sum_{j=1}^{m} \bar{c}_j z_j^2 - \sum_{j=2}^{m} \bar{\delta}_j \omega_j^2 + \frac{1}{2} z_{m+1}^2 + \frac{1}{2} \omega_{m+1}^2 - \frac{1}{r} \tilde{\theta} \dot{\hat{\theta}} \\ & + \sum_{j=1}^{m} \frac{1}{2b_j^2} z_j^2 \tilde{\theta} + (\bar{g} N(\zeta) + 1) \dot{\zeta} + \sum_{i=m+1}^{n} \sum_{j=m+1}^{i} e^{\tau_i} z_j^2 h_{i,j}^2 (Z_j) + \epsilon_m, \end{aligned} \tag{4.117}$$

where $\bar{c}_m = c_m - \frac{3}{2} - \frac{\kappa_{\varepsilon m}}{2}$, $\kappa_{\varepsilon_m} > 0$ and $\epsilon_m = \epsilon_{m-1} + \frac{1}{2\kappa_{\alpha m}} \dot{\alpha}_m^2 + \frac{1}{2} b_m^2 + \frac{1}{2\kappa_{\varepsilon m}} \bar{\varepsilon}_m^2$. Let α_m be the following first-order filter input

$$\delta_{m+1} \dot{\alpha}_{m+1}^f + \alpha_{m+1}^f = \alpha_m, \ \alpha_{m+1}^f(0) = \alpha_m(0), \tag{4.118}$$

where δ_m denotes a positive constant.

Step n: Consider the definition of z_n and ω_n. We can get the time derivative of z_n as

$$\dot{z}_n = u + M_n, \tag{4.119}$$

where $M_n = -k_n \hat{\xi}_1 + \frac{1}{\delta_m} \omega_m$. The actual control input and adaptive law have

$$v = -c_n z_n - M_n - \frac{1}{2b_n^2} z_n \hat{\theta}, \tag{4.120}$$

$$\dot{\hat{\theta}} = \sum_{j=1}^{n} \frac{r}{2b_j^2} z_j^2 - k_0 \hat{\theta}, \tag{4.121}$$

where c_n, b_n and k_0 are all appropriate positive constants. Then, we consider the Lyapunov function V_n

$$V_n = V_{n-1} + \frac{1}{2} z_n^2 + \frac{1}{2} \omega_n^2. \tag{4.122}$$

Its derivative can be developed as

$$
\begin{aligned}
\dot{V}_n \leq & -\lambda_0 \|\xi\|^2 - \bar{\lambda} V_{q_1} - \sum_{j=1}^{n-1} \bar{c}_j z_j^2 - c_n z_n^2 - \sum_{j=2}^{n} \bar{\delta}_j \omega_j^2 + \frac{3}{4} z_n^2 - \sum_{j=1}^{n} \frac{1}{2b_j^2} z_j^2 \tilde{\theta} \\
& + \frac{k_0}{r} \tilde{\theta}\hat{\theta} + z_n F_n - \frac{1}{2b_n^2} z_n^2 \hat{\theta} + \frac{1}{2\kappa_{\alpha_n}} \dot{\alpha}_{n-1}^2 + \sum_{j=1}^{n-1} \frac{1}{2b_j^2} z_j^2 \tilde{\theta} \\
& + \Delta v^2 + (\bar{g} N(\zeta) + 1)\dot{\zeta} + \epsilon_{n-1},
\end{aligned} \tag{4.123}
$$

where $\Delta v = |v - u|$, $\bar{\delta}_n = \frac{1}{\delta_n} - \frac{1}{2} - \frac{\kappa_{\alpha n}}{2}$, $\kappa_{\alpha_n} > 0$ and $F_n = e^{\tau_n} z_n h_{n,n}^2(Z_n)$. There exist fuzzy logic systems such that

$$
F_n = \theta_n^{*T} \varphi_n + \varepsilon_n, \; \|\varepsilon_n\| \leq \bar{\varepsilon}_n, \tag{4.124}
$$

where $\bar{\varepsilon}_n > 0$. Moreover, due to $\tilde{\theta} = \theta - \hat{\theta}$, we have

$$
\tilde{\theta}\hat{\theta} = \tilde{\theta}(\theta - \tilde{\theta}) \leq \frac{1}{2}\theta^2 - \frac{1}{2}\tilde{\theta}^2. \tag{4.125}
$$

Taking (4.125) and (4.126) into (4.123) yields

$$
\dot{V}_n \leq -\bar{\lambda} V_{q_1} - \lambda_0 \|\xi\|^2 - \sum_{j=1}^{n} \bar{c}_j z_j^2 - \sum_{j=2}^{n} \bar{\delta}_j \omega_j^2 - \frac{k_0}{2r} \tilde{\theta}^2 + (\bar{g} N(\zeta) + 1)\dot{\zeta} + \epsilon_n, \tag{4.126}
$$

where $\bar{c}_n = c_n - \frac{3}{4} - \frac{\kappa_{\varepsilon n}}{2}$, $\kappa_{\varepsilon_n} > 0$ and $\epsilon_n = \epsilon_{n-1} + \frac{b_n^2}{2} + \frac{1}{2\kappa_{\varepsilon n}} \varepsilon_n^2 + \frac{1}{2\kappa_{\alpha n}} \dot{\alpha}_{n-1}^2 + \frac{k_0}{2r}\theta^2$ $+ \Delta v^2$.

Let $Z = [z_1, \ldots, z_n, \omega_2, \ldots, \omega_n, \xi^T, \tilde{\theta}, \tau_1, \ldots, \tau_n]^T$. For any given positive constants $\bar{P}_z$, the set $\mathcal{Z} := \{Z : V_n(Z) \leq \bar{P}_z\}$ is a compact set. Then, the following theorem about the SPC design for the time-delay nonlinear system can be concluded.

Theorem 4.2 *Consider the nonlinear time-delay systems (4.77) under Assumptions 4.2–4.4, the input saturation (4.78) and the FTPP (4.80) with the input-driven observer (4.83), virtual control input (4.102), (4.107), (4.113), control input (4.120) and the adaptive law (4.121). For the given positive constant $\bar{P}_z$, if the initial states have $V_n(Z(0)) \leq \bar{P}_z$, there exist $c_i, b_i, \delta_i, k_0, \kappa_0, \kappa_{\eta_i}, \kappa_{\varepsilon_i}, \kappa_{\alpha_i}$ and $\bar{u}$ such that: $\frac{(\bar{g}N(\zeta)+1)\dot{\zeta}+\epsilon_n}{\lambda_1} \leq \bar{P}_z$ and $|u - v| \leq \bar{u}$ for $\forall t > 0$. And then all the closed-loop system signals are ensured to be bounded under the proposed Nussbaum-based Tunnel Prescribed Control and the tracking error e_1 can converge into the specified region in finite time, while the FTPP constraints are never violated despite the input saturation and unknown time delay.*

Proof From (4.126), we can have

$$
\begin{aligned}
\dot{V}_n \le & -\frac{\lambda_0}{\lambda_{\max}(P)}\xi^T P\xi - \sum_{j=1}^{n} 2\bar{c}_j\frac{1}{2}z_j^2 - \bar{\lambda}V_{q_1} - \frac{k_0}{2r}\tilde{\theta}^2 \\
& - \sum_{j=2}^{n} 2\bar{\delta}_j\frac{1}{2}\omega_j^2 + (\bar{g}N(\zeta)+1)\dot{\zeta} + \epsilon_n \\
\le & -\lambda_1 V_n + (\bar{g}N(\zeta)+1)\dot{\zeta} + \epsilon_n,
\end{aligned}
\tag{4.127}
$$

where $\lambda_1 = \min(\frac{\lambda_0}{\lambda_{\max}(P)}, 2\bar{c}_j, 2\bar{\delta}_j, \bar{\lambda}, k_0)$. According to the Lemma 2.5, the boundedness of $(\bar{g}N(\zeta)+1)\dot{\zeta}$ and $\zeta(t)$ can be guaranteed. We can appropriately select parameters $c_i, b_i, \delta_i, k_0, \kappa_0, \kappa_{\eta_i}, \kappa_{\varepsilon_i}$ and κ_{α_i} such that $\frac{(\bar{g}N(\zeta)+1)\dot{\zeta}+\epsilon_n}{\lambda_1} \le \bar{P}_z$. By the Assumption 4.3, these parameters can also make $|u-v| \le \bar{u}$ hold. From (4.127), it implies that $\dot{V}_n \le 0$ on $V_n = \bar{P}_z$. Then the set $\mathcal{P}$ is invariant. Therefore, we can have $Z \in \mathcal{Z}$ for all $t > 0$ under the initial condition of $V_n(Z(0)) \le \bar{P}_z$. Namely, all the closed-loop system states are ensured to be bounded, where the boundedness of z_1 indicates that the tracking signal $e_1(t)$ satisfies the FTPP constraints (4.80) despite of unknown time delay and input saturation. Moreover, owing to the distinctive merits of FTPP, we can obtain better tracking performance with smaller overshoot and the tracking signal e_1 is conducted into the specified region in finite time. Moreover, a block diagram of the control procedure and signals is depicted in Fig. 4.11 to clearly show the control scheme details. This completes the proof. □

Remark 4.6 By introducing the dynamic surface control, it lowers the computation level of the control design. By defining the maximum value of unknown parameters θ_i^* $(i = 0\ldots, n)$ as $\theta = \max\{\|\theta_i^*\|^2 : i = 0, \ldots, n\}$. Therefore, by updating

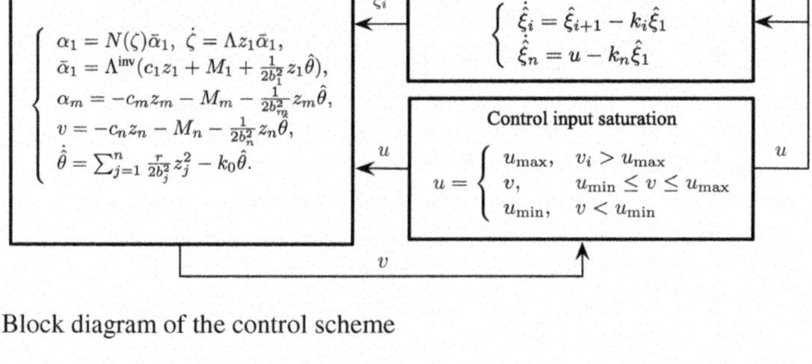

Fig. 4.11 Block diagram of the control scheme

(estimating) the single parameter θ, instead of each individual parameter for θ_i^*, the proposed control solves the underlying tracking control problem at undemanding computational expenses, leading to low-complexity control strategy.

Remark 4.7 We divide the control parameters into three categories and give more details about their selections. First, for the parameter c_i in the virtual and actual control input α_i, its value influences the convergence rate and tracking accuracy. With the increasing of c_i, the control input increases. But too large c_i may seriously degrade the tracking performance and even lead to instability due to the input saturation limiting the tracking ability. Second, larger k_0 and b_i lead to faster convergence rate of the adaptive law, but result in poor tracking accuracy. Third, for the parameters Γ_i in auxiliary system, too large value leads to smaller modified signal. In this condition, SPP cannot enlarge the boundaries to a large enough allowable set to guarantee both input and performance constraints. On the contrary, too small Γ_i increases the control design conservatism. Therefore, its value belongs to a balance product.

4.3.5 Numerical Simulations

To better illustrate the distinctive merits of the presented Finite-time Tunnel Prescribed Control, simulation results are given for a nonlinear time-delay system. Give the following second-order nonlinear system:

$$\begin{cases} \dot{x}_1 = g_1(x_1)x_2 + f_1(x_1) + q_1(x_1(t-\tau_1)) \\ \dot{x}_2 = g_2(\bar{x}_2)u + f_2(\bar{x}_2) + q_2(\bar{x}_2(t-\tau_2)) \\ y = x_1 \end{cases} \tag{4.128}$$

where $g_1(x_1) = 1 + 0.1\sin x_1^2$, $g_2(\bar{x}_2) = 1 + 0.1\cos x_1 x_2$, $f_1(x_1) = 0.01\sin(0.6x_1) + 1$, $f_2(\bar{x}_2) = 0.05x_1x_2$, $q_1(x_1(t-\tau_1)) = 0.05\sin(x_1^2(t-\tau_1))$, $q_2(\bar{x}_2) = 0.04\cos(2x_1(t-\tau_2)\cdot x_2(t-\tau_2)) + 0.02$, $\tau_1 = 1 + 0.5\sin t$ and $\tau_2 = 5 - 0.2\cos 0.5t$. We select the initial states as $x_1(0) = x_2(0) = 0$. And the reference trajectory is given by $y_r = 0.9\cos(0.2t - 0.3)$.

According to (4.83), the input-driven observer can be constructed as

$$\begin{cases} \dot{\hat{\xi}}_1 = \hat{\xi}_2 - k_1\hat{\xi}_1 \\ \dot{\hat{\xi}}_2 = u - k_2\hat{\xi}_1 \end{cases} \tag{4.129}$$

where $k_1 = 4.8$, $k_2 = 5$, and $\hat{\xi}_1(0) = \hat{\xi}_2(0) = 0$. Define the tracking error states as $e_1 = x_1 - y_r$, $z_1 = \ln(\frac{e_l + e_1}{e_u - e_1})$, $z_2 = \hat{\xi}_2 - \alpha_2^f$ and $\omega_2 = \alpha_2^f - \alpha_1$, where the control strategy is design as follows

$$\alpha_1 = N(\zeta)\bar{\alpha}_1, \ \dot{\zeta} = \Lambda z_1 \bar{\alpha}_1,$$

$$\bar{\alpha}_1 = \Lambda^{\text{inv}}(c_1 z_1 + M_1 + \frac{1}{2b_1^2} z_1 \hat{\theta}),$$

$$v = -c_2 z_2 - M_2 - \frac{1}{2b_2^2} z_2 \hat{\theta},$$

$$\dot{\hat{\theta}} = \sum_{j=1}^{2} \frac{r}{2b_j^2} z_j^2 - k_0 \hat{\theta},$$

where $N(\zeta) = \zeta^2 \cos \zeta$, $c_1 = 4.0$, $\zeta(0) = 1.5$, $c_2 = 40.0$, $b_1 = b_2 = 0.5$, $r = 0.5$, $k_0 = 1.0$, $\delta_2 = 0.1$, $\hat{\theta}(0) = 0.5$, $u_{\max} = 15.0$, and $u_{\min} = -15.0$. For the FTPP in (4.80), let $\delta_l = 0.2$, $\delta_u = 0.6$, $\rho_0 = 2.0$, $\rho_{t_f} = 0.3$, $l = 0.5$ and $t_f = 5.0$.

We show the simulation results in Figs. 4.12, 4.13, 4.14, 4.15, 4.16 and 4.15. The tracking error trajectory $e_1(t) = y(t) - y_r(t)$ is depicted in Fig. 4.12, where its trajectory is always evolved within the FTPP-constrained allowable region despite input saturation, unknown control directions and time delay nonlinearities. We first provide more descriptions about the salient features of FTPP. From Fig. 4.12, the performance boundaries $e_u(t)$ and $-e_l(t)$ are distributed at the same side. In this way, a narrow allowable region is established, which imposes more strict constraint on the tracking error trajectory. Namely, the overshoot performance can be effectively reduced under this condition, which leads to faster transient performance. Moreover, the tracking error trajectory is steered to the steady state set within user-specified finite time t_f. The control input signal $u(t)$ is depicted in Fig. 4.13, where $u(t)$ is limited within a feasible region set by $u_{\max}$ and $u_{\min}$. The boundedness of the adaptive law is illustrated in Fig. 4.14. Moreover, the estimation errors $\tilde{\xi}_1$ and $\tilde{\xi}_2$ are depicted in Fig. 4.15, and their values are guaranteed to be bounded.

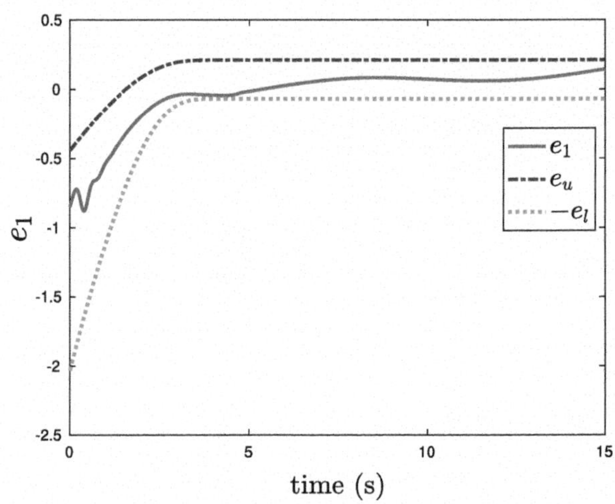

Fig. 4.12 The tracking error e_1, and constraints e_u and e_l under FTPP

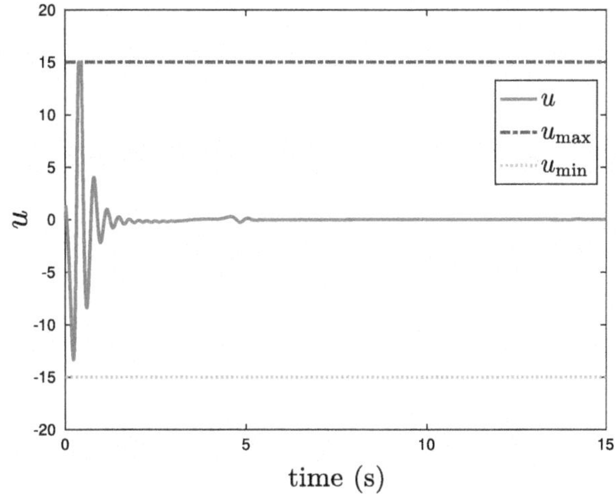

Fig. 4.13 The control input u and its constraints

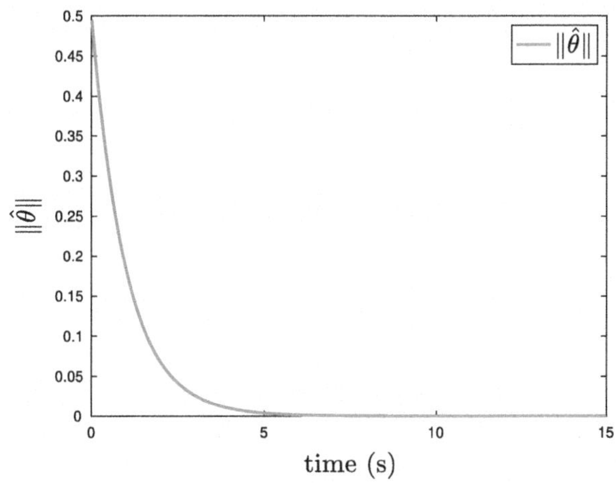

Fig. 4.14 The adaptive law $\hat{\theta}$

To clearly show the robustness of our proposed control scheme, we consider some measurement noise in simulation. The actual measurement state x_1' with measurement noise is expressed as $x_1' = 1.1x_1 + 0.08\sin(0.2\pi t)$, which will be utilized in the control design. The tracking error e_1 is depicted in Fig. 4.16, whose evolution is always within the allowable set. Therefore, it indicates the robustness of such control strategy against unknown nonlinearities, such as time delay and measurable noise. Moreover, we also provide comparative simulations of the classical PPC. According to (4.79), the PPFs are selected by $\bar{\rho}(t) = (\rho_0 - \rho_\infty)e^{-lt} + \rho_\infty$, with $\rho_0 = 2.0$, $\rho_\infty = 0.3$, $l = 0.5$ and $\delta = 0.6$. The PPC-based tracking error result is plotted in Fig. 4.17. Obviously, the loose performance boundaries e_u and e_l cannot impose heavy constraints on the maximum overshoot due to the large opening. Furthermore, considering the unknown control directions and unknown time-delay nonlinearities,

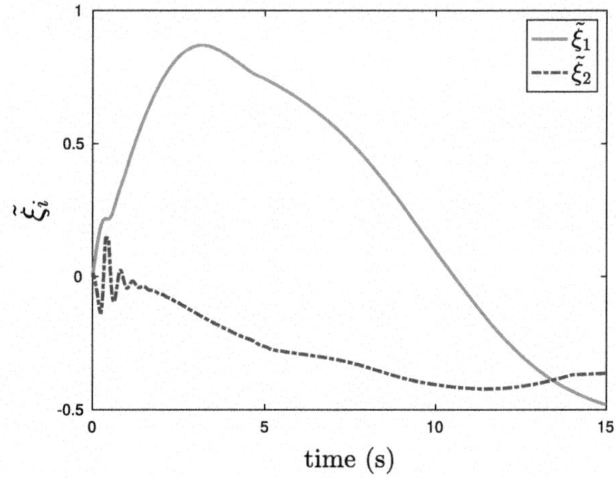

Fig. 4.15 The estimation error $\tilde{\xi}_i$

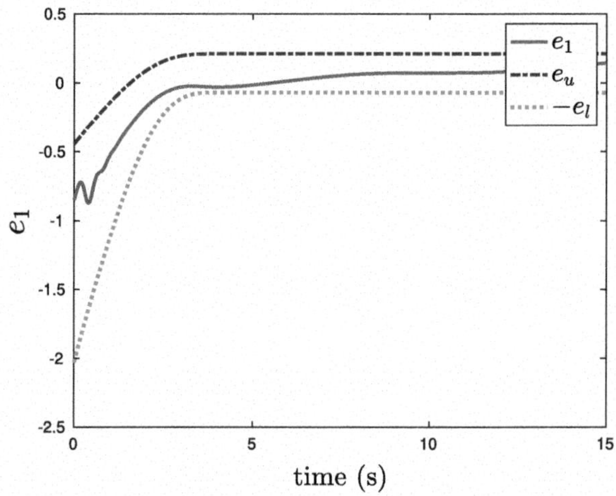

Fig. 4.16 The tracking error e_1 under measurement noise

the tracking result of e_1 shows poor transient performance and e_1 approaches to the boundary e_u. Unfortunately, under the input saturation, the PPC-based control scheme cannot successfully fulfill both input and output constraints for the given condition. The tracking error e_1 approaches to or even violates the performance constraint e_u at $2.96s$, which leads to control singularity problem. Overall, the distinctive merits of our proposed control scheme have been demonstrated by the above tracking results.

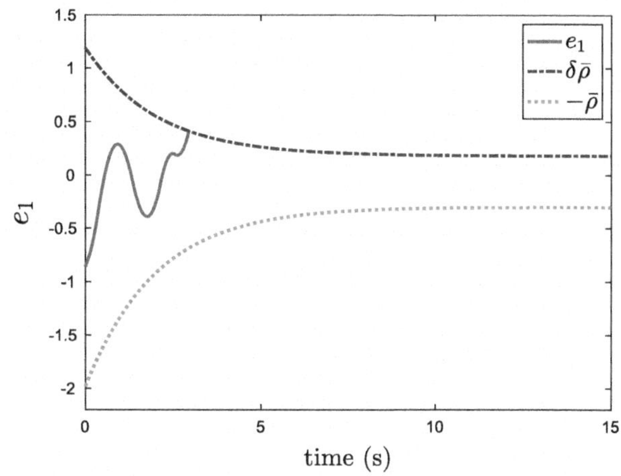

Fig. 4.17 The tracking error e_1, and constraints e_u and e_l under PPC

4.4 Conclusions

In this chapter, we have investigated two control strategies for a class of nonlinear systems with input saturation and performance constraints under different conditions, such as actuator faults, unknown time delay, unknown control directions, and unmeasurable states. The proposed concepts of Tunnel Prescribed Performance and Finite-time Tunnel Prescribed Performance have proven effective in reducing overshoot and achieving finite-time convergence. These novel performance constraints simplify both the control design and stability analysis, offering a more structured framework. Specifically, we developed the Tunnel Prescribed Control and Finite-time Tunnel Prescribed Control for non-strict feedback nonlinear systems and strict feedback time-delay nonlinear systems, respectively. In the first section, Tunnel Prescribed Control is designed such that all the closed-loop signals are guaranteed to be bounded, and the tracking error is evolved within the prescribed performance boundaries even in the presence of input saturation and actuator faults. In the second section, a Finite-time Tunnel Prescribed Control is proposed that drives the tracking error into the specified region within finite time under the unknown time delay and input saturation. The control scheme is designed with low complexity which can facilitate the implementation in applications. Simulation results have further demonstrated the effectiveness of our presented two control strategies under different conditions.

References

1. Bechlioulis, C.P., Rovithakis, G.A.: Robust adaptive control of feedback linearizable mimo nonlinear systems with prescribed performance. IEEE Transactions on Automatic Control **53**(9), 2090–2099 (2008)

2. Bechlioulis, C.P., Rovithakis, G.A.: Robust partial-state feedback prescribed performance control of cascade systems with unknown nonlinearities. IEEE Trans. Autom. Control **56**(9), 2224–2230 (2011)
3. Bikas, L.N., Rovithakis, G.A.: Tracking performance guarantees in the presence of quantization for uncertain nonlinear systems. IEEE Transactions on Automatic Control (2020)
4. Cao, Y., Cao, J., Song, Y.: Practical prescribed time tracking control over infinite time interval involving mismatched uncertainties and non-vanishing disturbances. Automatica **136**, 110050 (2022)
5. Chen, M., Tao, G., Jiang, B.: Dynamic surface control using neural networks for a class of uncertain nonlinear systems with input saturation. IEEE transactions on neural networks and learning systems **26**(9), 2086–2097 (2014)
6. Chen, W., Li, X., Ren, W., Wen, C.: Adaptive consensus of multi-agent systems with unknown identical control directions based on a novel nussbaum-type function. IEEE Transactions on Automatic Control **59**(7), 1887–1892 (2013)
7. Cheng, C., Zhang, Y., Liu, S.: Neural observer-based adaptive prescribed performance control for uncertain nonlinear systems with input saturation. Neurocomputing **370**, 94–103 (2019)
8. Cui, G., Yu, J., Shi, P.: Observer-based finite-time adaptive fuzzy control with prescribed performance for nonstrict-feedback nonlinear systems. IEEE Transactions on Fuzzy Systems (2020)
9. Guo, Q., Zhang, Y., Celler, B.G., Su, S.W.: Neural adaptive backstepping control of a robotic manipulator with prescribed performance constraint. IEEE Trans. Neural Netw. Learning Syst. **30**(12), 3572–3583 (2018)
10. Hu, C., Wang, Z., Qin, Y., Huang, Y., Wang, J., Wang, R.: Lane keeping control of autonomous vehicles with prescribed performance considering the rollover prevention and input saturation. IEEE transactions on intelligent transportation systems **21**(7), 3091–3103 (2019)
11. Huang, H., He, W., Li, J., Xu, B., Yang, C., Zhang, W.: Disturbance observer-based fault-tolerant control for robotic systems with guaranteed prescribed performance. IEEE Transactions on Cybernetics (2020)
12. Huang, J., Wang, W., Wen, C., Zhou, J.: Adaptive control of a class of strict-feedback time-varying nonlinear systems with unknown control coefficients. Automatica **93**, 98–105 (2018)
13. Ji, R., Ma, J., Li, D., Ge, S.S.: Finite-time adaptive output feedback control for mimo nonlinear systems with actuator faults and saturations. IEEE Transactions on Fuzzy Systems **29**(8), 2256–2270 (2020)
14. Jiang, Z.P., Mareels, I., Hill, D.J., Huang, J.: A unifying framework for global regulation via nonlinear output feedback: from iss to iiss. IEEE Transactions on Automatic Control **49**(4), 549–562 (2004)
15. Li, D., Ge, S.S., Lee, T.H.: Simultaneous arrival to origin convergence: sliding-mode control through the norm-normalized sign function. IEEE Transactions on Automatic Control **67**(4), 1966–1972 (2021)
16. Li, T., Li, Z., Wang, D., Chen, C.P.: Output-feedback adaptive neural control for stochastic nonlinear time-varying delay systems with unknown control directions. IEEE Transactions on Neural Networks and Learning Systems **26**(6), 1188–1201 (2014)
17. Li, Y., Li, K., Tong, S.: Adaptive neural network finite-time control for multi-input and multi-output nonlinear systems with positive powers of odd rational numbers. IEEE Trans. Neural Netw. Learning Syst. (2019)
18. Li, Y., Tong, S., Li, T.: Composite adaptive fuzzy output feedback control design for uncertain nonlinear strict-feedback systems with input saturation. IEEE Transactions on Cybernetics **45**(10), 2299–2308 (2014)
19. Mehdifar, F., Bechlioulis, C.P., Hashemzadeh, F., Baradarannia, M.: Prescribed performance distance-based formation control of multi-agent systems. Automatica **119**, 109086 (2020)
20. Mitchell, I.M., Bayen, A.M., Tomlin, C.J.: A time-dependent hamilton-jacobi formulation of reachable sets for continuous dynamic games. IEEE Transactions on automatic control **50**(7), 947–957 (2005)

21. Oliveira, T.R., Peixoto, A.J., Hsu, L.: Sliding mode control of uncertain multivariable nonlinear systems with unknown control direction via switching and monitoring function. IEEE Transactions on Automatic Control **55**(4), 1028–1034 (2010)
22. Qiu, J., Sun, K., Wang, T., Gao, H.: Observer-based fuzzy adaptive event-triggered control for pure-feedback nonlinear systems with prescribed performance. IEEE Trans. Fuzzy Syst. **27**(11), 2152–2162 (2019)
23. Shao, X., Hu, Q., Shi, Y.: Adaptive pose control for spacecraft proximity operations with prescribed performance under spatial motion constraints. IEEE Trans. Contr. Syst. Technol. pp. 1–15 (2020)
24. Shao, X., Hu, Q., Shi, Y., Jiang, B.: Fault-tolerant prescribed performance attitude tracking control for spacecraft under input saturation. IEEE Transactions on Control Systems Technology **28**(2), 574–582 (2018)
25. Sui, S., Tong, S., Li, Y.: Observer-based fuzzy adaptive prescribed performance tracking control for nonlinear stochastic systems with input saturation. Neurocomputing **158**, 100–108 (2015)
26. Sun, K., Qiu, J., Karimi, H.R., Fu, Y.: Event-triggered robust fuzzy adaptive finite-time control of nonlinear systems with prescribed performance. IEEE Transactions on Fuzzy Systems **29**(6), 1460–1471 (2020)
27. Sun, W., Wu, Y.Q., Sun, Z.Y.: Command filter-based finite-time adaptive fuzzy control for uncertain nonlinear systems with prescribed performance. IEEE Transactions on Fuzzy Systems **28**(12), 3161–3170 (2020)
28. Wang, M., Chen, B., Shi, P.: Adaptive neural control for a class of perturbed strict-feedback nonlinear time-delay systems. IEEE Transactions on Systems, Man, and Cybernetics, Part B (Cybernetics) **38**(3), 721–730 (2008)
29. Wang, W., Liang, H., Pan, Y., Li, T.: Prescribed performance adaptive fuzzy containment control for nonlinear multiagent systems using disturbance observer. IEEE Trans. Cybern. (2020)
30. Wang, W., Wang, D., Peng, Z., Li, T.: Prescribed performance consensus of uncertain nonlinear strict-feedback systems with unknown control directions. IEEE Transactions on Systems, Man, and Cybernetics: Systems **46**(9), 1279–1286 (2015)
31. Wei, C., Luo, J., Dai, H., Duan, G.: Learning-based adaptive attitude control of spacecraft formation with guaranteed prescribed performance. IEEE transactions on cybernetics **49**(11), 4004–4016 (2018)
32. Wen, C., Zhou, J., Liu, Z., Su, H.: Robust adaptive control of uncertain nonlinear systems in the presence of input saturation and external disturbance. IEEE Transactions on Automatic Control **56**(7), 1672–1678 (2011)
33. Xu, J.X., Yan, R.: Iterative learning control design without a priori knowledge of the control direction. Automatica **40**(10), 1803–1809 (2004)
34. Xudong, Y.: Asymptotic regulation of time-varying uncertain nonlinear systems with unknown control directions. Automatica **35**(5), 929–935 (1999)
35. Yong, K., Chen, M., Shi, Y., Wu, Q.: Flexible performance-based robust control for a class of nonlinear systems with input saturation. Automatica **122**, 109268 (2020)
36. Zhang, B., Jiang, L., Bai, J.: Velocity-free prescribed performance control for spacecraft hovering over an asteroid with input saturation. J. Franklin Institute (2020)
37. Zhang, C., Ma, G., Sun, Y., Li, C.: Observer-based prescribed performance attitude control for flexible spacecraft with actuator saturation. ISA Trans. **89**, 84–95 (2019)
38. Zhang, J.X., Wang, Q.G., Ding, W.: Global output-feedback prescribed performance control of nonlinear systems with unknown virtual control coefficients. IEEE Transactions on Automatic Control (2021)
39. Zhang, J.X., Yang, G.H.: Prescribed performance fault-tolerant control of uncertain nonlinear systems with unknown control directions. IEEE Trans. Autom. Control **62**(12), 6529–6535 (2017)
40. Zhang, T., Ge, S.S.: Adaptive neural control of mimo nonlinear state time-varying delay systems with unknown dead-zones and gain signs. Automatica **43**(6), 1021–1033 (2007)
41. Zhou, Q., Shi, P., Xu, S., Li, H.: Adaptive output feedback control for nonlinear time-delay systems by fuzzy approximation approach. IEEE Transactions on Fuzzy Systems **21**(2), 301–313 (2012)

Chapter 5
Tunnel Prescribed Control of Nonlinear System for Any Initial Conditions

5.1 Introduction

In the previous chapters, we have discussed classical prescribed performance control and our proposed tunnel prescribed control. However, there exists an implicit assumption in these performance control results, that is, the tracking error should be constrained from the beginning of the system operation, indicating that the knowledge of initial condition is completely known. If the initial condition violation occurs, these methods becomes invalid since they are not defined outside the allowable set. From a global perspective, it yields local tracking results. However, Entry Capture Problem (ECP) is frequently encountered in practical scenarios, where the tracking error may violate the performance constraints initially, then is regulated into an allowable set within a reasonable user-specified time [21]. A representative example is a manipulator system, starting from an unconstrained region for some time, then is expected to track and grasp an object in a specified performance. Such scenario is also common in many other practical applications (i.e., multi-agent formation under various initial condition, aerial refueling, etc.). Therefore, we, in this chapter, address a more general, attractive yet challenging ECP for uncertain nonlinear systems under any initial condition (including the initial constraint-violation) and asymmetric performance constraints. Two novel control strategies, namely, Error Self-tuning Function-based Tunnel Prescribed Control and Self-adjustable Performance Control, are proposed to solve the ECP. In lieu of the Nussbaum technique, an orientation function is introduced to deal with unknown control directions while such way is capable of reducing the control peaking problem. The resulted control designs show low complexity and efficiency in addressing ECP task. Simulation results are provided to demonstrate the effectiveness of the proposed control strategies.

R. Ji et al., *Constrained Nonlinear Control: Performance and Robustness*,
https://doi.org/10.1007/978-981-96-2971-8_5

5.2 Error Self-tuning Function-Based Tunnel Prescribed Control

5.2.1 State-Feedback Control of Nonlinear Systems with Known Control Directions

5.2.1.1 System Description

Consider the following strict-feedback nonlinear systems

$$\begin{cases} \dot{x}_i = f_i(\bar{x}_i) + g_i(\bar{x}_i)x_{i+1}, \ 1 \le i \le n-1 \\ \dot{x}_n = f_n(\bar{x}_n) + g_n(\bar{x}_n)u \\ y = x_1 \end{cases} \tag{5.1}$$

where $\bar{x}_i = [x_1, \dots, x_i]^T \in \mathbb{R}^i, i = 1, \dots, n$ denote measurable system state vectors, $f_i(\bar{x}_i) \in \mathbb{R}$ is an unknown smooth function, $g_i(\bar{x}_i) \in \mathbb{R}$ denotes an unknown control coefficient, $u \in \mathbb{R}$ and $y \in \mathbb{R}$ are the system input and output, respectively. In practice, many platforms can be represented or transferred into (5.1), such as manipulator systems, flight systems, etc. [13].

5.2.1.2 Entry Capture Problem

Define the tracking error $e_1(t) = x_1(t) - y_d(t)$ with $y_d(t)$ being the desired tracking trajectory. The control objective of this section is to develop a low-complex tracking control for the nonlinear system (5.1) under any initial tracking condition such that:

(i) All signals in the closed-loop systems are bounded; and
(ii) For any initial tracking condition (including the initial constraint-violation), the tracking error $e_1(t)$ can be regulated into the prescribed performance set within a user-specified settling time.

For clarity of the control objective, we formulate the following Entry Capture Problem (ECP) definition.

Definition 5.1 The tracking error $e_1(t)$ is free or released from any constraint during some initial time interval, then is strictly constrained by the prescribed performance right after a user-given settling time $T > 0$ as:

$$-e_l(t) < e_1(t) < e_u(t), \ \forall t \ge T, \tag{5.2}$$

where $e_u(t)$ and $-e_l(t)$ represent the upper and lower prescribed performance given by (5.6), respectively. Upon this condition, such error performance is said to be ECP.

Remark 5.1 Most existing studies on the PPC comply with an implicit error assumption that the performance constraints should be satisfied from the beginning of the system operation, yielding local property. The ECP focuses on a more general tracking condition in practice which involves the initial constraint violation but satisfies performance constraints within a given time. One typical example is a manipulator system, which starts from any unconstrained region for a period of time, then is required to track and grasp an object in a prescribed performance for obstacle avoidance and mission accomplishment. Therefore, such ECP is interesting yet challenging from both theoretical and practical significance.

For subsequently development, some standard assumptions are given on the nonlinear systems.

Assumption 5.1 The desired trajectory $y_d(t)$ and its derivatives up to $(n+1)$th-order are known, piece-wise continuous and bounded.

Assumption 5.2 The control coefficients $g_i(\bar{x}_i), i = 1, \ldots, n$, are unknown and bounded yet away from zero, that is, there exist positive constants $\underline{g}_i$ and $\bar{g}_i$ such that $0 < \underline{g}_i \leq |g_i(\bar{x}_i)| \leq \bar{g}_i < \infty$. Without loss of generality, the signs of $g_i(\bar{x}_i)$ are assumed to be positive.

Assumption 5.3 Certain crude structures of $f_i(\bar{x}_i)$ can be abstracted, where an unknown constant $b_i \geq 0$ and a known smooth function $\varphi_i(\bar{x}_i) \geq 0$ make the inequality $|f_i(\bar{x}_i)| \leq b_i \varphi_i(\bar{x}_i)$ hold.

Remark 5.2 Assumption 5.1 is widely used for the tracking control problem of the nonlinear systems [28, 49]. Assumption 5.2 is necessary for the nonlinear systems to be controllable and stability analysis, which is usually used in the existing studies [14, 44, 51]. For the Assumption 5.3, it is also extensively used in the control design as presented in [10, 36, 37], where $\varphi_i(\bar{x}_i)$ can be easily abstracted in practice by some crude system information, and please find more details in the simulation section.

5.2.1.3 Scaling Function

To deal with the second control objective, we first introduce the definition of a Scaling function.

Definition 5.2 The scaling function $s_f(t)$ possesses the following unique properties:

(i) $s_f(t)$ is C^{n+1} and bounded for $t \geq 0$;
(ii) $s_f(t)$ is monotonically increasing for $t \in [0, T)$ where T denotes the settling time given in (5.2); and
(iii) $s_f(0) = 0$, $s_f(t) = 1$ for $t \geq T$, $s_f(t) \in [0, 1]$ for all $t \geq 0$, and $\dot{s}_f(t) \to 0$ as $t \to T$.

Proof By utilizing L'Hospital's rule and some simple derivations, we omit the proof here due to the space limitation. □

Typically, there are many candidate functions that satisfy the above properties in Definition 5.1, for example,

$$s_f(t) = \begin{cases} 1 - e^{-\frac{lTt}{T-t}}, & 0 \leq t < T, \\ 1, & t \geq T, \end{cases} \tag{5.3}$$

where l denotes a positive constant.

In practice, nonlinear systems can be subject to a wide range of initial conditions, which may lead to singularity problem if they violate performance constraints, and then control design becomes invalid. To handle this limitation, we propose a scaling function as well as its properties. From Definition 5.1, if we multiply the tracking error by our scaling function, namely, $s_f(t)e_1(t)$, the resulted initial value is always equal to 0 due to $s_f(0) = 0$, which provide a unified solution to any initial condition. As performance constraints should be imposed on the tracking error right after a finite-time T, the result of their multiplication is $s_f(t)e_1(t) = e_1(t)$ with the aid of $s_f(t) = 1$ for all $t \geq T$ so that the ECP can be solved. For $0 < t < T$, the scaling function is required to increase from 0 to 1 smoothly and remains constant thereafter. Such a design mechanism could arouse extensive research on integrating various scaling functions with alternative control methods.

5.2.1.4 Error Self-tuning Function

For any initial tracking condition, the following Error Scaling Function (ESF) $e_s(t)$ is constructed by employing the scaling function $s_f(t)$ on the tracking error $e_1(t)$ as

$$e_s(t) = s_f(t)e_1(t) + (1 - s_f(t))\bar{e}_{ul}(t). \tag{5.4}$$

where $\bar{e}_{ul}(t)$ is given by

$$\bar{e}_{ul}(t) = \frac{e_u(t) - e_l(t)}{2}, \tag{5.5}$$

where $e_u(t)$ and $e_l(t)$ denote the Tunnel Prescribed Performance (TPP) with

$$\begin{aligned} e_l(t) &= \begin{cases} \delta\rho(t), & \text{if } e_{1,0} = 0 \\ (\delta - 1)\rho(t) + \bar{\rho}_\infty, & \text{if } e_{1,0} > 0 \\ (\delta + 1)\rho(t) - \bar{\rho}_\infty, & \text{if } e_{1,0} < 0 \end{cases} \\ e_u(t) &= \begin{cases} \delta\rho(t), & \text{if } e_{1,0} = 0 \\ (\delta + 1)\rho(t) - \bar{\rho}_\infty, & \text{if } e_{1,0} > 0 \\ (\delta - 1)\rho(t) + \bar{\rho}_\infty, & \text{if } e_{1,0} < 0 \end{cases} \end{aligned} \tag{5.6}$$

where $0 < \delta < 1$, $e_{1,0} = e_1(0)$ represents the initial value of the tracking error $e_1(t)$, and $\rho(t) = (\rho_0 - \rho_\infty)e^{-\ell t} + \rho_\infty$ with $0 < \rho_\infty < \rho_0$ and $\ell > 0$.

Invoking (5.3)–(5.5), we can derive two salient features of ESF, which is essential in the control design:

$$e_s(0) = s_f(0)e_1(0) + (1 - s_f(0))\bar{e}_{ul}(0) = \bar{e}_{ul}(0), \tag{5.7}$$

$$e_s(t) = s_f(t)e_1(t) + (1 - s_f(t))\bar{e}_{ul}(t) = e_1(t), \ t \geq T. \tag{5.8}$$

To better illustrate the mechanism behind such ESF (5.4), we take $e_{1,0} < 0$ in (5.6) as an example, where the trajectories of $e_1(t)$ and $e_s(t)$ are plotted in Fig. 5.1. By (5.4), the parameters are selected as: $\delta = 0.5$, $\ell = 1$, $\rho_0 = 0.6$, $\rho_\infty = 0.2$, $T = 3$, and $l = 1$. Given any initial tracking error $e_1(0)$ (including the initial constraint-violation), the initial ESF $e_s(0)$ in (5.7) is always guaranteed within the allowable set since $-e_l(0) < e_s(0) = \bar{e}_{ul}(0) < e_u(0)$. It can be concluded that $\bar{e}_{ul}(0)$ is served as a shifting function to avoid the singularity problem in the control design as conceptually illustrated in Fig. 5.1 at the initial time. To address ECP in (5.2), the tracking error $e_1(t)$ should be constrained by the TPP right after a time instant $t = T$. Therefore, with aid of $s_f(t)$, the effect of the shifting function $\bar{e}_{ul}(t)$ disappears for all $t \geq T$ as depicted in (5.8), and we have $e_s(t) = e_1(t)$ there, which is also clearly illustrated in Fig. 5.1. Then, the following lemma shows that such ESF does not alter the boundedness of $e_1(t)$, which is critical to achieve our control objective.

Lemma 5.1 *If $-e_l(t) < e_s(t) < e_u(t)$ is always satisfied for $t \in [0, T)$, then $e_1(t)$ is bounded there. If $-e_l(t) < e_s(t) < e_u(t)$ holds for all $t \geq 0$, the ECP in (5.2) is accomplished.*

Proof For the first claim, it can be verified by finding contradiction. Suppose that $-e_l(t) < e_s(t) < e_u(t)$ hold for $t \in [0, T)$, but a time instant $t^* \in [0, T)$ exists such that

$$\lim_{t \to t^*} |e_1(t)| = \infty. \tag{5.9}$$

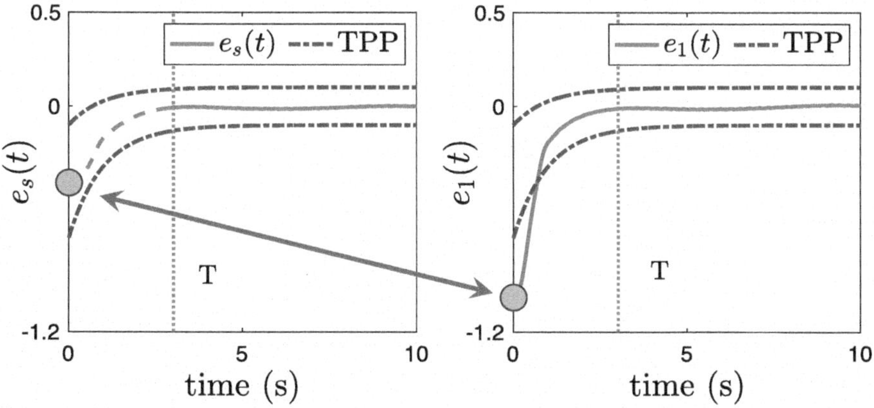

Fig. 5.1 The mechanism behind the time-varying scaling function

By (5.4), there has

$$-e_l(t) < s_f(t)e_1(t) + (1 - s_f(t))\bar{e}_{ul}(t) < e_u(t), \quad (5.10)$$

Inspecting (5.9), we can derive that if and only if $s_f(t^*) = 0$ such that (5.10) holds. From (5.3) and the properties of $s_f(t)$ in Definition 5.2, there is $t^* = 0$ and (5.9) is therefore rewritten as

$$\lim_{t \to 0} |e_1(t)| = \infty. \quad (5.11)$$

Note that all the initial tracking conditions are bounded, i.e., $x_1(0)$ and $y_d(0)$, which contradicts (5.11). This indicates that the hypothesis (5.9) does not exist, instead, the first claim is established. The second claim is observed from the first claim and (5.8). This completes the proof. □

Therefore, based on the above Lemma 5.1, it indicates that the control design is converted to guarantee the ESF $e_s(t)$ within the prescribed performance, namely

$$-e_l(t) < e_s(t) < e_u(t), \ \forall t \geq 0. \quad (5.12)$$

Remark 5.3 Considering the classical PPC methods in [2, 9, 30, 38, 57], they become invalid if the initial-condition violation occurs. Some recent studies have been investigated to remove the limitation on the initial condition: for example, global prescribed control [32, 48, 53] and funnel control [4, 7, 18]. Compared with these existing results, some salient features of our proposed ESF are observed. Different from these performance functions and funnel boundaries which are symmetrically distributed on both sides, our proposed ESF is applicable to both symmetric and asymmetric performance constraints. Moreover, such ESF can be easily extended to other control methods since it is introduced only on the tracking error signal to obtain global tracking results, instead of relying on specific performance functions. For example, it can be integrated in the iterative adaptive critic designs in [16] to eliminate the tracking error completely for better tracking performance.

5.2.1.5 Error-Dependent Transformation Function

In order to guarantee (5.12), we introduce an auxiliary error variable s_1 first

$$s_1 = e_s - \bar{e}_{ul} = s_f(e_1 - \bar{e}_{ul}). \quad (5.13)$$

In response to the asymmetric constraints by TPP, the following ETF z_1 based purely upon auxiliary error variable s_1:

$$z_1 = \frac{s_1}{[e_l + (\bar{e}_{ul} + s_1)][e_u - (\bar{e}_{ul} + s_1)]} = \frac{s_1}{\beta_1}, \tag{5.14}$$

where $\beta_1 = [e_l + (\bar{e}_{ul} + s_1)][e_u - (\bar{e}_{ul} + s_1)]$.

A key feature attained by such ETF is therefore given by the following lemma.

Lemma 5.2 *If the boundedness of z_1 is guaranteed, then $-e_l(t) < e_s(t) < e_u(t)$ holds.*

Proof For any initial tracking condition $e_1(0)$, there has $s_1(0) = 0$ due to $s_f(0) = 0$. According to the definitions of $\bar{e}_{ul}$, e_u and e_l, the inequalities $-e_l(0) < s_1(0) + \bar{e}_{ul}(0) < e_u(0)$ is always satisfied. Then, if z_1 is bounded, from (5.14), it can be derived that $-e_l < s_1 + \bar{e}_{ul} < e_u$. By (5.13), it yields $-e_l < e_s < e_u$. □

The novelty of the ETF as given in (5.14) is twofold:

(i) Our proposed ETF provides a unified solution for dealing with both symmetric and asymmetric constraints, by which the redevelopment of the transformation function is no longer needed. In contrast to the studies in [10, 48, 53, 57], which only address the case of symmetric constraints, the proposed ETF, therefore, enlarges the applicability to more general conditions.

(ii) The ETF depicts an explicit relationship between z_1 and s_1 as demonstrated in (5.14). This relationship plays a crucial role in the control design procedure to address the ECP. However, the commonly used logarithm and hyperbolic tangent transformation functions are difficult to obtain such clear relationship because of their inherently complex and coupled expressions.

5.2.1.6 Control Design

In what follows, the arguments in variables or functions are sometimes omitted if no confusion is caused. To proceed the control design, we introduce the following tracking errors z_1 and e_i $(i = 2, \ldots, n)$ as:

$$z_1 = \frac{s_1}{\beta_1}, \tag{5.15}$$

$$e_i = x_i - \alpha_{i-1},\ i = 2, \ldots, n, \tag{5.16}$$

where s_1 and β_1 can be found in (5.14); and α_i represents the virtual control input to be defined later. With (5.4), (5.13), and (5.14)–(5.16), the virtual control law is derived as

$$\alpha_1 = -\frac{e_1 - \bar{e}_{ul}}{\mu_1 \beta_1}(c_1 + \hat{\theta}_1 \Phi_1), \tag{5.17}$$

$$\alpha_i = -(c_i e_i + \hat{\theta}_i e_i \Phi_i),\ i = 2, \ldots, n-1, \tag{5.18}$$

$$\dot{\hat{\theta}}_1 = r_1 z_1^2 \Phi_1 - \sigma_1 \hat{\theta}_1, \tag{5.19}$$

$$\dot{\hat{\theta}}_i = r_i e_i^2 \Phi_i - \sigma_i \hat{\theta}_i,\ i = 2, \ldots, n-1, \tag{5.20}$$

where c_i, r_i and σ_i are all positive constants; μ_1 is given in (5.24); $\hat{\theta}_i$ denote the estimation values of θ_i defined in (5.31), (5.49) and (5.62); and Φ_i represent known, and computable functions depicted in (5.32), (5.50) and (5.63). The control law is given by

$$u = -(c_n e_n + \hat{\theta}_n e_n \Phi_n), \tag{5.21}$$

$$\dot{\hat{\theta}}_n = r_n e_n^2 \Phi_n - \sigma_n \hat{\theta}_n, \tag{5.22}$$

where $c_n, r_n, \sigma_n > 0$; $\hat{\theta}_n$ and Φ_n are the estimation of θ_n and known function defined in (5.76) and (5.77), respectively.

5.2.1.7 Stability Analysis

From the control design in (5.15)–(5.22), we follow the standard backstepping technique to analyze the system performance.

Step 1: Considering (5.4) and (5.16) with $i = 2$, we have

$$\dot{e}_s = \dot{s}_f e_1 - \bar{e}_{ul} \dot{s}_f + (1 - s_f)\dot{\bar{e}}_{ul} + s_f [f_1 - \dot{y}_d + g_1 (e_2 + \alpha_1)]. \tag{5.23}$$

Then, according to (5.13) and (5.15), the following dynamics is obtained

$$\dot{z}_1 = \mu_1 \dot{s}_1 - \mu_2 = \mu_1 (\dot{e}_s - \dot{\bar{e}}_{ul}) - \mu_2, \tag{5.24}$$

where $\mu_1 = \frac{1}{\beta_1^2}(e_l + \bar{e}_{ul})(e_u - \bar{e}_{ul}) + s_1^2$ and $\mu_2 = \frac{s_1}{\beta_1^2}[(\dot{e}_l + \dot{\bar{e}}_{ul})(e_u - \bar{e}_{ul} - s_1) + (\dot{e}_u - \dot{\bar{e}}_{ul})(e_l + \bar{e}_{ul} + s_1)]$. The derivative of $\frac{1}{2}z_1^2$ along (5.23) and (5.24) is given

$$z_1 \dot{z}_1 = z_1 g_1 \mu_1 s_f \alpha_1 + \Xi_1, \tag{5.25}$$

where $\Xi_1 = z_1 \mu_1 (l_1 + s_f f_1 + s_f g_1 e_2) - z_1 \mu_2$ with $l_1 = \dot{s}_f e_1 - \bar{e}_{ul} \dot{s}_f - s_f \dot{\bar{e}}_{ul} - s_f \dot{y}_d$. Bearing in mind Assumptions 5.1–5.3, and using Young's inequality, it yields

$$z_1 \mu_1 l_1 \le \underline{g}_1 z_1^2 \mu_1^2 l_1^2 + \frac{1}{4\underline{g}_1}, \tag{5.26}$$

$$z_1 \mu_1 s_f f_1 \le \underline{g}_1 z_1^2 \mu_1^2 s_f^2 b_1^2 \varphi_1^2 + \frac{1}{4\underline{g}_1}, \tag{5.27}$$

$$z_1\mu_1 s_f g_1 e_2 \le \underline{g}_2 z_1^2\mu_1^2 s_f^2 e_2^2 + \frac{\bar{g}_1^2}{4\underline{g}_2}, \tag{5.28}$$

$$z_1\mu_2 \le \underline{g}_1 z_1^2\mu_2^2 + \frac{1}{4\underline{g}_1}. \tag{5.29}$$

Therefore, we obtain

$$\Xi_1 \le \underline{g}_1\theta_1 z_1^2\Phi_1 + \underline{g}_2 z_1^2\mu_1^2 s_f^2 e_2^2 + \frac{3}{4\underline{g}_1} + \frac{\bar{g}_1^2}{4\underline{g}_2} \tag{5.30}$$

with

$$\theta_1 = \max\left\{1, b_1^2\right\}, \tag{5.31}$$

$$\Phi_1 = \mu_1^2 l_1^2 + \mu_1^2 s_f^2\varphi_1^2 + \mu_1^2 s_f^2 + \mu_2^2. \tag{5.32}$$

Moreover, based on (5.13), (5.14) and (5.17), there has

$$\begin{aligned} z_1 g_1\mu_1 s_f\alpha_1 =& - z_1 g_1\mu_1\frac{s_f(e_1 - \bar{e}_{ul})}{\mu_1\beta_1}(c_1 + \hat{\theta}_1\Phi_1) \\ =& - g_1 c_1 z_1^2 - g_1\hat{\theta}_1 z_1^2\Phi_1. \end{aligned} \tag{5.33}$$

Consider the following Lyapunov function:

$$V_1 = \frac{1}{2}z_1^2 + \frac{g_1}{2r_1}\tilde{\theta}_1^2, \tag{5.34}$$

where $\tilde{\theta}_1 = \theta_1 - \hat{\theta}_1$; $\hat{\theta}_1$ and r_1 are defined in (5.19). The time derivative of V_1 is derived as

$$\dot{V}_1 = z_1\dot{z}_1 - \frac{g_1}{r_1}\tilde{\theta}_1\dot{\hat{\theta}}_1. \tag{5.35}$$

By taking (5.19), (5.25), (5.30) and (5.33) into (5.35), one has

$$\dot{V}_1 \le - \underline{g}_1 c_1 z_1^2 + \underline{g}_2 z_1^2\mu_1^2 s_f^2 e_2^2 + \frac{3}{4\underline{g}_1} + \frac{\bar{g}_1^2}{4\underline{g}_2} + \frac{\underline{g}_1\sigma_i}{r_1}\tilde{\theta}_1\hat{\theta}_i. \tag{5.36}$$

Upon the definition $\tilde{\theta}_1 = \theta_1 - \hat{\theta}_1$, it follows that

$$\tilde{\theta}_1\hat{\theta}_1 = \tilde{\theta}_1(\theta_1 - \tilde{\theta}_1) \le -\frac{1}{2}\tilde{\theta}_1^2 + \frac{1}{2}\theta_1^2. \tag{5.37}$$

Substituting (5.37) into (5.36) yields

$$\dot{V}_1 \le -\underline{g}_1 c_1 z_1^2 - \bar{\sigma}_1 \tilde{\theta}_1^2 + \underline{g}_2 z_1^2 \mu_1^2 s_f^2 e_2^2 + \varepsilon_1, \tag{5.38}$$

where $\bar{\sigma}_1 = \frac{\underline{g}_1 \sigma_1}{2r_1}$ and $\varepsilon_1 = \frac{3}{4\underline{g}_1} + \frac{\bar{g}_1^2}{4\underline{g}_2} + \frac{\underline{g}_1 \sigma_i}{2r_1}\theta_1^2$ is a bounded signal. Note that $\underline{g}_2 z_1^2 \mu_1^2 s_f^2 e_2^2$ will be tackled later.

Step 2: Taking the time derivative of e_2 in (5.16) for $i = 2$, there has

$$\dot{e}_2 = \dot{x}_2 - \dot{\alpha}_1 = f_2 + g_2 e_3 + g_2 \alpha_2 - \dot{\alpha}_1. \tag{5.39}$$

where $\dot{\alpha}_1$ is derived as

$$\dot{\alpha}_1 = \frac{\partial \alpha_1}{\partial x_1}(f_1 + g_1 x_2) + \Delta\alpha_1, \tag{5.40}$$

where $\Delta\alpha_1 = \frac{\partial \alpha_1}{\partial y_d}\dot{y}_d + \frac{\partial \alpha_1}{\partial \hat{\theta}_1}\dot{\hat{\theta}}_1 + \frac{\partial \alpha_1}{\partial \Phi_1}\dot{\Phi}_1 + \frac{\partial \alpha_1}{\partial e_l}\dot{e}_l + \frac{\partial \alpha_1}{\partial e_u}\dot{e}_u + \frac{\partial \alpha_1}{\partial s_f}\dot{s}_f$, which is computable. Then, consider the Lyapunov function V_2 as

$$V_2 = V_1 + \frac{1}{2}e_2^2 + \frac{\underline{g}_2}{2r_2}\tilde{\theta}_2^2, \tag{5.41}$$

where $\tilde{\theta}_2 = \theta_2 - \hat{\theta}_2$ and r_2 is defined in (5.20). With aid of (5.38) and (5.39), the time derivative of V_2 is derived as

$$\begin{aligned} \dot{V}_2 \le& -\underline{g}_1 c_1 z_1^2 - \bar{\sigma}_1 \tilde{\theta}_1^2 + \underline{g}_2 z_1^2 \mu_1^2 s_f^2 e_2^2 + \varepsilon_1 - \frac{\underline{g}_2}{r_2}\tilde{\theta}_2 \dot{\hat{\theta}}_2 \\ &+ e_2(f_2 + g_2 e_3 + g_2 \alpha_2 - \dot{\alpha}_1) \\ \le& -\underline{g}_1 c_1 z_1^2 - \bar{\sigma}_1 \tilde{\theta}_1^2 + e_2 g_2 \alpha_2 - \frac{\underline{g}_2}{r_2}\tilde{\theta}_2 \dot{\hat{\theta}}_2 + \Xi_2 + \varepsilon_1. \end{aligned} \tag{5.42}$$

with $\Xi_2 = \underline{g}_2 z_1^2 \mu_1^2 s_f^2 e_2^2 + e_2(f_2 + g_2 e_3 - \frac{\partial \alpha_1}{\partial x_1}(f_1 + g_1 x_2) - \Delta\alpha_1)$. Using Young's inequality similar to (5.26)–(5.29), we have

$$e_2 f_2 \le \underline{g}_2 e_2^2 b_2^2 \varphi_2^2 + \frac{1}{4\underline{g}_2}, \tag{5.43}$$

$$e_2 g_2 e_3 \le \underline{g}_3 e_2^2 e_3^2 + \frac{\bar{g}_2^2}{4\underline{g}_3}, \tag{5.44}$$

$$-e_2 \frac{\partial \alpha_1}{\partial x_1} f_1 \le \underline{g}_2 e_2^2 \left(\frac{\partial \alpha_1}{\partial x_1}\right)^2 b_1^2 \varphi_1^2 + \frac{1}{4\underline{g}_2}, \tag{5.45}$$

$$-e_2 \frac{\partial \alpha_1}{\partial x_1} g_1 x_2 \le \underline{g}_2 e_2^2 \left(\frac{\partial \alpha_1}{\partial x_1} x_2\right)^2 + \frac{\bar{g}_1^2}{4\underline{g}_2}, \tag{5.46}$$

$$-e_2 \Delta\alpha_1 \le \underline{g}_2 e_2^2 (\Delta\alpha_1)^2 + \frac{1}{4\underline{g}_2}. \tag{5.47}$$

Therefore, we can obtain

$$\Xi_2 \leq \underline{g}_2\theta_2 e_2^2\Phi_2 + \underline{g}_3 e_2^2 e_3^2 + \frac{3}{4\underline{g}_2} + \frac{\bar{g}_2^2}{4\underline{g}_3} + \frac{\bar{g}_1^2}{4\underline{g}_2}, \tag{5.48}$$

with

$$\theta_2 = \max\left\{1, b_1^2, b_2^2\right\}, \tag{5.49}$$

$$\Phi_2 = \varphi_2^2 + \left(\frac{\partial\alpha_1}{\partial x_1}\varphi_1\right)^2 + \left(\frac{\partial\alpha_1}{\partial x_1}x_2\right)^2 + (\Delta\alpha_1)^2 + z_1^2\mu_1^2 s_f^2. \tag{5.50}$$

Substituting (5.18), (5.20) and (5.48) into (5.42) yields

$$\dot{V}_2 \leq -\underline{g}_1 c_1 z_1^2 - \bar{\sigma}_1\tilde{\theta}_1^2 - \underline{g}_2 c_2 e_2^2 - \bar{\sigma}_2\tilde{\theta}_2^2 + \underline{g}_3 e_2^2 e_3^2 + \varepsilon_2, \tag{5.51}$$

where $\bar{\sigma}_2 = \frac{\underline{g}_2\sigma_2}{2r_2}$ and $\varepsilon_2 = \frac{3}{4\underline{g}_2} + \frac{\bar{g}_2^2}{4\underline{g}_3} + \frac{\bar{g}_1^2}{4\underline{g}_2} + \frac{\underline{g}_2\sigma_2}{2r_2}\theta_2^2 + \varepsilon_1$.

Step i ($3 \leq i \leq n-1$)*:* From (5.16), we can have

$$\begin{aligned}\dot{e}_i &= f_i + g_i x_{i+1} - \dot{\alpha}_{i-1}\\ &= f_i + g_i e_{i+1} + g_i\alpha_i - \dot{\alpha}_{i-1},\end{aligned} \tag{5.52}$$

where $\dot{\alpha}_{i-1}$ is given by

$$\dot{\alpha}_{i-1} = \sum_{j=1}^{i-1}\frac{\partial\alpha_{i-1}}{\partial x_j}(f_j + g_j x_{j+1}) + \Delta\alpha_{i-1}, \tag{5.53}$$

where $\Delta\alpha_{i-1} = \sum_{j=0}^{i-1}\frac{\partial\alpha_{i-1}}{\partial y_d^{(j)}}y_d^{(j+1)} + \sum_{j=1}^{i-1}\frac{\partial\alpha_{i-1}}{\partial\hat{\theta}_j}\dot{\hat{\theta}}_j + \sum_{j=1}^{i-1}\frac{\partial\alpha_{i-1}}{\partial\Phi_j}\dot{\Phi}_j + \sum_{j=0}^{i-1}\frac{\partial\alpha_{i-1}}{\partial e_l^{(j)}}e_l^{(j+1)} + \sum_{j=0}^{i-1}\frac{\partial\alpha_{i-1}}{\partial e_u^{(j)}}e_u^{(j+1)} + \sum_{j=0}^{i-1}\frac{\partial\alpha_{i-1}}{\partial s_f^{(j)}}s_f^{(j+1)}$ which is computable.

Define the following Lyapunov function V_i as

$$V_i = V_{i-1} + \frac{1}{2}e_i^2 + \frac{\underline{g}_i}{2r_i}\tilde{\theta}_i^2. \tag{5.54}$$

Therefore, the time derivative of V_i along (5.52) is

$$\dot{V}_i = -\underline{g}_1 c_1 z_1^2 - \sum_{i=1}^{i-1}\bar{\sigma}_i\tilde{\theta}_i^2 - \sum_{i=2}^{i-1}\underline{g}_i c_i e_i^2 + \varepsilon_{i-1} + \Xi_i + e_i g_i\alpha_i - \frac{\underline{g}_i}{r_i}\tilde{\theta}_i\dot{\hat{\theta}}_i \tag{5.55}$$

where $\Xi_i = \underline{g}_i e_{i-1}^2 e_i^2 + e_i(f_i + g_i e_{i+1} - \dot{\alpha}_{i-1})$. By using Young's inequality for Ξ_i similar to (5.43)–(5.47), one has

$$e_i f_i \leq \underline{g}_i e_i^2 b_i^2 \varphi_i^2 + \frac{1}{4\underline{g}_i}, \tag{5.56}$$

$$e_i g_i e_{i+1} \leq \underline{g}_{i+1} e_i^2 e_{i+1}^2 + \frac{\bar{g}_i^2}{4\underline{g}_{i+1}}, \tag{5.57}$$

$$-e_i \sum_{j=1}^{i-1} \frac{\partial \alpha_{i-1}}{\partial x_j} f_j \leq \underline{g}_i e_i^2 \sum_{j=1}^{i-1} \left(\frac{\partial \alpha_{i-1}}{\partial x_j} \right)^2 b_j^2 \varphi_j^2 + \sum_{j=1}^{i-1} \frac{1}{4\underline{g}_i}, \tag{5.58}$$

$$-e_i \sum_{j=1}^{i-1} \frac{\partial \alpha_{i-1}}{\partial x_j} g_j x_{j+1} \leq \underline{g}_i e_i^2 \sum_{j=1}^{i-1} \left(\frac{\partial \alpha_{i-1}}{\partial x_j} x_{j+1} \right)^2 + \sum_{j=1}^{i-1} \frac{\bar{g}_j^2}{4\underline{g}_i}, \tag{5.59}$$

$$-e_i \Delta\alpha_{i-1} \leq \underline{g}_i e_i^2 (\Delta\alpha_{i-1})^2 + \frac{1}{4\underline{g}_i}. \tag{5.60}$$

It follows that

$$\Xi_i \leq \underline{g}_i \theta_i e_i^2 \Phi_i + \underline{g}_{i+1} e_i^2 e_{i+1}^2 + \frac{1}{2\underline{g}_i} + \frac{\bar{g}_i^2}{4\underline{g}_{i+1}} + \sum_{j=1}^{i-1} \frac{1}{4\underline{g}_i} + \sum_{j=1}^{i-1} \frac{\bar{g}_j^2}{4\underline{g}_i}, \tag{5.61}$$

where

$$\theta_i = \max\left\{1, b_1^2, b_2^2, \ldots, b_i^2\right\}, \tag{5.62}$$

$$\Phi_i = \sum_{j=1}^{i-1} \left(\frac{\partial \alpha_{i-1}}{\partial x_j} \right)^2 \varphi_j^2 + \sum_{j=1}^{i-1} \left(\frac{\partial \alpha_{i-1}}{\partial x_j} x_{j+1} \right)^2 + (\Delta\alpha_{i-1})^2 + \varphi_i^2 + e_{i-1}^2. \tag{5.63}$$

Then, substituting (5.18), (5.20), and (5.61) into (5.55), there is

$$\begin{aligned} \dot{V}_i = & -\underline{g}_1 c_1 z_1^2 - \sum_{i=1}^{i-1} \bar{\sigma}_i \tilde{\theta}_i^2 - \sum_{i=2}^{i} \underline{g}_i c_i e_i^2 + \underline{g}_{i+1} e_i^2 e_{i+1}^2 + \varepsilon_{i-1} \\ & + \frac{1}{2\underline{g}_i} + \frac{\bar{g}_i^2}{4\underline{g}_{i+1}} + \sum_{j=1}^{i-1} \frac{1}{4\underline{g}_i} + \sum_{j=1}^{i-1} \frac{\bar{g}_j^2}{4\underline{g}_i} + \frac{\underline{g}_i \sigma_i}{r_i} \tilde{\theta}_i \hat{\theta}_i. \end{aligned} \tag{5.64}$$

Based on the definition of $\tilde{\theta}_i = \theta_i - \hat{\theta}_i$, then it yields

$$\tilde{\theta}_i \hat{\theta}_i = \tilde{\theta}_i (\theta_i - \tilde{\theta}_i) \leq -\frac{1}{2} \tilde{\theta}_i^2 + \frac{1}{2} \theta_i^2. \tag{5.65}$$

Therefore, we can rewrite (5.64) as

$$\dot{V}_i = -\underline{g}_1 c_1 z_1^2 - \sum_{i=1}^{i} \bar{\sigma}_i \tilde{\theta}_i^2 - \sum_{i=2}^{i} \underline{g}_i c_i e_i^2 + \underline{g}_{i+1} e_i^2 e_{i+1}^2 + \varepsilon_i \tag{5.66}$$

where $\bar{\sigma}_i = \frac{\underline{g}_i \sigma_i}{2r_i}$ and $\varepsilon_i = \frac{1}{2\underline{g}_i} + \frac{\bar{g}_i^2}{4\underline{g}_{i+1}} + \sum_{j=1}^{i-1} \frac{1}{4\underline{g}_i} + \sum_{j=1}^{i-1} \frac{\bar{g}_j^2}{4\underline{g}_i} + \frac{\underline{g}_i \sigma_i}{2r_i} \theta_i^2 + \varepsilon_{i-1}$.

Step n: Taking the time derivative of e_n in (5.16) yields

$$\dot{e}_n = \dot{x}_n - \dot{\alpha}_{n-1} = f_n + g_n u - \dot{\alpha}_{n-1}, \tag{5.67}$$

where $\dot{\alpha}_{n-1}$ is given by

$$\dot{\alpha}_{n-1} = \sum_{j=1}^{n-1} \frac{\partial \alpha_{n-1}}{\partial x_j} (f_j + g_j x_{j+1}) + \Delta\alpha_{n-1}, \tag{5.68}$$

where $\Delta\alpha_{n-1} = \sum_{j=0}^{n-1} \frac{\partial \alpha_{n-1}}{\partial y_d^{(j)}} y_d^{(j+1)} + \sum_{j=1}^{n-1} \frac{\partial \alpha_{n-1}}{\partial \hat{\theta}_j} \dot{\hat{\theta}}_j + \sum_{j=1}^{n-1} \frac{\partial \alpha_{n-1}}{\partial \Phi_j} \dot{\Phi}_j + \sum_{j=0}^{n-1} \frac{\partial \alpha_{n-1}}{\partial e_l^{(j)}} e_l^{(j+1)} + \sum_{j=0}^{n-1} \frac{\partial \alpha_{n-1}}{\partial e_u^{(j)}} e_u^{(j+1)} + \sum_{j=0}^{n-1} \frac{\partial \alpha_{n-1}}{\partial s_f^{(j)}} s_f^{(j+1)}$ which is computable.

Given the Lyapunov function V_n as

$$V_n = V_{n-1} + \frac{1}{2} e_n^2 + \frac{\underline{g}_n}{2r_n} \tilde{\theta}_n^2. \tag{5.69}$$

Its time derivative can be derived as

$$\dot{V}_n \le -\underline{g}_1 c_1 z_1^2 - \sum_{i=2}^{n} \underline{g}_i c_i e_i^2 - \sum_{i=1}^{n-1} \bar{\sigma}_i \tilde{\theta}_i^2 - \frac{\underline{g}_n}{r_n} \tilde{\theta}_n \dot{\hat{\theta}}_n - g_n \hat{\theta}_n e_n^2 \Phi_n + \Xi_n + \varepsilon_{n-1} \tag{5.70}$$

where $\bar{\sigma}_n = \frac{\underline{g}_n \sigma_n}{2r_n}$ and $\Xi_n = \underline{g}_n e_{n-1}^2 e_n^2 + e_n f_n - e_n \dot{\alpha}_{n-1}$. Similar to (5.43)–(5.47), there is

$$e_n f_n \le \underline{g}_n e_n^2 b_n^2 \varphi_n^2 + \frac{1}{4\underline{g}_n}, \tag{5.71}$$

$$-e_i \sum_{j=1}^{n-1} \frac{\partial \alpha_{n-1}}{\partial x_j} f_j \le \underline{g}_n e_n^2 \sum_{j=1}^{n-1} \left(\frac{\partial \alpha_{n-1}}{\partial x_j} \right)^2 b_j^2 \varphi_j^2 + \sum_{j=1}^{n-1} \frac{1}{4\underline{g}_n}, \tag{5.72}$$

$$-e_n \sum_{j=1}^{n-1} \frac{\partial \alpha_{n-1}}{\partial x_j} g_j x_{j+1} \le \underline{g}_n e_n^2 \sum_{j=1}^{n-1} \left(\frac{\partial \alpha_{n-1}}{\partial x_j} x_{j+1} \right)^2 + \sum_{j=1}^{n-1} \frac{\bar{g}_j^2}{4\underline{g}_n}, \tag{5.73}$$

$$-e_n \Delta\alpha_{n-1} \le \underline{g}_n e_n^2 (\Delta\alpha_{n-1})^2 + \frac{1}{4\underline{g}_n}, \tag{5.74}$$

Therefore, it yields

$$\Xi_n \leq \underline{g}_n \theta_n e_n^2 \Phi_n \upsilon, \tag{5.75}$$

with

$$\theta_n = \max\left\{b_1^2, b_2^2, \ldots, b_n^2\right\}, \tag{5.76}$$

$$\Phi_n = \sum_{j=1}^{n-1}\left(\frac{\partial \alpha_{n-1}}{\partial x_j}\varphi_j\right)^2 + \sum_{j=1}^{n-1}\left(\frac{\partial \alpha_{n-1}}{\partial x_j}x_{j+1}\right)^2 + (\Delta\alpha_{n-1})^2 + e_{n-1}^2 + \varphi_n^2. \tag{5.77}$$

Substituting (5.22) and (5.75) into (5.70), we can have

$$\dot{V}_n \leq -\underline{g}_1 c_1 z_1^2 - \sum_{i=2}^{n}\underline{g}_i c_i e_i^2 - \sum_{i=1}^{n}\bar{\sigma}_i\tilde{\theta}_i^2 + \varepsilon_n, \tag{5.78}$$

where $\bar{\sigma}_n = \frac{\underline{g}_n \sigma_n}{2r_n}$ and $\varepsilon_n = \frac{\underline{g}_n \sigma_n}{2r_n}\theta_n^2 + \frac{1}{2\underline{g}_n} + \sum_{j=1}^{n-1}\frac{1}{4\underline{g}_n} + \sum_{j=1}^{n-1}\frac{\bar{g}_j^2}{4\underline{g}_n} + \varepsilon_{n-1}$.

To summarize, we establish the following theorem.

Theorem 5.1 *Consider the nonlinear system* (5.1) *with Assumptions 5.1–5.3. The virtual control inputs* $\alpha_i, i = 1, \ldots, n-1$ *is developed in* (5.17) *and* (5.18)*, the actual control law u is established in* (5.21)*, and the adaptive laws is given by* (5.19), (5.20) *and* (5.21)*. Then, the proposed control strategy guarantees the following.*

(i) The tracking is guaranteed to be ultimately uniformly bounded regardless of initial conditions; and
(ii) The nonlinear system is said to be EPC as the tracking signal $e_1(t)$ *is strictly constrained by the TPP right after the finite time T as presented in Definition 5.1.*

Proof According to (5.78), there has

$$\dot{V}_n \leq -\underline{g}_1 c_1 z_1^2 - \sum_{i=2}^{n}\underline{g}_i c_i e_i^2 - \sum_{i=1}^{n}\bar{\sigma}_i\tilde{\theta}_i^2 + \varepsilon_n \leq -\pi V_n + \varepsilon_n, \tag{5.79}$$

where $\pi = \min\{2\underline{g}_1 c_1, 2\underline{g}_i c_i, 2\frac{r_i\bar{\sigma}_i}{\underline{g}_i}\} > 0$. Then, we obtain

$$0 \leq V_n(t) \leq \frac{\varepsilon_n}{\pi} + \left(V_n(0) - \frac{\varepsilon_n}{\pi}\right)e^{-\pi t}. \tag{5.80}$$

It can be concluded that $V_n(t) \in L_\infty$, which indicates that $z_1, e_2, \ldots, e_n, \tilde{\theta}_1, \ldots, \tilde{\theta}_n \in L_\infty$ for any initial conditions. According to the transformation (5.14), the boundedness of z_1 implies that $-e_l - \bar{e}_{ul} < s_1 < e_u - \bar{e}_{ul}$ holds. From Lemma 5.2, there has $-e_l < e_s < e_u$ for $t \geq 0$. Then, according to Lemma 5.1, we can derive that $e_1 \in L_\infty$, and the output constraints are strictly imposed on e_1 right after a given time T, i.e., $-e_l < e_1 < e_u$, $\forall t \geq T$. Therefore, the nonlinear system satisfies ECP as depicted in (5.2). The boundedness of $\hat{\theta}_i, \dot{\hat{\theta}}_i, \alpha_i$ and v is guaranteed as shown in (5.17)–(5.22) since $y_d, \dot{y}_d, s_f, z_1, e_i, \varphi_i \in L_\infty$ with $i = 1 \ldots, n$. Moreover, based on (5.16), it can be derived that $x_i \in L_\infty, i = 2, \ldots, n$ as $\alpha_i \in L_\infty$. Consequently, all the signals in the closed-loop system are ultimately uniformly bounded.

□

Remark 5.4 As the system order increases, our proposed tracking control would encounter the computational explosion problem. This issue arises from the recursive derivations of virtual control inputs when following backstepping procedures. Dynamic Surface Control (DSC), proposed in [43], is regarded as an effective technique to address the complexity arising due to these explosion of terms. In this way, we can integrate such DSC into our future control design to make it more applicable in practical implementation.

5.2.1.8 Numerical Simulations

To illustrate the effectiveness of the developed control strategy, we consider the following single-link robot, which represents a typical example of practical systems:

$$M\ddot{q} + N\dot{q} + \frac{1}{2}mg\gamma \sin q = u, \tag{5.81}$$

where $q, \dot{q}$ and $\ddot{q}$ denote the angular displacement, velocity and acceleration, respectively, M represents the moment of inertia, N is the damping coefficient, g is the gravity acceleration, m and γ are the link mass and length, and u denotes the system input. Let $x_1 = q$ and $x_2 = \dot{q}$, then (5.81) is rewritten as

$$\begin{cases} \dot{x}_1 = f_1 + g_1 x_2 \\ \dot{x}_2 = f_2 + g_2 u \end{cases} \tag{5.82}$$

where $f_1 = 0$, $g_1 = 1$, $f_2 = -\frac{1}{2M}mg\gamma \sin x_1 - \frac{N}{M}x_2$, and $g_2 = \frac{1}{M}$. Upon Assumption 5.3, it can be derived that $|f_1| \leq 0$ and $|f_2| \leq \frac{1}{2M}mg\gamma + \frac{N^2}{4M^2} + x_2^2$, which yields $\varphi_1 = 0$, $b_1 = 0$, $\varphi_2 = 1 + x_2^2$ and $b_2 = \max\{\frac{1}{2M}mg\gamma + \frac{N^2}{4M^2}, 1\}$. To conduct simulation, the system parameters are set as $M = 0.5$, $m = 1$, $\gamma = 1$, $N = 1$, and $g = 9.8$. The initial system states are chosen as $x_1(0) = x_2(0) = 0$. The specifications of TPP (5.6) are given by $\delta = 0.3$, $\rho_\infty = 1$, $\rho_0 = 2.3$, and $\ell = 0.5$. The control parameters are selected as $c_1 = 4$, $c_2 = 2$, $r_1 = r_2 = 0.01$, $\sigma_1 = \sigma_2 = 0.01$, $T = 3.0, 5.0, 7.0$, and $l = 1$. The desired trajectory are $y_d = 0.2\cos(0.5t)$. The sample time is set as 0.01 and the system runs 20s.

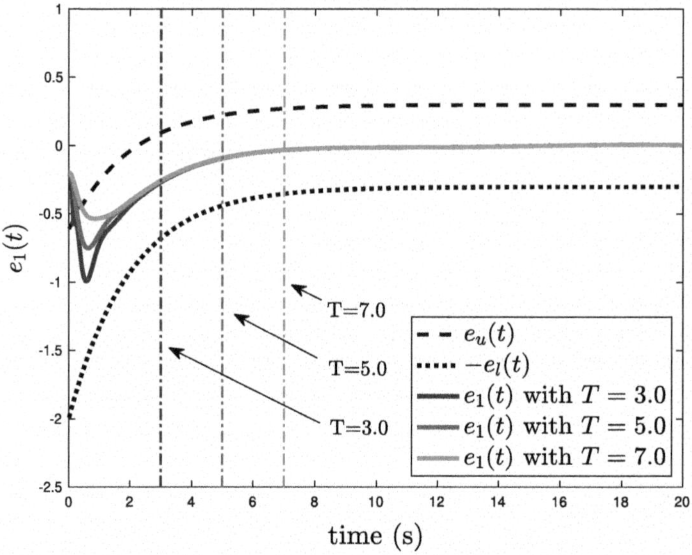

Fig. 5.2 The tracking trajectory of $e_1(t)$

The simulation results are shown in Figs. 5.2, 5.3 and 5.4. The trajectories of $e_1(t) = x_1(t) - y_d(t)$ are depicted in Fig. 5.2 under different settling times T. It can be seen that although the initial condition of $e_1(0)$ is outside the allowable set, the tracking error $e_1(t)$ satisfies performance constraints after the settling time chosen by designers.

With the aid of ESF in (5.3)–(5.5), the initial value of $e_s(0)$ is adjusted to the nominal position between upper and lower boundaries as shown in Fig 5.3, and $e_s(t)$ is always within the allowable set to ensure the boundedness of $e_1(t) = x_1(t) - y_d(t)$ as presented in Lemma 5.1. The control signal $u(t)$ under different settling times is plotted in Fig. 5.4.

To further illustrate the effectiveness of our method, we conduct comparison simulation with the funnel control in [19], which follows

$$\begin{aligned} u &= -\bar{\mu}[(\cos(\chi_2) - \chi_2\sin(\chi_2))^2 e^2]\chi_2^4(1+\bar{\omega}^2)\vartheta \\ &\quad - \bar{\mu}N_2 e - \bar{\mu}N_2^2\chi_2^4(1+\bar{\omega}^2)\vartheta, \\ N_2 &= \chi_2\cos(\chi_2),\ \chi_2 = \frac{1}{1-\mathcal{F}^2 e^2}, \\ \vartheta &= \bar{\omega} - N_2 e,\ \dot{\bar{\omega}} = -\bar{\mu}\bar{\omega} + u,\ e = x_1 - y_d, \end{aligned} \tag{5.83}$$

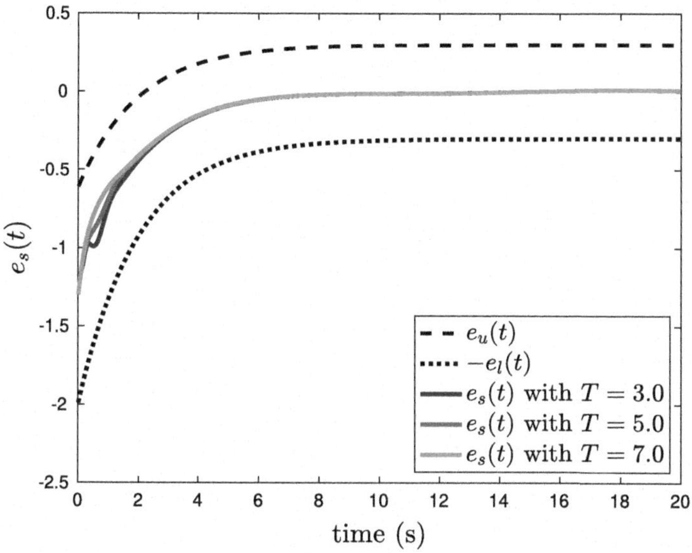

Fig. 5.3 The tracking trajectory of $e_s(t)$

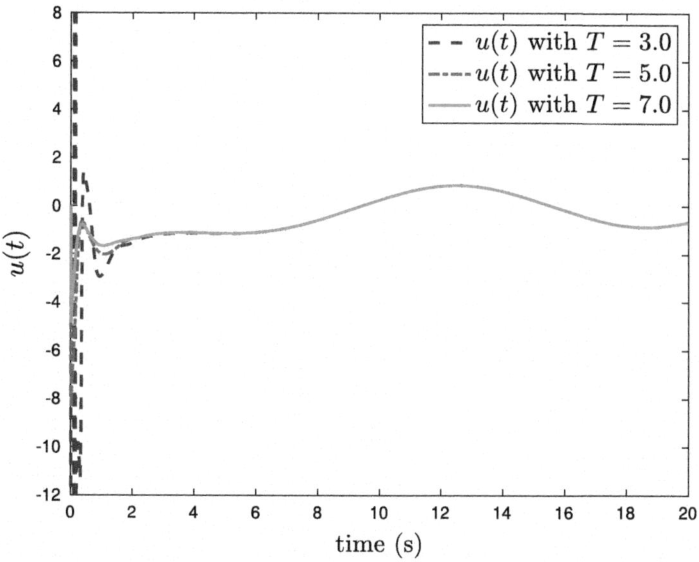

Fig. 5.4 The control input signal $u(t)$

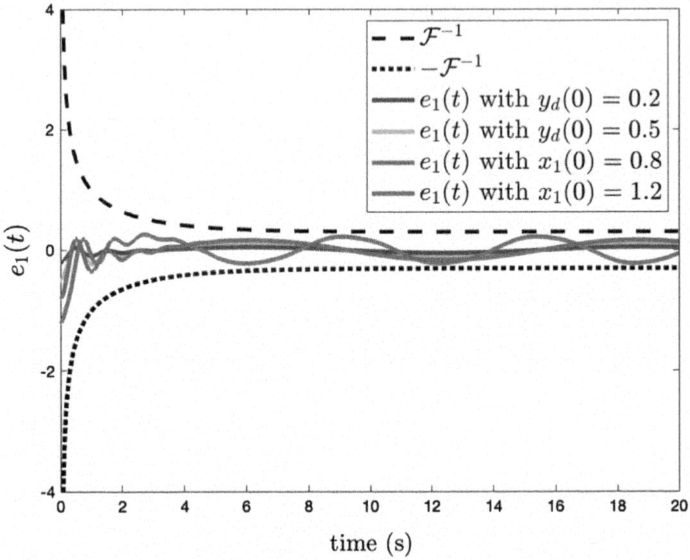

Fig. 5.5 The tracking error $e_1(t)$ with funnel control

where $\bar{\mu} = 4$ and $\mathcal{F} = \frac{\sqrt{1-\psi^2}}{\sqrt{\ell\psi}}$ with $\psi = (1 - 0.3)e^{-0.5t} + 0.3$ and $\ell = 1 - 0.3^2$. The initial conditions are $x_1(0) = x_2(0) = 0$. Besides, the desired trajectories are with different settings: $y_d(t) = 0.2\cos(0.5t),\ 0.5\cos(0.5t), 0.8\cos(0.5t),$ $1.2\cos(0.5t)$. The simulation results are shown in Figs. 5.5 and 5.6. It can be seen that the tracking error $e_1(t)$ in Fig. 5.5 is guaranteed within the allowable set. However, obvious overshoot and oscillate are introduced, especially during the transient period. The control inputs are plotted in Fig. 5.6.

Remark 5.5 In addition to the example in (5.81), our approach can be adopted in many others practical systems, for example, m-joint robot manipulator [52], actuator hysteresis and failures [50], and spring-mass-damper system [41]. For the convenience of readers, we here present one second-order nonlinear system subject to input hysteresis in [46]:

$$\begin{aligned} \dot{x}_1 &= x_2 + 1 - \sin x_1^2, \\ \dot{x}_2 &= 1 - \sin x_1 \cos x_2 + x_2 e^{-x_1^2} u. \end{aligned} \tag{5.84}$$

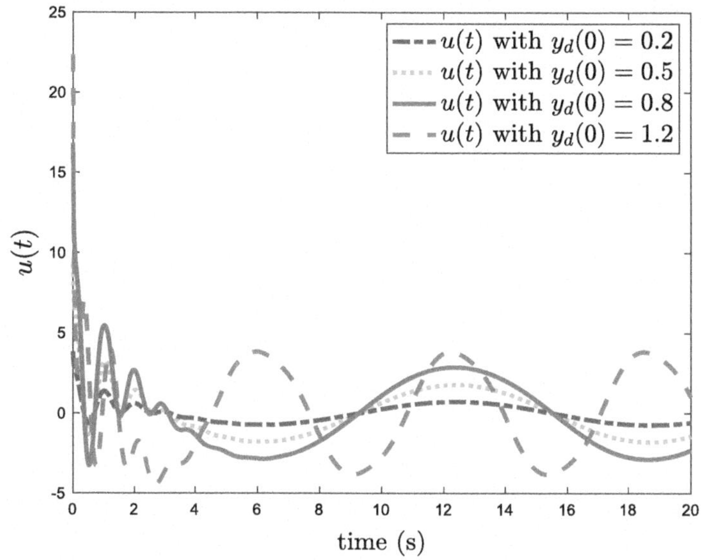

Fig. 5.6 The funnel control input under different desired trajectories

5.2.2 *Output-Feedback Control of Nonlinear Systems with Unknown Control Directions and External Disturbances*

5.2.2.1 System Descriptions

We consider the following strict-feedback nonlinear systems

$$\begin{cases} \dot{x}_i = f_i(\bar{x}_i) + g_i x_{i+1} + d_i(t), \ i \in \mathbb{I}_{1:n-1} \\ \dot{x}_n = f_n(\bar{x}_n) + g_n u + d_n(t) \\ y = x_1 \end{cases} \tag{5.85}$$

where $i \in \mathbb{I}_{1:n-1}$ represents the set $\{1, \ldots, n-1\}$, $\bar{x}_i = [x_1, \ldots, x_i]^T \in \mathbb{R}^i$ stands for the state vector; $f_i(\bar{x}_i) : \mathbb{R}^i \to \mathbb{R}$ is an unknown continuous function and Lipschitz in $\bar{x}_i \in \mathbb{R}^i$; $g_i \in \mathbb{R}$ represents the unknown control gain; $d_i(t) \in \mathbb{R}$ denotes the unknown external disturbance; $u \in \mathbb{R}$ and $y \in \mathbb{R}$ are the system input and output, respectively, under the assumption that only y is measurable. It is worth noting that the structures and parameters of the nonlinearities, the control directions and their magnitudes, and the bounds of external disturbances in system (5.85) are completely unknown.

5.2.2.2 Problem Formulation

Define the tracking error $e_1(t) = y(t) - y_r(t)$ where $y_r(t)$ denotes the reference trajectory. The control objective is to design a low-complexity semi-global control strategy for the nonlinear systems under unknown control directions and symmetric/asymmetric performance constraints such that:

(i) All the signals of the closed-loop systems remain bounded.

(ii) For any initial condition (including the initial constraint-violation), the tracking error $e_1(t)$ is regulated into the Tunnel Prescribed Performance constraints before the user-given time.

(iii) Construct an orientation function to handle the unknown control directions which can effectively avoid large initial control input.

Before moving forward, we formally define the Entry Capture Problem (ECP) as follows.

Definition 5.3 Consider the nonlinear system (5.85), if the tracking error $e_1(t)$ is free or released from performance constraints during some initial time interval, then the above constraints are imposed right after a given time as:

$$-e_{1,l}(t) < e_1(t) < e_{1,u}(t), \ \forall t \geq T, \tag{5.86}$$

where $e_{1,u}(t)$ and $-e_{1,l}(t)$ are TPP boundaries, and T denotes the user-defined settling time. If the error performance satisfies the above tracking condition, it is said to be ECP.

Remark 5.6 In most PPC methods [11, 34, 55], the prescribed performance in (5.86) is assumed to be satisfied from the beginning of system operation. However, in this work, we focus on a more frequently encountered ECP that has not been adequately addressed by the existing studies. By ECP, we mean that no constraints are imposed on the tracking error during the initial time interval, but symmetric/asymmetric performance constraints are imposed right after a given time. One typical example is a manipulator system which often operates from any initial condition and then tracks/grasps an object in a prescribed performance for successful task completion. Therefore, addressing such ECP has been a task of practical interest and theoretical significance.

For the subsequent development, some widely used assumptions are made.

Assumption 5.4 ([15]) The signs of control coefficients are unknown, and there exist unknown positive constants $\underline{g}$ and $\bar{g}$ such that $0 < \underline{g} \leq |g_i| \leq \bar{g}$ for $i \in \mathbb{I}_{1:n}$.

Assumption 5.5 ([8]) The unknown disturbance is bounded by $|d_i(t)| \leq \bar{d}$, where $\bar{d}$ is an unknown positive constant.

Assumption 5.6 ([54]) The desired trajectory $y_r(t)$ and its time derivative $\dot{y}_r(t)$ are bounded.

Remark 5.7 Compared with the previous literature, less knowledge of the nonlinear system (5.85) is required. In the existing works: (i) The upper bound $\bar{d}$ of disturbances is assumed to be known [12]; (ii) $f_i(\bar{x}_i)$ or its bounding function is partially known [10, 56]. Note that only Assumptions 5.4–5.6 are necessary in this section without extra knowledge of these system nonlinearities, which are commonly used in the existing studies [8, 33].

5.2.2.3 Fuzzy Neural Network Observer

In practical applications, most system states are not completely measurable. As only y in (5.85) is measurable, we, in this section, introduce a fuzzy neural network observer before developing the control design. To move forward, the following new state variable ξ_i is defined as shown in [31]:

$$\xi_i = \frac{x_i}{\bar{g}_i},\ \bar{g}_i = \prod_{j=i}^{n} g_j. \tag{5.87}$$

Accordingly, the system (5.85) can be rewritten as

$$\begin{cases} \dot{\xi}_i = \bar{f}_i(\bar{\xi}_i) + \xi_{i+1} + \bar{d}_i(t),\ i \in \mathbb{I}_{1:n-1} \\ \dot{\xi}_n = \bar{f}_n(\bar{\xi}_n) + u + \bar{d}_n(t) \end{cases} \tag{5.88}$$

where $\bar{\xi}_i = [\xi_1, \ldots, \xi_i]^T \in \mathbb{R}^i$, $\bar{f}_i(\bar{\xi}_i) = \frac{f_i(\bar{x}_i)}{\bar{g}_i} \in \mathbb{R}$ and $\bar{d}_i(t) = \frac{d_i(t)}{\bar{g}_i} \in \mathbb{R}$. According to the universal approximation ability of the fuzzy logic systems in Lemma 3.2, one has

$$\bar{f}_i(\bar{\xi}_i) = \theta_i^T S_i(\bar{\xi}_i) + \varepsilon_i,\ \forall \bar{\xi}_i \in U_i, \tag{5.89}$$

where θ_i denotes the optimal parameter vector, $S_i(\bar{\xi}_i)$ is the fuzzy membership function, $|\varepsilon_i| \leq \bar{\varepsilon}$ with $\bar{\varepsilon} > 0$ and U_i is a compact set.

We then develop the following fuzzy neural network observer to approximate the unmeasurable states:

$$\begin{cases} \dot{\hat{\xi}}_i = -l_i\hat{\xi}_1 + \hat{\xi}_{i+1} + \hat{\theta}_i^T S_i(\hat{\bar{\xi}}_i),\ i \in \mathbb{I}_{1:n-1} \\ \dot{\hat{\xi}}_n = -l_n\hat{\xi}_1 + u + \hat{\theta}_n^T S_n(\hat{\bar{\xi}}_n) \end{cases} \tag{5.90}$$

where $\hat{\xi}_i$ denotes the estimation of ξ_i whose initial value is set as 0, and l_i is a positive constant such that

$$A = \begin{bmatrix} -l_1 & & \\ \vdots & I_{n-1} & \\ -l_n & 0 & \cdots & 0 \end{bmatrix}, \tag{5.91}$$

is a Hurwitz matrix. For a positive definite matrix $Q = Q^T$, there is a positive definite matrix $P = P^T$ such that $A^T P + PA = -Q$. Moreover, the adaptive law $\hat{\theta}_i$ is designed by

$$\dot{\hat{\theta}}_i = -\sigma_i \hat{\theta}_i + r_i z_i \Lambda_i S_i(\hat{\bar{\xi}}_i),\ i \in \mathbb{I}_{1:n}, \tag{5.92}$$

where σ_i and $r_i > 0$ are positive constants; z_i and Λ_i are defined in (5.114), (5.119) and (5.124).

Define the estimation error $\tilde{\xi}_i$ and vector $\tilde{\xi}$ as

$$\tilde{\xi}_i = \xi_i - \hat{\xi}_i,\ i \in \mathbb{I}_{1:n}, \tag{5.93}$$

$$\tilde{\xi} = [\tilde{\xi}_1, \ldots, \tilde{\xi}_n]^T. \tag{5.94}$$

Therefore, we can have

$$\dot{\tilde{\xi}} = A\tilde{\xi} + \sum_{i=1}^{n} B_i(\tilde{\theta}_i^T \hat{S}_i(\hat{\bar{\xi}}_i) + l_i \xi_1 + \bar{d}_i) + \sum_{i=1}^{n} B_i(\theta_i^T \Delta s_i + \varepsilon_i),$$

where $\Delta s_i = S_i(\bar{\xi}_i) - S_i(\hat{\bar{\xi}}_i)$, $\tilde{\theta}_i = \theta_i - \hat{\theta}_i$, $B_i = [0, \ldots, 1, \ldots, 0]^T$, and $B_n = [0, \ldots, 1]^T$.

Consider a Lyapunov candidate function $V_0 = \tilde{\xi}^T P \tilde{\xi}$ and its time derivative of V_0 is given by

$$\begin{aligned} \dot{V}_0 = &- \tilde{\xi}^T Q \tilde{\xi} + 2\tilde{\xi}^T P \sum_{i=1}^{n} B_i(\tilde{\theta}_i^T S_i(\hat{\bar{\xi}}_i) + l_i \xi_1 + \theta_i^T \Delta s_i) \\ &+ 2\tilde{\xi}^T P \sum_{i=1}^{n} B_i(\bar{d}_i(t) + \varepsilon_i). \end{aligned} \tag{5.95}$$

From Assumption 5.5, there is

$$2\tilde{\xi}^T P \sum_{i=1}^{n} B_i(\bar{d}_i(t) + \varepsilon_i) \le \frac{n}{\kappa_1}\|\tilde{\xi}\|^2 + \kappa_1 \|P\|^2 (\bar{d} + \bar{\varepsilon})^2 \tag{5.96}$$

$$\begin{aligned} &2\tilde{\xi}^T P \sum_{i=1}^{n} B_i(\tilde{\theta}_i^T S_i(\hat{\bar{\xi}}_i) + l_i \xi_1 + \theta_i^T \Delta s_i) \\ &\le \frac{n}{\kappa_2}\|e\|^2 + 4\kappa_2 \|P\|^2 \sum_{i=1}^{n} \|\theta_i\|^2 + \frac{n}{\kappa_3}\|P\|^2 \|e\|^2 \\ &+ \kappa_3 \sum_{i=1}^{n} \|\tilde{\theta}_i\|^2 + \frac{n}{\kappa_4}\|e\|^2 + \kappa_4 \|P\|^2 \sum_{i=1}^{n} l_i^2 \xi_1^2, \end{aligned} \tag{5.97}$$

where κ_i is a positive constant. Substituting (5.96) and (5.97) into (5.95), it yields

$$\begin{aligned}\dot{V}_0 \leq & -c_0\|\tilde{\xi}\|^2+\kappa_1\|P\|^2(\bar{d}+\bar{\varepsilon})+4\kappa_2\|P\|^2\sum_{i=1}^{n}\|\theta_i\|^2 \\ & +\kappa_3\sum_{i=1}^{n}\|\tilde{\theta}_i\|^2+\kappa_4\|P\|^2\sum_{i=1}^{n}l_i^2\xi_1^2, \end{aligned} \tag{5.98}$$

where $c_0=\lambda_{\min}(Q)-\frac{n}{\kappa_1}-\frac{n}{\kappa_2}-\frac{n}{\kappa_3}\|P\|^2-\frac{n}{\kappa_4}$ with $\lambda_{\min}(\cdot)$ represents the minimum eigenvalue of Q, and c_0 is a positive constant if we appropriately select parameter l_i.

Lemma 5.3 *If $z_i, i \in \mathbb{I}_{1:n}$ defined in* (5.114) *and* (5.119) *are bounded, so is $\tilde{\xi}_i, i \in \mathbb{I}_{1:n}$. In addition, if the boundedness of $\hat{\xi}_i, i \in \mathbb{I}_{1:n}$ is guaranteed, so are ξ_i and x_i.*

Proof Based on Assumption 5.5 and Lemma 3.2, we can have $\bar{d}, \bar{\varepsilon}, \theta_i \in L^{\infty}$. Hence, by the definition of V_0 and its time derivative in (5.98), if $\tilde{\theta}_i \in L^{\infty}$ and $\xi_1 \in L^{\infty}$, the first claim $\tilde{\xi}_i \in L^{\infty}$ can be proved.

Therefore, the boundedness of $\tilde{\theta}_i$ and ξ_1 is analyzed as follows. If $z_i, i \in \mathbb{I}_{1:n}$ defined in (5.114) and (5.119) are bounded, the boundedness of $\hat{\theta}_i, i \in \mathbb{I}_{1:n}$ in (5.92) are guaranteed on the basis of bounded-input bounded-output stability since $z_i \in L^{\infty}$ and $S_i^T(\hat{\bar{\xi}}_i)S_i(\hat{\bar{\xi}}_i) \leq 1$. Then, from $\tilde{\theta}_i=\theta_i-\hat{\theta}_i$, it can be derived that $\tilde{\theta}_i \in L^{\infty}$ due to $\theta_i \in L^{\infty}$ and $\hat{\theta}_i \in L^{\infty}$. By (5.114), if z_1 is bounded, we can have $\bar{e}_1 \in L^{\infty}$. Then, it can be observed from Lemma 5.4 that the tracking error $e_1(t)=y(t)-y_r(t)$ is bounded. Due to $y_r(t) \in L^{\infty}$ as stated in Assumption 5.6, one has $y(t) \in L^{\infty}$. From (5.87), the boundedness of ξ_1 is thus guaranteed. This completes the first claim.

For the second claim, from the definition of $\tilde{\xi}_i$ in (5.93), it can be obtained that $\xi_i \in L^{\infty}$ if $\tilde{\xi}_i \in L^{\infty}$ and $\hat{\xi}_i \in L^{\infty}$ hold, which further guarantees the boundedness of x_i based on (5.87). □

Remark 5.8 Considering unmeasurable system states $x_i, i \in \mathbb{I}_{2:n}$ and unknown control direction $g_i, i \in \mathbb{I}_{1:n}$ of nonlinear systems (5.85), new state variables $\xi_i, i \in \mathbb{I}_{1:n}$ are thus introduced to simplify the control design. It should be noted that the fuzzy neural network observer (5.90) is designed to estimate the unmeasurable state variable ξ_i in (5.88) instead of the system state x_i since only $\hat{\xi}_i$ is used for the later control design. Moreover, the control objective of this section is to address the ECP in (5.86) and guarantee the boundedness of all tracking signals. If $z_i \in L^{\infty}$ and $\hat{\xi}_i \in L^{\infty}$ hold for $i \in \mathbb{I}_{1:n}$, from Lemma 5.3, we can establish the boundedness of system states $x_i, i \in \mathbb{I}_{1:n}$. Therefore, the estimations of $\hat{\xi}_i, i \in \mathbb{I}_{1:n}$ are served as auxiliary states for the control design and the stability analysis, and the estimations of $x_i, i \in \mathbb{I}_{1:n}$ with high accuracy are not considered in this work.

5.2.2.4 Scaling Function

To make the control method applicable to any initial condition, we introduce the following Scaling Function $w(t)$ as

$$w(t) = \begin{cases} \sin \frac{\pi t}{2T} & \text{if } t < T \\ 1 & \text{otherwise.} \end{cases} \tag{5.99}$$

where T represents the user-given time in (5.86). Some salient features of the scaling function are summarized as follows, which are significant for handling the ECP:

$$\text{(i) } w(0) = 0, \text{ and } w(t) = 1 \text{ for } t \geq T; \tag{5.100}$$

$$\text{(ii) } w(t) \text{ is monotonically increasing for } t \in [0, T); \text{ and} \tag{5.101}$$

$$\text{(iii) } 0 < \dot{w}(t) < \infty,\ 0 < t < T, \text{ and } \dot{w}(t) = 0,\ t \geq T. \tag{5.102}$$

We pause here to reveal motivations of such scaling function and demonstrate its scalability to other control methods. To obtain semi-global tracking ability, it is necessary to propose a simple yet effective mechanism for handling various initial condition. It is intuitive to adjust the initial tracking error signal to a fixed point for any cases. By the first property in (5.100), if we multiply $w(t)$ with $e_1(t)$, we can always have $w(t)e_1(t) = 0$ at $t = 0$ because of $w(0) = 0$. In this way, it provides a unified processing for any initial conditions, which facilitates later control design. From the ECP in (5.86), performance constraints are involved right after a finite time T. It indicates that $w(t)$ should monotonically increase from 0 to 1 during the initial time interval. Specifically, their multiplication becomes $w(t)e_1(t) = e_1(t)$ for $t \geq T$, which makes the performance constraints be imposed on the tracking error hereafter. In addition, other properties of such scaling function are used in the stability analysis. Generally, these scaling function properties in (5.100)–(5.102) offer a basic mechanism to address the underlying ECP, which can stimulate fruitful expressions of such scaling function not limited to (5.99).

5.2.2.5 Error Self-tuning Functions

To address the ECP, we introduce the following Error Self-tuning Functions (ESFs):

$$\bar{e}_1(t) = w(t)e_1(t) + h_1(t), \tag{5.103}$$

$$\bar{e}_i(t) = e_i(t) + h_i(t),\ i \in \mathbb{I}_{2:n}, \tag{5.104}$$

where $e_i(t) = \hat{\xi}_i(t) - \alpha_{i-1}(t)$ for $i \in \mathbb{I}_{2:n}$, and $\alpha_{i-1}(t)$ represents the virtual control input defined in (5.120). As discussed in fuzzy neural network observer (5.90) and Remark 5.10, there are always $\hat{\xi}_i(0) = 0$ and $\alpha_{i-1}(0) = 0$, leading to $e_i(0) = 0$ for $i \in \mathbb{I}_{2:n}$. Therefore, it is unnecessary to introduce $w(t)$ in the definition of $\bar{e}_i(t)$ for $i \in \mathbb{I}_{2:n}$. Moreover, $h_i(t), i \in \mathbb{I}_{1:n}$ are designed as

$$h_i(t) = \begin{cases} h_{i,0}\exp(-\frac{\mu_i T t}{T-t}) & \text{if } t < T \\ 0 & \text{otherwise.} \end{cases} \tag{5.105}$$

where μ_i is a positive constant, $h_{i,0} = \frac{e_{i,u}(0)-e_{i,l}(0)}{2}$, $e_{i,l}(0)$ and $e_{i,u}(0)$ denote the initial value of the TPP boundaries defined in (5.113).

It should be noted that two important properties on $\bar{e}_1(t)$ can be concluded:

$$\bar{e}_1(0) = w(0)e_1(0) + h_1(0) = h_{1,0}, \tag{5.106}$$

$$\bar{e}_1(t) = w(t)e_1(t) + h_1(t) = e_1(t), \ t \geq T. \tag{5.107}$$

To better illustrate how we address the ECP by using $\bar{e}_1(t)$, we take $\bar{e}_1(t)$ and $e_1(t)$ as an example, where the performance constraints (5.113) are selected with $i = 1$ and $e_{1,0} < 0$. First, from (5.106), we always have $-e_{1,l}(0) < \bar{e}_1(0) = h_{1,0} < e_{1,u}(0)$. It indicates that given any initial tracking error $e_1(0)$ (including the initial constraint-violation as shown in the right figure of Fig. 5.7), $\bar{e}_1(0)$ always satisfies the initial performance constraints as shown in the left one. When $t \geq T$, $w(t) = 1$ and $h_1(t) = 0$ lead to $\bar{e}_1(t) = e_1(t)$, namely, they are the same trajectories, which is also depicted in Fig. 5.7. Therefore, if $\bar{e}_1(t)$ satisfies the performance constraints for $t \geq T$, so does $e_1(t)$. The following lemma is given to show that such $\bar{e}_1(t)$ does not affect the boundedness of $e_1(t)$.

Lemma 5.4 *If the tuned tracking error $\bar{e}_1(t)$ satisfies $-e_{1,l}(t) < \bar{e}_1(t) < e_{1,u}(t)$ for $t \in [0, T)$, then the tracking error $e_1(t)$ is bounded during this interval. Moreover, if $-e_{1,l}(t) < \bar{e}_1(t) < e_{1,u}(t)$ holds for $\forall t \geq 0$, then the ECP in (5.86) is accomplished.*

Proof The first claim can be obtained by contradiction. Suppose that $-e_{1,l}(t) < \bar{e}_1(t) < e_{1,u}(t)$ holds for $t \in [0, T)$, but there is a time instant $t_1 \in [0, T)$ such that

$$\lim_{t \to t_1} |e_1(t)| = \lim_{t \to t_1} |y(t) - y_r(t)| = \infty. \tag{5.108}$$

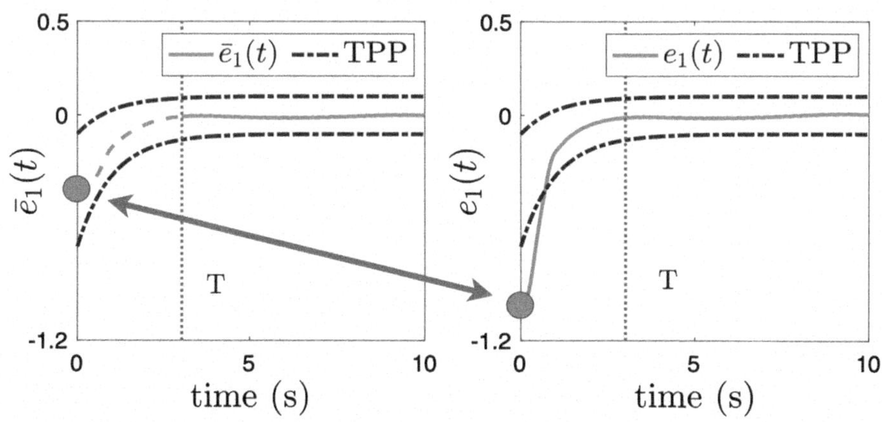

Fig. 5.7 The mechanism behind the time-varying scaling function

From (5.103), there exists

$$-e_{1,l}(t) < w(t)(y(t) - y_r(t)) + h_1(t) < e_{1,u}(t), \ t \in [0, T). \tag{5.109}$$

By the definition of $h_1(t)$ in (5.105) with $i = 1$, together with (5.108) and (5.109), we can have $\lim_{t \to t_1} w(t) = 0$, which holds if and only if $t_1 = 0$ from the definition of $w(t)$ in (5.99). Then, (5.108) can be rewritten as $\lim_{t \to 0} |y(0) - y_r(0)| = \infty$. Nevertheless, the bounded initial states $y(0)$ and $y_r(t)$ indicate the boundedness of $|y(0) - y_r(0)|$. Thus, (5.108) is false, and the first part is proved. It is easy to derive the second claim according to (5.107), and we omit here. This completes the proof. □

Based on Lemma 5.4, it indicates that the ECP in (5.86) can be solved as long as $\bar{e}_1(t)$ is guaranteed within the performance constraints:

$$-e_{1,l}(t) < \bar{e}_1(t) < e_{1,u}(t), \ t \geq 0. \tag{5.110}$$

Moreover, we, in this work, also consider the performance constraints on $\bar{e}_i(t)$ as:

$$-e_{i,l}(t) < \bar{e}_i(t) < e_{i,u}(t), \ t \geq 0, \ i = 2, \ldots, n. \tag{5.111}$$

It can be derived that the boundedness of $e_i(t)$, $i = 2, \ldots, n$ is guaranteed if (5.111) holds. The analysis is the same as presented in Lemma 5.4 and we omit here for brevity.

Remark 5.9 The proposed ESFs in (5.103)–(5.104) make the control design applicable to any initial condition although it may violate the initial performance constraints. The advantages of such ESFs can be observed from the following two aspects: (i) Different from the funnel control [3, 6, 18, 20] and global prescribed performance [17, 35, 53], ESFs introduce additional scaling functions on tracking errors instead of relying on specific performance functions. In this way, such ESFs can be easily extended to more general performance cases and other control methods. (ii) It should be noted that these existing results mainly focus on symmetric performance constraints but asymmetric ones have not been adequately addressed. The proposed ESFs can handle both symmetric and asymmetric performance constraints, which is more desirable in practice.

5.2.2.6 Tunnel Prescribed Performance

From (5.110) and (5.111), the control objective is converted to guarantee $\bar{e}_i(t)$ within the performance constraints,

$$-e_{i,l}(t) < \bar{e}_i(t) < e_{i,u}(t), \ i \in \mathbb{I}_{1:n}, \tag{5.112}$$

where $e_{i,u}(t)$ and $-e_{i,l}(t)$ are the Tunnel Prescribed Performance (TPP) with

$$
e_{i,l}(t) = \begin{cases} \delta_{i,l}\rho_i(t), & \text{if } e_{i,0} = 0 \\ (\delta_{i,l} - 1)\rho_i(t) + \bar{\rho}_{i,\infty}, & \text{if } e_{i,0} > 0 \\ (\delta_{i,l} + 1)\rho_i(t) - \bar{\rho}_{i,\infty}, & \text{if } e_{i,0} < 0 \end{cases}
$$

$$
e_{i,u}(t) = \begin{cases} \delta_{i,u}\rho_i(t), & \text{if } e_{i,0} = 0 \\ (\delta_{i,u} + 1)\rho_i(t) - \bar{\rho}_{i,\infty}, & \text{if } e_{i,0} > 0 \\ (\delta_{i,u} - 1)\rho_i(t) + \bar{\rho}_{i,\infty}, & \text{if } e_{i,0} < 0 \end{cases} \tag{5.113}
$$

where $e_{i,0} = e_i(0)$, $0 < \delta_{i,l}, \delta_{i,u} < 1$, $\rho_i(t) = (\rho_{i,0} - \rho_{i,\infty})e^{-\ell_i t} + \rho_{i,\infty}$, $\rho_{i,0} > \rho_{i,\infty} > 0$, $\ell_i > 0$, and $0 < \bar{\rho}_{i,\infty} \leq \min((1 - \delta_{i,l}\text{sign}(e_{i,0}))\rho_{i,0}, (1 + \delta_{i,u}\text{sign}(e_{i,0}))\rho_{i,0})$.

Compared with the existing prescribed performance, the typical feature of such TPP lies in the following aspects.

(i) Unlike the performance design in [1, 2, 26], the brief expressions of $e_{i,l}(t)$ and $e_{i,u}(t)$ avoid the tedious design procedure with respect to the different initial error condition, such as performance selection, control design development and stability analysis.

(ii) The initial performance boundaries of TPP are distributed on the same side as shown in Fig. 5.8a and b, rendering a tight allowable set. In this way, TPP can accomplish more strict tracking requirements of practical systems, such as tight allowable set and smaller overshoot performance. Notice that if select $\bar{\rho}_{i,\infty} = (1 - \delta_{i,l})\rho_{i,\infty}$ when $e_{i,0} > 0$ or $\bar{\rho}_{i,\infty} = (1 + \delta_{i,u})\rho_{i,\infty}$ when $e_{i,0} < 0$, even zero overshoot can be imposed as shown in Fig. 5.8c.

(iii) Finally, the TPP becomes $e_{i,l}(t) = \delta_{i,l}\rho_i(t)$ and $e_{i,u}(t) = \delta_{i,u}\rho_i(t)$ in the case of $e_{i,0} = 0$. Then, (5.112) can be rewritten as $-\delta_{i,l}\rho_i(t) < e_i(t) < \delta_{i,u}\rho_i(t)$, which is the same as the common prescribed performance as shown in Fig. 5.8d.

We utilize the sign function in TPP to unify different initial conditions into one concise expression. Namely, the TPP has the ability of flexibly adjusting its orientation of performance constraints as shown in Fig. 5.8a and b, instead of redeveloping or reprogramming the control design. This simple design attribute is desirable in most practical engineering, which makes the control design easy and reliable to be implemented. In the existing literature, much effort is dedicated to confine the tracking error within relative loose allowable set, where the constraint boundaries are distributed on both sides. With the increasing requirements from mission tasks, strict performance constraints deserve special attention. One typical example is the flight tracking systems whose tracking trajectory is required in a small region around the target to avoid target loss due to the limited perception range. Therefore, the proposed TPP establishes a tunnel-shape performance constraints to provide better transient performance. Moreover, such TPP can be further improved in the future by combining with existing approaches in [42, 47] to obtain finite time or flexible characteristics.

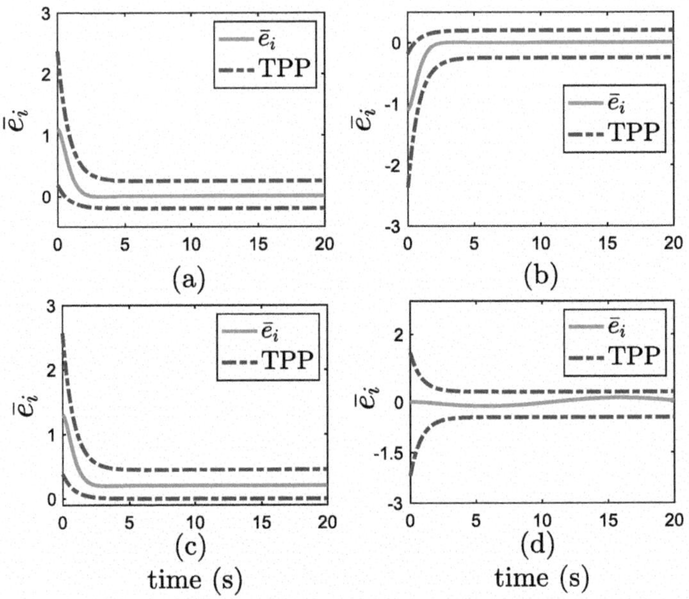

Fig. 5.8 The tracking error $e_i(t)$ under TPP. **a** and **b** denotes the case of $e_i(0) > 0$ and $e_i(0) < 0$. **c** denotes the case of no overshoot performance, and **d** denotes the case of $e_i(0) = 0$

5.2.2.7 Control Design

To address the performance constraints in (5.112), we define the following error transformation function:

$$z_1(\bar{e}_1) = \ln\left(\frac{e_{1,l} + \bar{e}_1}{e_{1,u} - \bar{e}_1}\right). \tag{5.114}$$

It is adopted on the one hand as a transformation mechanism, that is $\bar{e}_1(t)$ into an equivalent unconstrained variable $z_1(\bar{e}_1)$. On the other hand, z_1 is used to develop an orientation function $\phi(z_1)$ to handle the unknown control directions, which has the following properties:

$$|\phi(z_1)| < \infty, \ |d\phi/dz_1| < \infty, \ \forall |z_1| < \infty, \tag{5.115}$$

and

$$\begin{cases} \lim_{|z_1|\to\infty} \sup z_1(\bar{e}_1)\phi(z_1) = +\infty, \\ \lim_{|z_1|\to\infty} \inf z_1(\bar{e}_1)\phi(z_1) = -\infty. \end{cases} \tag{5.116}$$

There are many candidate functions satisfying the above properties, which indicates many expressions of such orientation function can be chosen, for example, $\phi(\cdot)$ can be selected as $\sin(\cdot)$ or $\cos(\cdot)$. Throughout this section, let $\phi(z_1) = \sin(z_1)$.

Then, we provide the following lemma, which indicates that the ESFs (5.103) does not affect the above-mentioned properties.

Lemma 5.5 *Consider the scaling function $w(t)$ defined in* (5.99). *There exists a constant $\zeta \neq 0$ such that*

$$\begin{cases} \lim_{|z_1|\to\infty} \sup \zeta w(t) z_1(\bar{e}_1)\phi(z_1) = +\infty, \\ \lim_{|z_1|\to\infty} \inf \zeta w(t) z_1(\bar{e}_1)\phi(z_1) = -\infty. \end{cases} \tag{5.117}$$

Proof These properties can be established by finding the contradictions. We omit it here for clarity and conciseness. □

Based on the above discussions, Tunnel Prescribed Control (TPC) is designed recursively. We first have

$$\alpha_1 = c_1 z_1 \phi(z_1), \tag{5.118}$$

where c_1 is a positive constant. Identically, the following error transformation function and control input signal are introduced to handle the performance constraints on $\bar{e}_i(t)$ in (5.111):

$$z_i = \ln\left(\frac{e_{i,l} + \bar{e}_i}{e_{i,u} - \bar{e}_i}\right), \tag{5.119}$$

$$\alpha_i = -c_i z_i, \tag{5.120}$$

where $i \in \mathbb{I}_{2:n}$ and note that for $i = n$, we have the actual control input $u = \alpha_n$.

One salient feature associated with the ESFs is given by the following lemma.

Lemma 5.6 *Given any initial condition of the system states, there is*

$$z_i(0) = 0, \ \forall i = \mathbb{I}_{1:n}. \tag{5.121}$$

Proof From (5.106), we have $\bar{e}_1(0) = h_{1,0}$. According to the definition of $h_{1,0}$ in (5.105), it represents the middle position of the performance boundaries at the initial time. Substituting $\bar{e}_1(0) = h_{1,0}$ into the error transformation function (5.114), we have $z_1(\bar{e}_1(0)) = \ln(\frac{e_{1,l}+h_{1,0}}{e_{1,u}-h_{1,0}}) = 0$. Then, it yields $\alpha_1(0) = 0$ from (5.118). Based on the definition of $\bar{e}_2(t)$ in (5.104), it follows $\bar{e}_2(0) = e_2(0) + h_2(0) = h_{2,0}$ due to $e_2(0) = 0$. Therefore, one has $z_2(\bar{e}_2(0)) = \ln(\frac{e_{2,l}+h_{2,0}}{e_{2,u}-h_{2,0}}) = 0$ leading to $\alpha_2(0) = 0$ defined in (5.120). Identically, we can also establish $e_i(0) = h_i(0) = h_{i,0}$, $z_i(0) = 0$ and $\alpha_i = 0$ for $i \in \mathbb{I}_{3:n}$ according to (5.119)–(5.120). This completes the proof. □

Remark 5.10 Compared with the control design in [15, 45], we stress that no intermediate control signal derivatives, dynamic surface control or command filters are required by reviewing the control design in (5.118) and (5.120). It makes a simple and less computation consumption control design possible. In lieu of Nussbaum-gain techniques, switching rules and parameter estimation methods, the function $\phi(z_1)$ is utilized to tackle the unknown control direction problem. Moreover, the control peaking phenomenon possibly introduced by the Nussbaum techniques can be significantly avoided or alleviated. Lemma 5.6 has illustrated that the control input $\alpha_i(0) = 0$, $i \in \mathbb{I}_{1:n}$, which indicates that the large control input is evaded at the initial time, which is desirable in most practical applications, such as flight systems and AC motors.

5.2.2.8 Performance Analysis

We give the following sufficient condition about the derivative of the virtual control input α_i, $i \in \mathbb{I}_{1:n}$.

Lemma 5.7 *If* $-e_{i,l} < \bar{e}_i < e_{i,u}$, $\dot{\bar{e}}_i$ *and* z_i *are bounded, so does* $\dot{\alpha}_i$ *for* $i \in \mathbb{I}_{1:n}$.

Proof Differentiating (5.118) and (5.120) yields

$$\dot{\alpha}_1 = c_1 \dot{z}_1 \left(\phi(z_1) + z_1 \frac{d\phi}{dz_1} \right), \tag{5.122}$$

$$\dot{\alpha}_i = -c_i \dot{z}_i, \ i \in \mathbb{I}_{2:n} \tag{5.123}$$

Let $\Lambda_i = \partial z_i / \partial \bar{e}_i$, $\Lambda_{i,u} = \partial z_i / \partial e_{i,u}$ and $\Lambda_{i,l} = \partial z_i / \partial e_{i,l}$. Recalling (5.114) and (5.119), we have

$$\dot{z}_i = \Lambda_i \dot{\bar{e}}_i + \Lambda_{i,u} \dot{e}_{i,u} + \Lambda_{i,l} \dot{e}_{i,l}, \tag{5.124}$$

Suppose that $-e_{i,l} < \bar{e}_i < e_{i,u}$ for $i \in \mathbb{I}_{1:n}$ and thus Λ_i, $\Lambda_{i,l}$ and $\Lambda_{i,u}$ are bounded. Upon the assumptions in Lemma 5.7 and boundaries (5.113), the boundedness of $\dot{\alpha}_i$, $i \in \mathbb{I}_{1:n}$ can thus be guaranteed. □

We now in a position to present the main results.

Theorem 5.2 *For the nonlinear system in* (5.85) *under Assumptions 5.4–5.6, TPC scheme with* (5.90), (5.114), *and* (5.118)–(5.120) *can guarantee: (i) First, all the closed-loop system signals are bounded for any initial condition, and the ESFs* $\bar{e}_i(t)$, $i \in \mathbb{I}_{1:n}$ *in* (5.103) *and* (5.104) *are always within TPP* (5.112). *(ii) Second, the tracking error* $e_1(t)$ *satisfies ECP as depicted in* (5.86), *and semi-global ultimately uniformly bounded tracking is ensured regardless of the initial conditions.*

Proof For clarity and conciseness, we omit the time or state dependence of functions in the following analysis. From (5.87), (5.93), (5.103), and (5.118)–(5.120), the time derivatives of the tuned tracking errors e_1 and e_i are

$$\dot{\bar{e}}_1 = \sigma_1 + w \bar{g}_1 c_1 z_1 \phi_1(z_1), \tag{5.125}$$

$$\dot{\bar{e}}_i = \sigma_i - c_i z_i,\ i \in \mathbb{I}_{2:n-1}, \tag{5.126}$$

$$\dot{\bar{e}}_n = \sigma_n + u, \tag{5.127}$$

where $\sigma_1 = \dot{w}(x_1 - y_r) + w(f_1 + d_1 - \dot{y}_r) + \dot{h}_1 + w\bar{g}_1(\bar{e}_2 + \tilde{\xi}_2 - h_2)$, $\sigma_i = -l_i\hat{\xi}_1 + \hat{\theta}_i^T S_i - \dot{\alpha}_{i-1} + \dot{h}_i + \bar{e}_{i+1} - h_{i+1}$ for $i \in \mathbb{I}_{2:n-1}$, and $\sigma_n = -l_n\hat{\xi}_1 + \hat{\theta}_n^T S_n - \dot{\alpha}_{n-1} + \dot{h}_n$.

We claim that the tuned tracking error satisfies

$$-e_{i,l} < \bar{e}_i < e_{i,u},\ i \in \mathbb{I}_{1:n},\ \forall t \geq 0, \tag{5.128}$$

and it is to be proved by seeking a contradiction. According to Lemma 5.6, we know that (5.128) is satisfied at $t = 0$. As $\bar{e}_1$ is continuous, so is z_1 in (5.114) if $-e_{1,l} < \bar{e}_1 < e_{1,u}$. Then, we also have α_1 in (5.118) to be continuous in time. From (5.90) and (5.104), we can get the continuity of $\bar{e}_2$ in time. Identically, the continuity of $\bar{e}_i$ in time can be inductively established by following the same analysis, as long as $-e_{i-1,l} < \bar{e}_{i-1} < e_{i-1,u}$.

The above discussion indicates that if the tuned tracking error violates the allowable set (5.128) on some $j \in \{1, \ldots, n\}$, there exists the first time instant t^* such that

$$-e_{i,l} < \bar{e}_i < e_{i,u},\ i \in \mathbb{I}_{1:n},\ t < t^*, \tag{5.129}$$

$$\lim_{t\to t^*} \bar{e}_j = e_{j,u},\ \text{or}\ \lim_{t\to t^*} \bar{e}_j = -e_{j,l}. \tag{5.130}$$

Suppose (5.130) and each case of $j = 1, \ldots, n$ will be analyzed one-by-one. First, we consider the following case:

$$\lim_{t\to t^*} \bar{e}_1 = e_{1,u},\ \text{or}\ \lim_{t\to t^*} \bar{e}_1 = -e_{1,l}. \tag{5.131}$$

And from the definition of z_1 in (5.114), we can further have

$$\lim_{t\to t^*} |z_1| = \infty. \tag{5.132}$$

By recalling the TPP design in (5.112), it is naturally divided into three conditions, i.e., $e_1(0) > 0$, $e_1(0) < 0$, and $e_1(0) = 0$. Each condition includes the tuned tracking error increases or decreases to the upper boundary $e_{1,u}(t)$ or the lower boundary $-e_{1,l}(t)$. No matter which trend is, there should exist constants w_1 and w_2 such that

$$\liminf_{t\to t^*} \dot{\bar{e}}_1 \geq w_1, \tag{5.133}$$

$$\limsup_{t\to t^*} \dot{\bar{e}}_1 \leq w_2. \tag{5.134}$$

According to (5.99)–(5.102), and (5.105), one has $w, \dot{w}, \dot{h}_1, h_2 \in L^\infty$ when $t < t^*$. Assumptions 5.5–5.6 indicate that $d_1, y_r, \dot{y}_r \in L^\infty$. Based on (5.129) and

Lemma 5.3, $\bar{e}_2$ and $\tilde{\xi}_2$ are bounded over $t \in [0, t^*)$. From Lemma 5.4 and Assumption 5.6, note that x_1 is further bounded, which implies $f_1 \in L^\infty$ when $t < t^*$. In view of the above discussion, it consequently implies that σ_1 in (5.125) are bounded on $t \in [0, t^*)$.

Referring to (5.132) and Lemma 5.5 by applying $\zeta = c_1 \bar{g}_1$ yields

$$\liminf_{t \to t^*} \dot{\bar{e}}_1 = \liminf_{t \to t^*} \sigma_1 + \liminf_{t \to t^*} \zeta w z_1 \phi_1 = -\infty, \tag{5.135}$$

$$\limsup_{t \to t^*} \dot{\bar{e}}_1 = \limsup_{t \to t^*} \sigma_1 + \limsup_{t \to t^*} \zeta w z_1 \phi_1 = +\infty, \tag{5.136}$$

which apparently contradicts (5.133) and (5.134), respectively. Therefore, the supposition (5.131) is false, and it indicates that the tuned tracking error $\bar{e}_1$ cannot approach the upper or the lower boundaries of TPP over $t \in [0, t^*)$. Apparently, we have a positive constant λ_1 satisfying

$$-e_{1,l} + \lambda_1 < \bar{e}_1 < e_{1,u} - \lambda_1, \ t < t^*. \tag{5.137}$$

See from (5.114) that z_1 is thus bounded over $t \in [0, t^*)$. Upon (5.125), $\dot{\bar{e}}_1$ is bounded there. By Lemma 5.7, one has $\dot{\alpha}_1 \in L^\infty$ on $[0, t^*)$.

Next, we analyze the tuned tracking error $\bar{e}_2$. From (5.124) and (5.126), we have

$$\dot{z}_2 = -\Lambda_2 c_2 z_2 + \bar{\sigma}_2 \tag{5.138}$$

where $\bar{\sigma}_2 = \Lambda_2 \sigma_2 + \Lambda_{2,u} \dot{e}_{2,u}(t) + \Lambda_{2,l} \dot{e}_{2,l}(t)$. Consider the following Lyapunov candidate function as $V_2 = \frac{1}{2} z_2^2$ and its time derivative is given by

$$\dot{V}_2 = z_2(-\Lambda_2 c_2 z_2 + \bar{\sigma}_2) \leq |z_2|(-\Lambda_2 c_2 |z_2| + |\bar{\sigma}_2|). \tag{5.139}$$

We discuss the boundedness of $\bar{\sigma}_2$ as follows. From (5.93), we have $\hat{\xi}_1 = \xi_1 - \tilde{\xi}_1$. Based on Lemma 5.3, $\tilde{\xi}_1$ is bounded over $t < t^*$, which indicates $\hat{\xi}_1 \in L^\infty$ for $t \in [0, t^*)$. Referring to (5.105), $\dot{h}_2$ and h_3 are also bounded there. By (5.92), (5.104) and (5.129), there is $\bar{e}_3, \hat{\theta}_2 \in L^\infty$ for $t < t^*$. By (5.113), we have $\dot{e}_{2,u}, \dot{e}_{2,l} \in L^\infty$. As the above analysis, there exist a positive constant b_2 such that $|\bar{\sigma}_2| < b_2$ for $t < t^*$. Therefore, it yields

$$\dot{V}_2 \leq |z_2|(-\Lambda_2 c_2 |z_2| + b_2), \ t < t^*. \tag{5.140}$$

From the definition of Λ_2 in (5.124), there exists a positive constant ς_2 such that $\Lambda_2 \geq \frac{4}{e_{2,u} + e_{2,l}} \geq \varsigma_2 > 0$. From Lemma 5.6, we have $z_2(0) = 0$ and obtain that z_2 is bounded by $|z_2| \leq \frac{b_2}{\varsigma_2 c_2}$ since $\dot{V}_2 < 0$ if $|z_2| > \frac{b_2}{\varsigma_2 c_2}$, for $t < t^*$. Therefore, based on (5.119), there is a positive constant λ_2 with

$$-e_{2,l} + \lambda_2 < \bar{e}_2 < e_{2,u} - \lambda_2, \ t < t^*. \tag{5.141}$$

Namely, the tuned tracking error $\bar{e}_2$ is always within TPP for $t < t^*$ as depicted in (5.129), which indicates that the supposition (5.130) for $j = 2$ is false and (5.128) is true. Due to $\sigma_2, z_2 \in L^\infty$ on $[0, t^*)$, there is $\dot{\bar{e}}_2 \in L^\infty$. By recalling Lemma 5.7 and the above discussion, one has that $\dot{\alpha}_2$ is bounded.

Identically, we can also conclude the following result for $\bar{e}_i$, $i \in \mathbb{I}_{3:n-1}$ by changing the above argument $\bar{e}_2$ to $\bar{e}_i$. There also holds

$$- e_{i,l} + \lambda_i < \bar{e}_i < e_{i,u} - \lambda_i, \; t < t^*. \tag{5.142}$$

with λ_i denoting a positive constant. In fact, the above discussion shows that the tuned tracking error $\bar{e}_i$, $i \in \mathbb{I}_{3:n-1}$ is always within the performance constraints on $t \in [0, t^*)$. It reveals that (5.142) contradict (5.130), and therefore, (5.128) is true for $i \in \mathbb{I}_{1:n-1}$. Invoking Lemma 5.7, there also has $\dot{\alpha}_{n-1} \in L^\infty$.

By (5.124) and (5.127), the time derivative of z_n is

$$\dot{z}_n = \bar{\sigma}_n + \Lambda_n u, \tag{5.143}$$

where $\bar{\sigma}_n = \Lambda_n \sigma_n + \Lambda_{n,u} \dot{e}_{n,u} + \Lambda_{n,l} \dot{e}_{n,l}$. Substituting the actual control input u in (5.120) into (5.143) yields

$$\dot{z}_n = -\Lambda_n c_n z_n + \bar{\sigma}_n. \tag{5.144}$$

As $\hat{\xi}_1, \dot{\alpha}_{n-1}, \dot{h}_n, \hat{\theta}_n \in L^\infty$ on $t < t^*$, we have σ_n to be bounded there. By (5.113), we further get $\bar{\sigma}_n \in L^\infty$. It means that there has a positive constant b_n such that $|\bar{\sigma}_n| < b_n$ over $[0, t^*)$. Similarly, construct the Lyapunov function $V_n = \frac{1}{2} z_n^2$. Taking (5.144) into the time derivative of V_n yields

$$\dot{V}_n \leq |z_n|(-\Lambda_n c_n |z_n| + b_n), \; t < t^*. \tag{5.145}$$

It can be also derived that z_n is bounded during $t < t^*$ as shown in (5.140). By (5.119), we can have a positive constant λ_n such that

$$- e_{n,l} + \lambda_n < \bar{e}_n < e_{n,u} - \lambda_n, \; t < t^*. \tag{5.146}$$

Clearly, (5.146) contradict (5.130). Therefore, (5.130) is false and (5.128) is true for $i = n$. We, consequently, have that all the tracking error $\bar{e}_i$ is within the TPP allowable set despite the unknown control directions and unknown external disturbances. Moreover, due to Lemma 5.4 and (5.128), we have that the tracking error $e_1(t) = y(t) - y_r(t)$ follows the ECP depicted in (5.86).

We then show the boundedness of other system states in the closed-loop system. Since z_i and $\bar{e}_i$ are bounded, one sees from (5.104), (5.118), and (5.120) that α_i and $\hat{\xi}_i$ are bounded. Due to $\tilde{\xi}_i \in L^\infty$ by Lemma 5.3, it follows $\xi_i \in L^\infty$ from (5.93). Note from (5.87) that x_i is bounded for $i \in \mathbb{I}_{1:n}$. As discussed above, all the closed-loop system states are semi-global ultimately uniformly bounded regardless of the initial conditions. This completes the proof. □

Remark 5.11 As Fuzzy Neural Network is just limited to a compact set, $\Omega \subset \mathbb{R}^n$ stated in Lemma 5.4, only semi-global ultimately uniformly boundedness result is thereby achieved. Note that we can further extend our method to obtain global stability if we avoid using Fuzzy Neural Network, but employ an input-driven filter to construct state observer as shown in [24, 29, 48], where only the control signal is as the input variable of the filter.

Remark 5.12 About the settling time T, it determines when the performance functions are strictly imposed on the tracking error. Note that the value of finite settling time T should satisfy $0 < T_m < T$, where T_m represents the minimum amount of time necessary for state observer and signal processing. Moreover, such assumption is widely used in the control design [39, 53]. If T is chosen too small, although faster convergence can be obtained, it might result in larger control input, especially during the initial time interval. For control parameters, only c_i, $i \in \mathbb{I}_{1:n}$ should be selected, whose value affects the tracking accuracy. A large c_i also leads to fast convergence rate, and high tracking accuracy since it decides the ultimate boundary of the tracking error. If the gain of control is too high, it makes the system more vulnerable to external disturbances, leading to instability or other undesirable behavior. For the performance functions, parameters $\delta_{i,l}$, $\delta_{i,u}$, $\rho_{i,0}$, $\rho_{i,\infty}$ and ℓ_i determine the performance attributes. Smaller values of $\delta_{i,u}$ and $\delta_{i,l}$ result in a tighter allowable set. However, it is essential to choose appropriate values for the robustness to uncertainties. Hence, both the selections of settling time, control parameters, and performance functions should be a trade-off results in practice.

5.2.2.9 Simulation Examples

Example 1: In this section, we have conducted simulation studies to illustrate the superiority and effectiveness of our proposed TPC scheme. Consider the following dynamics of the jet engine compressor given by [27]

$$\begin{aligned} \dot{x}_1 &= -x_2 - \frac{3}{2}x_1^2 - \frac{1}{2}x_1^3 \\ \dot{x}_2 &= u + d_2(t) \\ y &= x_1 \end{aligned} \tag{5.147}$$

where $d_2(t) = 0.2\sin 0.2t$ is the external disturbance. All the initial states are set as zero, namely, $x_1(0) = x_2(0) = 0$, and the desired tracking trajectory is $y_r(t) = \cos(0.5t)$.

The parameters of TPP in (5.112) are selected as: $\delta_{1,l} = 0.6, \delta_{1,u} = 0.5, \delta_{2,l} = 0.7$, $\delta_{2,u} = 0.6, \bar{\rho}_{1,\infty} = 0.2$ and $\bar{\rho}_{2,\infty} = 0.2$, $\rho_1 = 0.98e^{-0.5t} + 0.22$, and $\rho_2 = 1.55e^{-t} + 0.45$. From (5.99) and (5.105), let $\mu_1 = 1.6$ and $\mu_2 = 0.5$. By the State Observer in (5.90), we have

$$\begin{cases} \dot{\hat{\xi}}_1 = -l_1\hat{\xi}_1 + \hat{\xi}_2 + \hat{\theta}_1^T S_1(\hat{\xi}_1) \\ \dot{\hat{\xi}}_2 = -l_2\hat{\xi}_1 + u \end{cases} \tag{5.148}$$

where $l_1 = 20$, $l_2 = 5$ and the adaptive law for $\hat{\theta}_1$ is

$$\dot{\hat{\theta}}_1 = -\sigma_1\hat{\theta}_1 + r_1 z_1 \Lambda_1 S_1(\hat{\xi}_1), \tag{5.149}$$

where $\sigma_1 = 2.0$ and $r_1 = 1.0$. The membership function is designed as

$$\mu_{1,j} = \exp\left(-\frac{(\hat{\xi}_1 - 1.5 + 0.5j)^2}{2}\right), \tag{5.150}$$

where $j = 1 \ldots 5$. Recalling (5.118) and (5.120), we have

$$\alpha_1 = c_1 z_1 \phi_1(z_1), \tag{5.151}$$

$$u = -c_2 z_2, \tag{5.152}$$

with $c_1 = 2$, $c_2 = 10$ and $\phi_1(z_1) = \cos z_1^2$.

To illustrate the effectiveness of TPC, we choose different settling time as $T = 1.0, 3.0, 5.0, 7.0$ for simulation. The tracking results are shown in Figs. 5.9, 5.10 and 5.11. From Fig. 5.9, the tracking error $e_1(t)$ violates the allowable set, initially. By the proposed TPC, it can be steered into there within finite time T. Specifically, $e_1(t)$ is bounded over $0 \leq t < T$ despite it may not meet performance constraints during this initial interval. Then, $e_1(t)$ satisfies the TPP: $-e_{1,l}(t) < e_1(t) < e_{1,u}(t)$ for $t \geq T$, where the tracking performance under different settling time T is plotted in Fig. 5.9. Therefore, the performance constraints are strictly imposed on the tracking error $e_1(t)$ right after the chosen settling time T. In this way, it solves the ECP. Notice that smaller overshoot performance can be achieved by selecting appropriate control parameters due to the tunnel-shape allowable set. Upon employing ESFs, the tuned tracking signal $\bar{e}_1(t)$ always evolves within the performance constraints from the beginning of system operation as shown in Fig. 5.10. These results indicate that the selections of prescribed performance are independent of the initial conditions, making the control design easy to be implemented. For the control input $u(t)$, it is depicted in Fig. 5.11, whose initial value is equal to zero since $z_1(0) = z_2(0) = 0$ as discussed in Lemma 5.6. And this control strategy indeed avoids control peaking phenomenon at the initial time, which is of significance in most practical applications as high-magnitude control input may shorten the service life of components.

Example 2: We further consider the following uncertain nonlinear system:

$$\begin{cases} \dot{x}_1 = f_1(\bar{x}_1) + g_1 x_2 + d_1(t), \\ \dot{x}_2 = f_2(\bar{x}_2) + g_2 u + d_2(t), \end{cases} \tag{5.153}$$

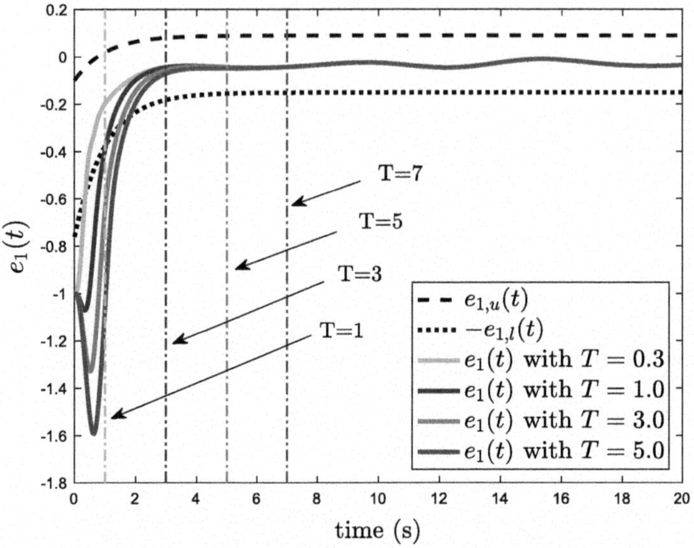

Fig. 5.9 The trajectory of the tracking error $e_1(t)$

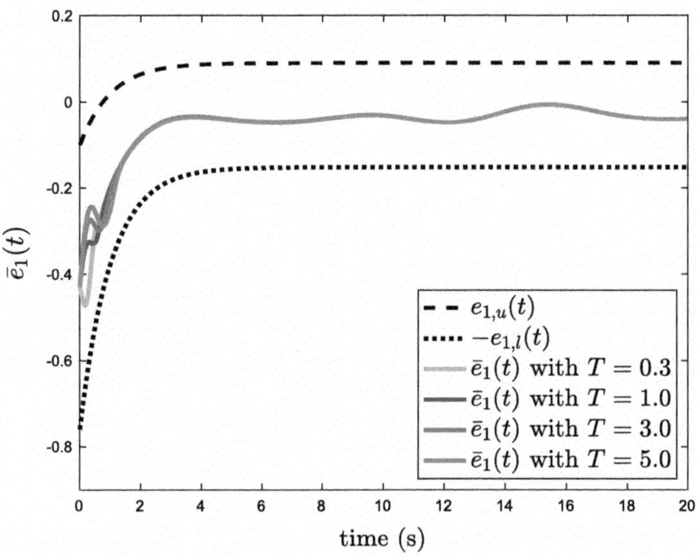

Fig. 5.10 The trajectory of the ESF $\bar{e}_1(t)$

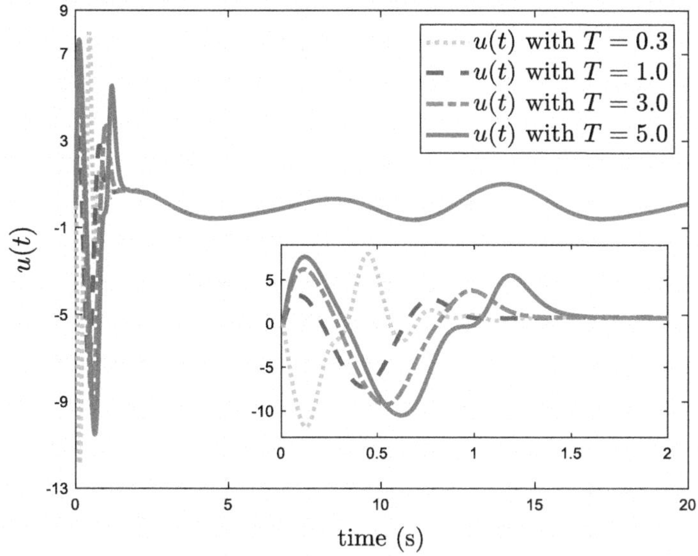

Fig. 5.11 The control input signal u

where $f_1(\bar{x}_1) = 0$, $f_2(\bar{x}_2) = -2x_2 - 5\sin x_1$, $g_1 = 1$, $g_2 = 3$, $d_1(t) = 0.1\sin t$, and $d_2(t) = 0.2\sin 0.5t$, The initial system states are $x_1(0) = 2$ and $x_2(0) = 0$. For simulation, let $y_r(t) = \sin 0.5t$, $\delta_{1,l} = \delta_{1,u} = 0.5$, $\rho_1 = 0.4e^{-0.5t} + 0.6$, $\bar{\rho}_{1,\infty} = 0.6$, $\delta_{2,l} = \delta_{2,u} = 0.5$, $\rho_2 = 1.9e^{-0.5t} + 0.6$, $\mu_1 = 5$, $l_1 = 10$, $l_2 = 5$, $\sigma_2 = 2.0$, $r_2 = 1.0$, $c_1 = 3.5$, and $c_2 = 8.5$. To simulate the ECP under different settling time, we select $T = 0.3, 1.0, 3.0$, respectively.

To verify the merit of the proposed method, we also conduct a comparison between the proposed TPC and the funnel control [19]. Therefore, the funnel control is given by

$$\begin{aligned}
u &= -\bar{\mu}[(\cos(\chi_2) - \chi_2\sin(\chi_2))^2 e^2]\chi_2^4(1+\bar{\omega}^2)\vartheta \\
&\quad - \bar{\mu}N_2 e - \bar{\mu}N_2^2\chi_2^4(1+\bar{\omega}^2)\vartheta \\
N_2 &= \chi_2\cos(\chi_2),\ \chi_2 = \frac{1}{1-\mathcal{F}^2e^2}, \\
\vartheta &= \bar{\omega} - N_2 e,\ \dot{\bar{\omega}} = -\bar{\mu}\bar{\omega} + u,\ e = x_1 - y_d,
\end{aligned} \tag{5.154}$$

where $\bar{\mu} = 4$ and $\mathcal{F} = \frac{\sqrt{1-\psi^2}}{\sqrt{\ell}\psi}$ with $\psi = (1-0.3)e^{-0.5t} + 0.3$ and $\ell = 1 - 0.3^2$. For the funnel control, the initial conditions are $x_1(0) = 0.5, 1.0, 1.5, 2.0$ and $x_2(0) = 0$. The simulation results are shown as follows.

From Fig. 5.12, it can be observed that the tracking error signal $e_1(t)$ is regulated into the allowable set before the different settling time T chosen by the designer,

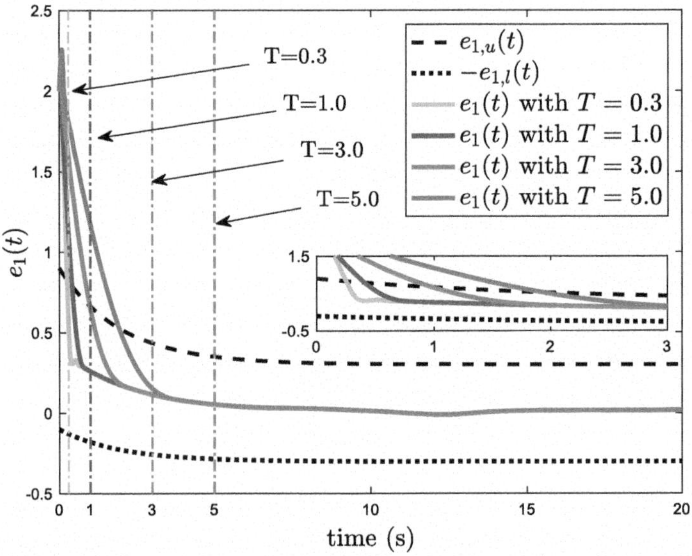

Fig. 5.12 The tracking error $e_1(t)$ under different settling time

regardless of the initial condition. Although $e_1(t)$ may violate the prescribed performance initially, with the aid of ESFs, the tuned tracking error signal $\bar{e}_1(t)$ always satisfies the allowable set from the beginning of system operation as shown in Fig. 5.13. Moreover, due to the tight allowable set involved by the tunnel prescribed performance, better transient performance (i.e., smaller overshoot performance) is achieved under the proposed TPC. In Fig. 5.14, the control input $u(t)$ is plotted under different settling time. Remarkably, all the initial control inputs are zero which avoids control peaking problem at the initial time. By Fig. 5.15, it shows the tracking error $e_1(t)$ under different initial conditions by employing funnel control. Although such funnel control can be applied to any initial conditions, obvious overshoot performance may be encountered since symmetric performance boundaries lead to loose performance constraints during the initial time interval. The funnel control signals are depicted in Fig. 5.16, whose initial values are not equal to zero. Compared with the funnel control, our proposed TPC provides a simple method to address ECP for any initial conditions without introducing any additional filters or relying on specific performance functions. Such TPC can be also applied for either symmetric or asymmetric performance constraints. In lieu of the Nussbaum function, the orientation function is an effective way to handle unknown control directions, which completes the avoidance of control peaking problem.

In order to show our proposed control method suitable to any initial condition, we further select extra large initial conditions with $x_1(0) = 100$ and $x_2(0) = 0$. In simulation, we have $y_r(t) = \sin 0.5t$, $\delta_{1,l} = \delta_{1,u} = 0.6$, $\rho_1 = 30e^{-0.8t} + 20$, $\bar{\rho}_{1,\infty} = 20$, $\mu_1 = 0.5$, $c_1 = 10.5$, and $c_2 = 3.5$. To simulate the ECP under different settling time, we select $T = 2.0, 5.0, 8.0$, respectively. From Fig. 5.17, the tracking error can be

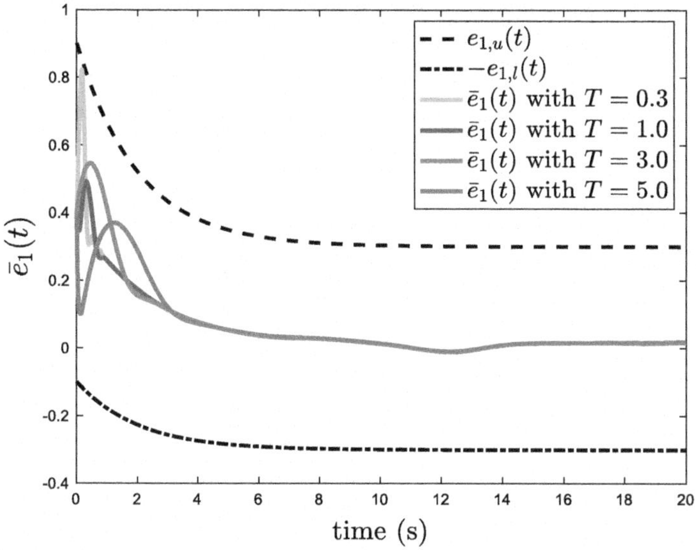

Fig. 5.13 The ESF $\bar{e}_1(t)$ under different settling time

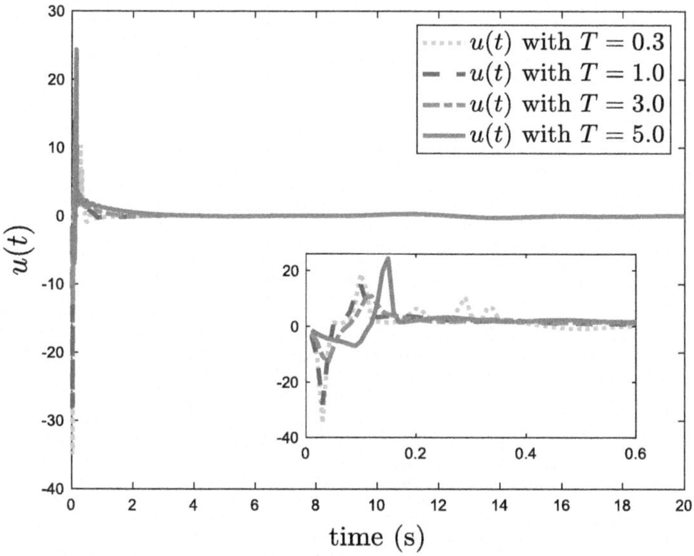

Fig. 5.14 The control input signal u under different settling time

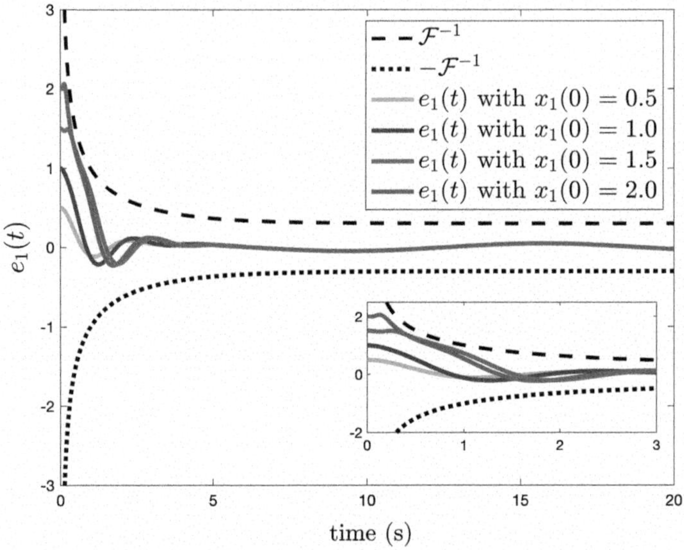

Fig. 5.15 The tracking error $e_1(t)$ under funnel control

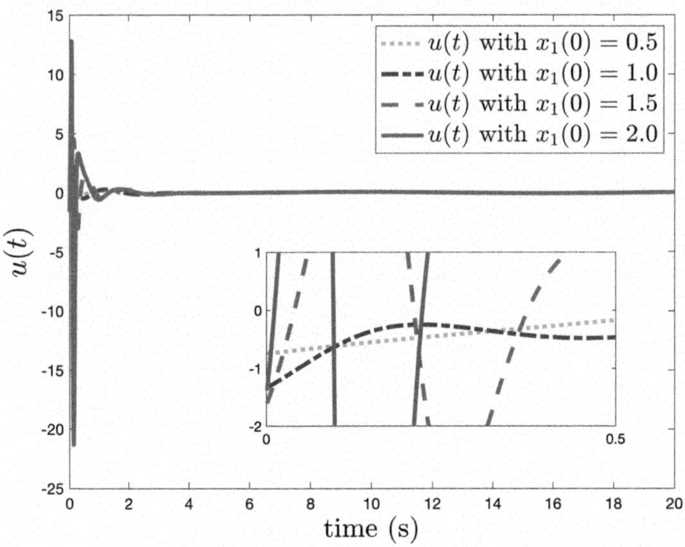

Fig. 5.16 The tracking error $e_1(t)$ under funnel control

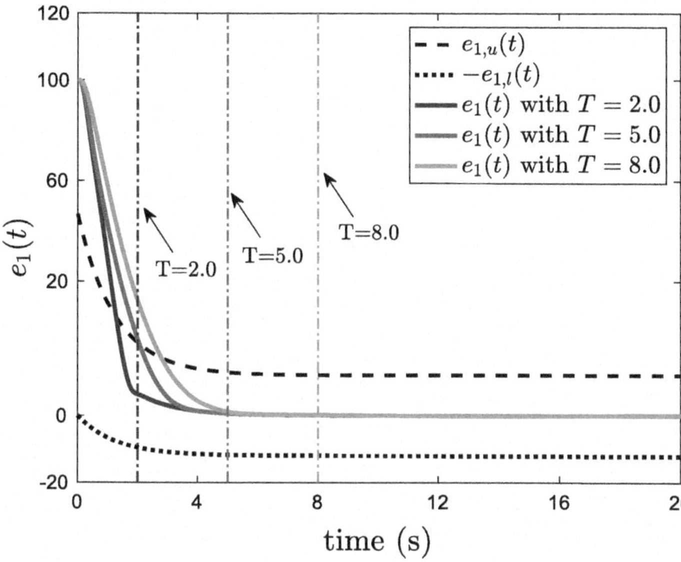

Fig. 5.17 The tracking error $e_1(t)$ under different settling time

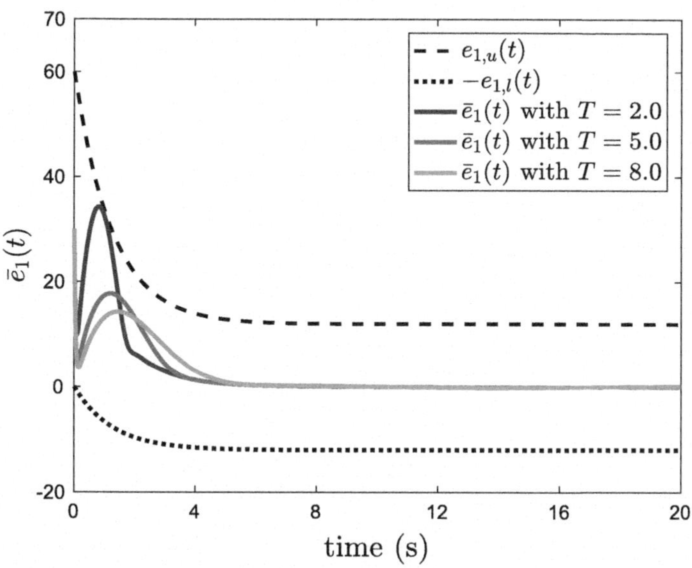

Fig. 5.18 The ESF $\bar{e}_1(t)$ under different settling time

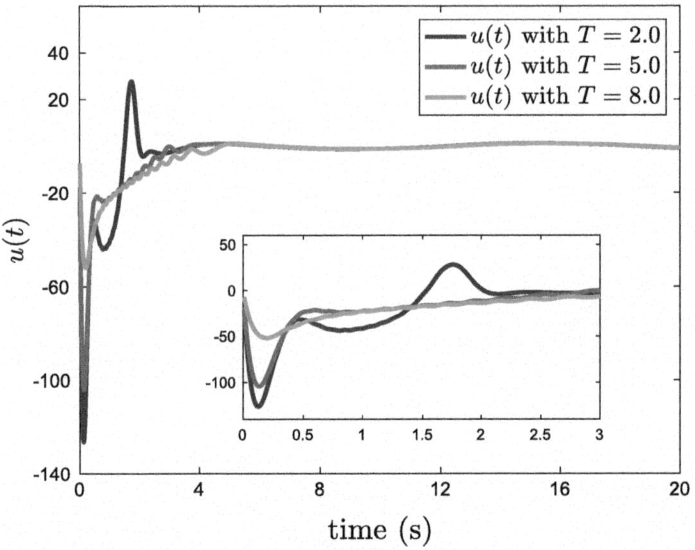

Fig. 5.19 The control input signal u under different settling time

regulated into the predefined region within given settling time T, its initial condition violates the allowable set. By ESF, the tuned tracking error $\bar{e}_1(t)$ is exhibited in Fig. 5.18, which satisfies the performance constraints from beginning to end. Additionally, the control inputs $u(t)$ under different settling times are depicted in Fig. 5.19, whose values are always equal to zero, initially.

5.3 Self-adjustable Performance-Based Tunnel Prescribed Control

5.3.1 Problem Formulation

Consider a class of strict-feedback nonlinear systems

$$\begin{cases} \dot{x}_i = f_i(\bar{x}_i) + g_i(\bar{x}_i)x_{i+1},\ i = 1, \ldots, n-1 \\ \dot{x}_n = f_n(\bar{x}_n) + g_n(\bar{x}_n)u \\ y = x_1 \end{cases} \tag{5.155}$$

where $x_i \in \mathbb{R}$, $i = 1, \ldots, n$, is the system state with $\bar{x}_i = [x_1, \ldots, x_i]^T \in \mathbb{R}^i$, $f_i(\cdot) : \mathbb{R}^i \to \mathbb{R}$ is an unknown but continuous function, $g_i(\cdot) : \mathbb{R}^i \to \mathbb{R}$ denotes an unknown time-varying control coefficient, $u \in \mathbb{R}$ and $y \in \mathbb{R}$ are system input and output, respectively.

Define the tracking error $e_1(t) = y(t) - y_d(t)$ with $y_d(t)$ being the desired trajectory. The control objective is to design a control scheme for the nonlinear systems (5.155) under any initial condition such that:

(i) All signals in the closed-loop systems are bounded; and
(ii) For any initial condition, the tracking error $e_1(t)$ can be regulated into the prescribed allowable set in a customizable manner before a given time and then be constrained within there.

Definition 5.4 For any initial tracking condition (including initial constraint violation), if the tracking error $e_1(t)$ is fully constrained by the prescribed performance boundaries right after a user-given settling time $T > 0$:

$$-e_l(t) < e_1(t) < e_u(t), \ \forall t \geq T, \tag{5.156}$$

then, such error tracking performance is said to be Entry Capture Problem, where $-e_l(t)$ and $e_u(t)$ denote lower and upper prescribed boundaries, respectively.

Remark 5.13 The studied Entry Capture Problem represents a frequently encountered tracking situation that the prescribed performance is involved right after a given time T, whereas, in most existing works, these performance constraints should be satisfied from the beginning of system operation. One typical example is a flight system which is often released from any condition (including initial-constraint violation), but is required to track and interact with a target in prescribed performance for successful task completion. Therefore, addressing such Entry Capture Problem is interesting yet more challenging since it needs the control design applicable for more general cases.

Assumption 5.7 The desired trajectory $y_d(t)$ and its time derivatives up to $(n+1)$th-order are known, bounded, and piece-wise continuous.

Assumption 5.8 The nonlinear function $f_i(\bar{x}_i)$, $i = 1, \ldots, n$, is unknown but certain crude information is available so that $|f_i(\bar{x}_i)| \leq b_i \phi_i(\bar{x}_i)$, $\forall t \geq 0$, where $b_i \geq 0$ is an unknown constant and $\phi_i(\bar{x}_i) \geq 0$ is a known smooth function.

Assumption 5.9 The control coefficient $g_i(\bar{x}_i)$, $i = 1, \ldots, n$, is unknown and time-varying, but away from zero, that is, there exist positive constants $\underline{g}_i$ and $\bar{g}_i$ such that $0 < \underline{g}_i \leq |g_i(\bar{x}_i)| \leq \bar{g}_i$. Without loss of generality, we assume that the signs of $g_i(\bar{x}_i)$ are known and all positive.

Remark 5.14 Assumption 5.7 is widely adopted in tracking control of nonlinear systems [28]. Assumption 5.8 indicates that some core functions can be easily extracted only based on some crude system information, which is reasonable and in line with practice [36]. Assumption 5.9 is necessary to guarantee the controllable condition of nonlinear system (5.155), which is made in most control design [25].

5.3.2 Auxiliary Functions

To deal with the third control objective of this article, we first introduce the definition of auxiliary functions.

Definition 5.5 Auxiliary functions $\eta_u(t)$ and $\eta_l(t)$ are scalar functions which satisfies the following properties:

(i) $\eta_u^{(k)}$ and $\eta_l^{(k)}$, $k = 0, \ldots, n+1$, are known, continuous, and bounded;
(ii) $\eta_u(0) > e_1(0) - e_u(0)$ and $\eta_l(0) < e_1(0) + e_l(0)$;
(iii) $\eta_u(t), \eta_l(t), \dot{\eta}_u(t), \dot{\eta}_l(t) \to 0$ as $t \to T$, where T is the settling time in (5.156); and
(iv) $\eta_u(t) \geq \eta_l(t), \forall t \geq 0$ and $\eta_u(t) = \eta_l(t) = 0, \forall t \geq T$.

Obviously, there exist many candidates satisfying these properties, for example,

$$\eta_u(t) = \begin{cases} (e_1(0) + \frac{(\lambda-2)e_u(0)+\lambda e_l(0)}{2})e^{-\frac{lTt}{T-t}}, & 0 \leq t < T \\ 0, & t \geq T \end{cases}$$

$$\eta_l(t) = \begin{cases} (e_1(0) - \frac{\lambda e_u(0)+(\lambda-2)e_l(0)}{2})e^{-\frac{lTt}{T-t}}, & 0 \leq t < T \\ 0, & t \geq T \end{cases}$$

where l and $\lambda \geq 1$ are positive constants, $e_l(0)$ and $e_u(0)$ denote the initial values of performance constraints $e_l(t)$ and $e_u(t)$ which will be defined in the next section. Throughout this section, we use the above functions as the auxiliary functions.

Remark 5.15 More details about the motivation of such auxiliary functions are provided here. From the third objective in Sect. 5.3.1, the control scheme should be adapted to any initial condition. However, in practice, the initial condition may violate the performance constraints initially, rendering most existing PPC methods inapplicable since they are not defined and suffer from singularity problem in such scenario. To handle this problem, a straightforward approach is to utilize these auxiliary functions to adjust the performance boundaries in accordance with the initial condition. By leveraging the second property in Definition 5.5, any given initial condition would remain within an updated and allowable set, thereby ensuring the applicability of the control method. Other properties are also crucial for system analysis under addressing the Entry Capture Problem (5.156). Moreover, there would be fruitful expressions of such additional auxiliary functions which facilitate their extension to other control approaches.

5.3.3 Self-adjustable Prescribed Performance

We design the following Self-adjustable Prescribed Performance (SPP) on the tracking error $e_1(t)$ by introducing the above auxiliary functions into the performance boundaries:

$$-E_l(t) < e_1(t) < E_u(t), \ \forall t \geq 0, \tag{5.157}$$

with

$$E_u(t) = e_u(t) + \eta_u(t), \tag{5.158}$$
$$E_l(t) = e_l(t) - \eta_l(t), \tag{5.159}$$

where $e_u(t)$ and $-e_l(t)$ represent the original upper and lower Tunnel Prescribed Performance (TPP) [22], which are given by:

$$e_u(t) = (\delta + \text{sign}(e_{1,0}))\rho(t) - \rho_\infty \text{sign}(e_{1,0}), \tag{5.160}$$
$$e_l(t) = (\delta - \text{sign}(e_{1,0}))\rho(t) + \rho_\infty \text{sign}(e_{1,0}), \tag{5.161}$$

where $0 < \delta < 1, e_{1,0} = e_1(0), \rho(t) = (\rho_0 - \rho_\infty)e^{-\varsigma t} + \rho_\infty$ with $\rho_0 > \rho_\infty > 0$ and $\varsigma > 0$.

To better illustrate the mechanism behind such SPP, we take $e_1(0) > 0$ as an example. The parameters of SPP (5.158)–(5.159) are selected as: $\delta = 0.6$, $\rho_0 = 0.5$, $\rho_\infty = 0.2$, $l = 1$, $\varsigma = 1.2$, $\lambda = 3$, $T = 4$, and the initial tracking condition is $e_1(0) = 2$. As shown in the left plot of Fig. 5.20, the initial tracking error $e_1(0)$ violates the original allowable set established by $e_u(t)$ and $-e_l(t)$, which is colored in brown. It renders the previous methods inapplicable due to the singularity problem encountered. With the aid of the auxiliary functions, SPP is capable of re-adjusting the performance boundaries according to different initial conditions such that $e_1(t)$ is always within the updated allowable set initially (i.e., $-E_l(0) < e_1(0) < E_u(0)$) as shown in the right plot of Fig. 5.20 in green color. During the initial time interval ($0 \leq t \leq T$), our SPP not only ensures the control method applicable for any initial conditions but also provides temporary performance constraints on the tracking error. When $t \geq T$, SPP is equivalent to the original performance boundaries, that is, $E_u(t) = e_u(t)$ and $-E_l(t) = -e_l(t)$ since $\eta_u(t) = \eta_l(t) = 0$ as depicted in the right figure. Therefore, SPP provides an effective solution for the underlying Entry Capture Problem (5.156) as stated in the following lemma.

Lemma 5.8 *If* $-E_l(t) < e_1(t) < E_u(t)$ *holds for all* $t \geq 0$, *then the Entry Capture Problem* (5.156) *is obtained.*

Proof When $0 \leq t < T$, the boundedness of $\eta_u(t)$ and $\eta_l(t)$ is guaranteed according to Definition 5.5. By (5.160)–(5.161), the original TPP is also bounded there. It ensures that $e_1(t)$ is bounded on $[0, T)$. When $t \geq T$, we have $-E_l(t) < e_1(t) < E_u(t)$ which can be rewritten as $-e_l(t) < e_1(t) < e_u(t)$. If the above statement holds, the Entry Capture Problem is, therefore, addressed by invoking Definition 5.4. □

From the above discussion, the control objective is successfully transferred to guarantee $e_1(t)$ evolving within the SPP envelope (5.157) to deal with the issue of Entry Capture Problem.

Remark 5.16 In comparison with funnel boundary [5], global prescribed performance [48, 53] and tuning functions [23, 48], some salient features of the proposed SPP (5.157) are observed as follows:

(i) **(Improved transient performance)** The funnel boundary and global prescribed performance often encounter an overshoot problem, as their initial performance constraints tend to infinity leading to a loose allowable set. Similarly, by tuning functions, there are no constraints on tracking errors during the initial interval, which is undesirable in practice due to safety concerns and task demands. Different from these methods, our SPP is capable of adjusting performance boundaries with the help of auxiliary functions such that the proposed control scheme can be applied to any initial condition even if the initial constraints are violated. It provides bounded yet customizable virtual performance boundaries during the initial interval as depicted in Fig. 5.20, effectively avoiding the aforementioned severe overshoot performance and unconstrained behaviors. The tracking error is regulated into the allowable set within finite time and Entry Capture Problem is therefore solved.

(ii) **(Flexibility and Extensibility)** Our SPP allows for an easy extension to other control methods by integrating the introduced auxiliary functions into the performance boundaries as formulated in (5.157)–(5.159). However, the funnel boundary and global prescribed performance rely on specific performance functions, which limits their adaptability to other control methods. Moreover, the previous methods are limited to symmetric performance distributions but the proposed SPP is applicable to both symmetric and asymmetric cases, thereby enhancing its applicability to more general cases. In this section, we employ asymmetric TPP as the baseline, which not only provides a tighter allowable set but also limits overshoot performance during the initial interval of the Entry Capture Problem.

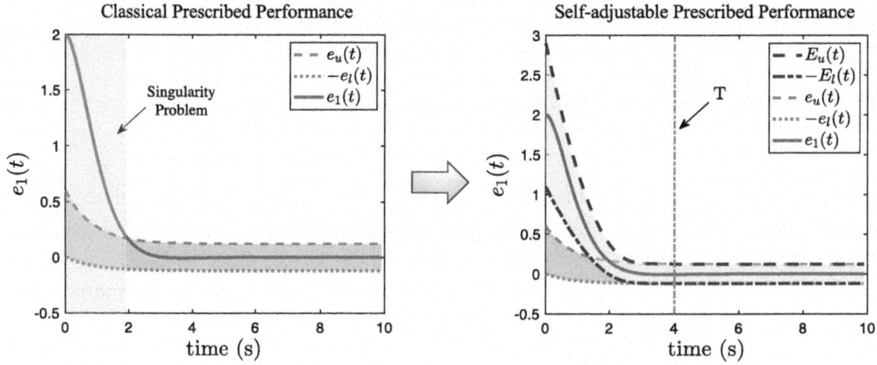

Fig. 5.20 The mechanism behind the Self-adjustable Prescribed Performance

5.3.4 Control Design for Second-Order Nonlinear Systems

To clearly illustrate our design methodology, we first consider the following second-order nonlinear system:

$$\begin{cases} \dot{x}_1 = f_1(\bar{x}_1) + g_1(\bar{x}_1)x_2, \\ \dot{x}_2 = f_2(\bar{x}_2) + g_2(\bar{x}_2)u, \end{cases} \tag{5.162}$$

where x_1 and x_2 are system states, $f_i(\bar{x}_i)$ and $g_i(\bar{x}_i), i = 1, 2$, are unknown yet smooth nonlinear functions satisfying Assumptions 5.8–5.9, and u denotes the system control input.

Step 1: We first define the tracking errors:

$$e_1 = x_1 - y_d \tag{5.163}$$

$$e_2 = x_2 - \alpha_1, \tag{5.164}$$

where α_1 represents the virtual control input to be defined shortly. In order to guarantee e_1 satisfying SPP (5.157) for all $t \geq 0$, we introduce the following error transformation function

$$z_1 = \ln\left(\frac{E_l + e_1}{E_u - e_1}\right) \tag{5.165}$$

then, by invoking (5.158)–(5.159) and (5.164), its time derivative is

$$\dot{z}_1 = \mu_1\dot{e}_1 + \mu_2 = \mu_1(f_1 + g_1e_2 - \dot{y}_d + g_1\alpha_1) + \mu_2 \tag{5.166}$$

where $\mu_1 = \ell(E_u + E_l)$ and $\mu_2 = \ell((\dot{e}_l - \dot{\eta}_l)(E_u - e_1) - (\dot{e}_u + \dot{\eta}_u)(E_l + e_1))$ and $\ell = \frac{1}{(e_1+E_l)(E_u-e_1)}$. From (5.158)–(5.161) and the property (iv) in Definition 5.5, it can be derived that $E_u(t) + E_l(t) > 0$ and $(E_u(t) - e_1(t))(E_l(t) + e_l(t)) > 0$ are bounded functions for $e_1(t)$ in the compact set $\Omega_{e_1} = \{e_1(t) \in \mathbb{R} : -E_l(t) < e_1(t) < E_u(t)\}$. Therefore, we have $\mu_1 \neq 0$ and $\mu_1 \in L_\infty$. Then, the time derivative of $\frac{1}{2}z_1^2$ along (5.166) is:

$$\begin{aligned} z_1\dot{z}_1 &= z_1\mu_1(f_1 + g_1e_2 - \dot{y}_d) + z_1\mu_1g_1\alpha_1 + z_1u_2 \\ &= z_1\mu_1g_1\alpha_1 + \Xi_1, \end{aligned} \tag{5.167}$$

where $\Xi_1 = z_1\mu_1(f_1 + g_1e_2 - \dot{y}_d) + z_1\mu_2$. Upon using Assumptions 5.7–5.9 and Young's inequality, we have

$$-z_1\mu_1\dot{y}_d \leq \underline{g}_1 z_1^2\mu_1^2\dot{y}_d^2 + \frac{1}{4\underline{g}_1}, \tag{5.168}$$

$$z_1\mu_1 f_1 \leq \underline{g}_1 z_1^2\mu_1^2 b_1^2\phi_1^2 + \frac{1}{4\underline{g}_1}, \tag{5.169}$$

$$z_1\mu_1 g_1 e_2 \leq \underline{g}_2 z_1^2\mu_1^2 e_2^2 + \frac{\bar{g}_1^2}{4\underline{g}_2}, \tag{5.170}$$

$$z_1\mu_2 \leq \underline{g}_1 z_1^2\mu_2^2 + \frac{1}{4\underline{g}_1}. \tag{5.171}$$

Therefore, Ξ_1 in (5.167) is bounded by

$$\Xi_1 \leq \underline{g}_1\theta_1 z_1^2\Phi_1 + \underline{g}_2 z_1^2\mu_1^2 e_2^2 + \frac{3}{4\underline{g}_1} + \frac{\bar{g}_1^2}{4\underline{g}_2} \tag{5.172}$$

with

$$\theta_1 = \max\left\{1, b_1^2\right\}, \tag{5.173}$$

$$\Phi_1 = \mu_1^2\dot{y}_d^2 + \mu_1^2\phi_1^2 + \mu_2^2. \tag{5.174}$$

We develop the virtual control input α_1 as

$$\alpha_1 = -\frac{1}{\mu_1}(c_1 z_1 + z_1\hat{\theta}_1\Phi_1), \tag{5.175}$$

$$\dot{\hat{\theta}}_1 = r_1 z_1^2\Phi_1 - \sigma_1\hat{\theta}_1, \;\; \hat{\theta}_1(0) \geq 0 \tag{5.176}$$

where c_1, r_1, and σ_1 are positive constants, and $\hat{\theta}_1$ is the estimation of θ_1.

Consider the following Lyapunov function

$$V_1 = \frac{1}{2}z_1^2 + \frac{\underline{g}_1}{2r_1}\tilde{\theta}_1^2,$$

where $\tilde{\theta}_1 = \theta_1 - \hat{\theta}_1$. Then, from (5.167), (5.172), (5.175) and (5.176), its time derivative can be derived as

$$\begin{aligned}\dot{V}_1 =& z_1\mu_1 g_1\alpha_1 + \Xi_1 - \frac{\underline{g}_1}{r_1}\tilde{\theta}_1\dot{\hat{\theta}}_1\\ \leq& -c_1\underline{g}_1 z_1^2 + \underline{g}_2 z_1^2\mu_1^2 e_2^2 + \frac{3}{4\underline{g}_1} + \frac{\bar{g}_1^2}{4\underline{g}_2} + \frac{\underline{g}_1\sigma_1}{r_1}\tilde{\theta}_1\hat{\theta}_1.\end{aligned}$$

Based on the definition of $\tilde{\theta}_1$, it yields

$$\tilde{\theta}_1\hat{\theta}_1 = \tilde{\theta}_1(\theta_1 - \tilde{\theta}_1) \leq -\frac{1}{2}\tilde{\theta}_1^2 + \frac{1}{2}\theta_1^2.$$

We then have

$$\dot{V}_1 \leq -c_1\underline{g}_1 z_1^2 - \bar{\sigma}_1\tilde{\theta}_1^2 + \underline{g}_2 z_1^2\mu_1^2 e_2^2 + \varepsilon_1, \tag{5.177}$$

where $\bar{\sigma}_1 = \frac{\underline{g}_1 \sigma_1}{2r_1}$ and $\varepsilon_1 = \frac{3}{4\underline{g}_1} + \frac{\bar{g}_1^2}{4\underline{g}_2} + \frac{\underline{g}_1 \sigma_i}{2r_1}\theta_1^2$ is a bounded signal. Notice that the item $\underline{g}_2 z_1^2 \mu_1^2 e_2^2$ will be tackled in the next step.

Step 2: From (5.162) and (5.164), the time derivative of e_2 is

$$\dot{e}_2 = \dot{x}_2 - \dot{\alpha}_1 = f_2 + g_2 u - \dot{\alpha}_1, \tag{5.178}$$

with

$$\dot{\alpha}_1 = \frac{\partial \alpha_1}{\partial x_1}(f_1 + g_1 x_2) + \Delta\alpha_1, \tag{5.179}$$

where $\Delta\alpha_1 = \sum_{k=0}^{1} \frac{\partial \alpha_1}{\partial y_d^{(k)}} y_d^{(k+1)} + \sum_{k=0}^{1} \frac{\partial \alpha_1}{\partial \rho^{(k)}} \rho^{(k+1)} + \sum_{k=0}^{1} \frac{\partial \alpha_1}{\partial \eta_u^{(k)}} \dot{\eta}_u^{(k+1)} + \sum_{k=0}^{1} \frac{\partial \alpha_1}{\partial \eta_l^{(k)}} \dot{\eta}_l^{(k+1)} + \frac{\partial \alpha_1}{\partial \hat{\theta}_1}\dot{\hat{\theta}}_1$, which is computable. We then consider the Lyapunov function V_2:

$$V_2 = V_1 + \frac{1}{2}e_2^2 + \frac{\underline{g}_2}{2r_2}\tilde{\theta}_2^2, \tag{5.180}$$

where $\tilde{\theta}_2 = \theta_2 - \hat{\theta}_2$, $\hat{\theta}_2$ is the estimation of θ_2 in (5.183), and r_2 is a positive constant. The time derivative of V_2, along (5.178), is derived

$$\begin{aligned} \dot{V}_2 =& \dot{V}_1 + e_2\dot{e}_2 - \frac{\underline{g}_2}{r_2}\tilde{\theta}_2\dot{\hat{\theta}}_2 \\ \leq& -c_1\underline{g}_1 z_1^2 - \bar{\sigma}_1\tilde{\theta}_1^2 + e_2 g_2 u - \frac{\underline{g}_2}{r_2}\tilde{\theta}_2\dot{\hat{\theta}}_2 + \Xi_2 + \varepsilon_1 \end{aligned} \tag{5.181}$$

where $\Xi_2 = e_2(f_2 - \frac{\partial \alpha_1}{\partial x_1}(f_1 + g_1 x_2) - \Delta\alpha_1) + \underline{g}_2 z_1^2 \mu_1^2 e_2^2$. Using Young's inequality, we can also expand Ξ_2 as inequalities in (5.168)–(5.171). It yields that

$$\Xi_2 \leq \underline{g}_2 \theta_2 e_2^2 \Phi_2 + \frac{3}{4\underline{g}_2} + \frac{\bar{g}_1^2}{4\underline{g}_2}, \tag{5.182}$$

where

$$\theta_2 = \max\left\{b_1^2, b_2^2\right\}, \tag{5.183}$$

$$\Phi_2 = \left(\frac{\partial \alpha_1}{\partial x_1}\phi_1\right)^2 + \left(\frac{\partial \alpha_1}{\partial x_1}x_2\right)^2 + (\Delta\alpha_1)^2 + \mu_1^2 z_1^2 + \phi_2^2. \tag{5.184}$$

To move forward, the actual control is developed as

$$u(t) = -(c_2 e_2 + \hat{\theta}_2 e_2 \Phi_2), \tag{5.185}$$

$$\dot{\hat{\theta}}_2 = r_2 e_2^2 \Phi_2 - \sigma_2 \hat{\theta}_2, \quad \hat{\theta}_2(0) \geq 0 \tag{5.186}$$

where $c_2, r_2, \sigma_2 > 0$ and $\hat{\theta}_2$ is the estimation of θ_2. By (5.182)–(5.186), the inequality (5.181) can be rewritten as

$$\dot{V}_2 \leq -c_1 \underline{g}_1 z_1^2 - \bar{\sigma}_1 \tilde{\theta}_1^2 - \underline{g}_2 c_2 e_2^2 + \frac{\sigma_2 \underline{g}_2}{r_2} \tilde{\theta}_2 \hat{\theta}_2 + \frac{3}{4\underline{g}_2} + \frac{\bar{g}_1^2}{4\underline{g}_2} + \varepsilon_1. \tag{5.187}$$

Based on the definition of $\tilde{\theta}_2$, we obtain

$$\tilde{\theta}_2 \hat{\theta}_2 = \tilde{\theta}_2 (\theta_2 - \tilde{\theta}_2) \leq -\frac{1}{2} \tilde{\theta}_2^2 + \frac{1}{2} \theta_2^2. \tag{5.188}$$

By substituting (5.188) into (5.187) yields

$$\dot{V}_2 \leq -c_1 \underline{g}_1 z_1^2 - \bar{\sigma}_1 \tilde{\theta}_1^2 - \underline{g}_2 c_2 e_2^2 - \bar{\sigma}_2 \tilde{\theta}_2^2 + \varepsilon_2, \tag{5.189}$$

where $\bar{\sigma}_2 = \frac{\underline{g}_2 \sigma_2}{2r_2}$ and $\varepsilon_2 = \frac{\underline{g}_2 \sigma_2}{2r_2} \theta_2^2 + \frac{3}{4\underline{g}_2} + \frac{\bar{g}_1^2}{4\underline{g}_2} + \varepsilon_1$.

To summarize, we establish the following theorem.

Theorem 5.3 *Consider the second-order nonlinear system* (5.162) *under Assumptions 5.7–5.9. The virtual control α_1 and the actual control input $v(t)$ are developed in* (5.175) *and* (5.185)*. The adaptive laws are given by* (5.176) *and* (5.186)*. The proposed control method guarantees the following.*

(i) The tracking error e_1 satisfies Entry Capture Problem property since it is strictly constrained by the prescribed performance: $-e_l(t) < e_1(t) < e_u(t)$ right after a user-given finite time T as presented in Definition 5.4; and

(ii) All signals in the closed-loop system are guaranteed to be bounded regardless of initial conditions.

Proof By revisiting (5.189), there is

$$\begin{aligned}\dot{V}_2 \leq & -c_1 \underline{g}_1 z_1^2 - \bar{\sigma}_1 \tilde{\theta}_1^2 - \underline{g}_2 c_2 e_2^2 - \bar{\sigma}_2 \tilde{\theta}_2^2 + \varepsilon_2 \\ \leq & -\pi V_2 + \varepsilon_2,\end{aligned}$$

where $\pi = \min\{2c_1\underline{g}_1, 2c_2\underline{g}_2, 2\frac{r_1\bar{\sigma}_1}{\underline{g}_1}, 2\frac{r_2\bar{\sigma}_2}{\underline{g}_2}\} > 0$. Subsequently, we can have

$$0 \leq V_2(t) \leq \frac{\varepsilon_2}{\pi} + (V_2(0) - \frac{\varepsilon_2}{\pi}) e^{-\pi t}.$$

It can be concluded that V_2 is bounded which guarantees the boundedness of z_1, e_2, $\tilde{\theta}_1$, and $\tilde{\theta}_2$. From the error transformation function in (5.165), it is certain that $-E_l(t) < e_1(t) < E_u(t)$ holds due to $z_1 \in L_\infty$. According to Lemma 5.8, e_1 satisfies $-e_l(t) < e_1(t) < e_u(t)$ for $t \geq T$, which indicates that e_1 follows the Entry Capture Problem in Definition 5.4.

Next, other tracking signals in the closed-loop system are proved to be bounded. Considering the adaptive laws (5.176) and (5.186), θ_1, θ_2, $\dot{\hat{\theta}}_1$ and $\dot{\hat{\theta}}_2$ remain bounded. The virtual control input α_1 defined in (5.175) belong to L_∞. Moreover, by (5.185), the boundedness of actual control signal $v(t)$ is guaranteed. From (5.163) and (5.164), it can be derived that x_1 and x_2 also keep bounded. All the closed-loop signals are therefore bounded for the second-order nonlinear system under any initial condition. □

5.3.5 Control Design for High-Order Nonlinear Systems

In this section, the proposed control scheme will be extended to nth-order systems (5.155). Instead of step-by-step procedure, its control design is omitted here for conciseness since its stability analysis is identical to the second-order system (5.162).

We first define the tracking error variables:

$$e_1 = x_1 - y_d \tag{5.190}$$

$$e_i = x_i - \alpha_{i-1},\ i = 2, \ldots, n, \tag{5.191}$$

where α_{i-1} represent virtual control inputs. In order to guarantee e_1 fulfilling the Self-adjustable Prescribed Performance (5.157), we introduce the following error transformation function:

$$z_1 = \ln\left(\frac{E_l + e_1}{E_u - e_1}\right) \tag{5.192}$$

Then, the virtual control inputs are designed as

$$\alpha_1 = -\frac{1}{\mu_1}(c_1 z_1 + z_1 \hat{\theta}_1 \Phi_1), \tag{5.193}$$

$$\alpha_i = -(c_i e_i + \hat{\theta}_i e_i \Phi_i),\ i = 2, \ldots, n-1, \tag{5.194}$$

and the adaptive laws are updated by

$$\dot{\hat{\theta}}_1 = r_1 z_1^2 \Phi_1 - \sigma_1 \hat{\theta}_1,\ \ \hat{\theta}_1(0) \geq 0, \tag{5.195}$$

$$\dot{\hat{\theta}}_i = r_i e_i^2 \Phi_i - \sigma_i \hat{\theta}_i,\ \ \hat{\theta}_i(0) \geq 0, \tag{5.196}$$

for $i = 2, \ldots, n-1$, where c_i, r_i and σ_i, $i = 1, \ldots, n-1$, are positive constants. $\hat{\theta}_i$ is the estimation of θ_i, and Φ_i is a computable function, where θ_i and Φ_i are

$$\theta_i = \max\{1, b_1^2, b_2^2, \ldots, b_i^2\},\ i = 1, \ldots, n-1, \tag{5.197}$$

$$\Phi_1 = \mu_1^2 \dot{y}_d^2 + \mu_1^2 \phi_1^2 + \mu_2^2, \tag{5.198}$$

$$\Phi_2 = \left(\frac{\partial \alpha_1}{\partial x_1}\phi_1\right)^2 + \left(\frac{\partial \alpha_1}{\partial x_1}x_2\right)^2 + (\Delta\alpha_1)^2 + \mu_1^2 z_1^2 + \phi_2^2, \tag{5.199}$$

$$\Phi_i = \sum_{k=1}^{i-1}\left(\frac{\partial \alpha_{i-1}}{\partial x_k}\phi_k\right)^2 + \sum_{k=1}^{i-1}\left(\frac{\partial \alpha_{i-1}}{\partial x_k}x_{k+1}\right)^2 + (\Delta\alpha_{i-1})^2 + e_{i-1}^2 + \phi_i^2,\ i = 3, \ldots, n-1, \tag{5.200}$$

where $\mu_1 = \ell(E_u + E_l)$ and $\mu_2 = \ell(\dot{e}_l - \dot{\eta}_l)(E_u - e_1) - (\dot{e}_u + \dot{\eta}_u)(E_l + e_1)$ and $\ell = \frac{1}{(e_1+E_l)(E_u-e_1)}$. We can also derive that $E_u(t) + E_l(t) > 0$ and $(E_u(t) - e_1(t))(E_l(t) + e_l(t)) > 0$ are bounded functions if $e_1(t)$ belongs to a compact set $\Omega_{e_1} = \{e_1(t) \in \mathbb{R} : -E_l(t) < e_1(t) < E_u(t)\}$. Therefore, we have $\mu_1 \neq 0$ and $\mu_1 \in L_\infty$. Identically, the actual control input and the adaptive law are developed:

$$u(t) = -(c_n e_n + \hat{\theta}_n e_n \Phi_n), \tag{5.201}$$

$$\dot{\hat{\theta}}_n = r_n e_n^2 \Phi_n - \sigma_n \hat{\theta}_n,\ \ \hat{\theta}_n(0) \geq 0, \tag{5.202}$$

where $r_n, r_n, \sigma_n > 0, \hat{\theta}_n$ denotes the estimation of θ_n and Φ_n is a computable function. Similar to (5.197)–(5.200), we have

$$\theta_n = \max\{b_1^2, b_2^2, \ldots, b_n^2\}, \tag{5.203}$$

$$\Phi_n = \sum_{k=1}^{n-1}\left(\frac{\partial \alpha_{i-1}}{\partial x_k}\phi_k\right)^2 + \sum_{k=1}^{n-1}\left(\frac{\partial \alpha_{i-1}}{\partial x_k}x_{k+1}\right)^2 + (\Delta\alpha_{n-1})^2 + e_{n-1}^2 + \phi_n^2. \tag{5.204}$$

In the following theorem, we summarize the result on the Coded Event-triggered Control for the nth-order nonlinear systems.

Theorem 5.4 *Consider nth-order nonlinear systems* (5.155) *under Assumptions* 5.7–5.9. *If the control inputs and the adaptive laws are designed as* (5.195)–(5.202), *we can guarantee the following properties:*

(i) Given any initial condition, the tracking error $e_1(t)$ is constrained by the prescribed performance: $-e_l(t) < e_1(t) < e_u(t)$, for $t \geq T$ with T being a preassigned finite time, that is, e_1 fulfills Entry Capture Problem properties in Definition 5.4*; and*

(ii) All signals in the closed-loop system are guaranteed to be bounded regardless of initial conditions.

Proof Consider the following Lyapunov candidates:

$$V_1 = \frac{1}{2}z_1^2 + \frac{g_1}{2r_1}\tilde{\theta}_1^2,$$

$$V_i = V_{i-1} + \frac{1}{2}e_n^2 + \frac{g_n}{2r_n}\tilde{\theta}_n^2.$$

where $\tilde{\theta}_i = \theta_i - \hat{\theta}_i, i = 1, \ldots, n$. Similar to the stability analysis in Theorem 5.3, it can be also derived:

$$
\begin{aligned}
\dot{V}_n \le & - c_1 \underline{g}_1 z_1^2 - \sum_{k=2}^{n} c_k \underline{g}_k e_k^2 - \sum_{k=1}^{n} \bar{\sigma}_k \tilde{\theta}_k^2 + \varepsilon_n \\
\le & - \pi V_n + \varepsilon_n,
\end{aligned}
$$

where $\pi = \min\{2c_1\underline{g}_1, \ldots, 2c_n\underline{g}_n, 2\frac{r_1\bar{\sigma}_1}{\underline{g}_1}, \ldots, 2\frac{r_n\bar{\sigma}_n}{\underline{g}_n}\} > 0$, $\varepsilon_n = \frac{\underline{g}_n \sigma_n}{2r_n}\theta_n^2 + \frac{3}{4\underline{g}_n} + \frac{\bar{g}_{n-1}^2}{4\underline{g}_n} + \varepsilon_1$. Then, we obtain

$$
0 \le V_n(t) \le \frac{\varepsilon_n}{\pi} + (V_n(0) - \frac{\varepsilon_n}{\pi})e^{-\pi t}.
$$

It indicates that $V_n(t)$ is bounded, leading to $z_1, e_2, \ldots, e_n, \tilde{\theta}_1, \ldots, \tilde{\theta}_n \in L_\infty$. According to the error transformation function (5.192), we can derive that e_1 is evolving within the SPP, namely, $-E_l(t) < e_1(t) < E_u(t)$. Based on Lemma 5.8, the Entry Capture Problem is achieved, that is, $-e_l(t) < e_1(t) < e_u(t)$ for all $t \ge T$.

Next, the boundedness of other signals in the closed-loop system is proved. From (5.195)–(5.197) and (5.202)–(5.203), and the definition of $\tilde{\theta}_i$, it leads to $\hat{\theta}_i$ and $\dot{\hat{\theta}}_i \in L_\infty$, for $i = 1, \ldots, n$. By (5.193)–(5.194) and (5.201), the boundedness of control input signals $\alpha_1, \ldots, \alpha_{n-1}, v$ is guaranteed. Considering (5.190)–(5.191), $x_1, \ldots, x_n$ belong to L_∞. Therefore, all signals in the closed-loop system are bounded. This completes the proof.

□

Remark 5.17 We provide a guideline for the parameter selections. For the settling time T in (5.156), its value should satisfy $0 < T_{\min} < T$ where $T_{\min}$ is the minimum time necessary for signal processing. This assumption is widely used in existing results [40, 53]. If T is chosen too small, although faster convergence, it could lead to large control signals. Considering the control parameters c_i and σ_i, their values will determine the convergence rate and tracking accuracy. Note that their extreme values may lead to large control inputs, making systems sensitive to external disturbances. Therefore, these parameters' selections should consider the balance between tracking performance and system ability.

5.3.6 Numerical Simulations

To illustrate the effectiveness of our proposed control scheme, simulations are conducted in this section. Consider the following second-order nonlinear system:

$$
\begin{cases}
\dot{x}_1 = f_1(\bar{x}_1) + g_1(\bar{x}_1)x_2, \\
\dot{x}_2 = f_2(\bar{x}_2) + g_2(\bar{x}_2)u,
\end{cases}
\tag{5.205}
$$

where $f_1(\bar{x}_1) = x_1^2 + 0.1\cos(0.5x_1)$, $f_2(\bar{x}_2) = 4x_1x_2 + x_1e^{-|x_2|} + 0.05\sin(x_1x_2)$, $g_1(\bar{x}_1) = 5 + 0.5\sin(x_1)$, and $g_2(\bar{x}_2) = 3 + 0.2\cos(x_1x_2)$. By Assumption 5.8, it can be derived that $|f_1(\bar{x}_1)| \le x_1^2 + 0.1$ with $b_1 = 1$ and $\phi_1 = 1 + x_1^2$, and $|f_2(\bar{x}_2)| \le 3x_1^2 + 2x_2^2 + 0.3$ with $b_2 = 3$ and $\phi_2 = x_1^2 + x_2^2 + 1$. The initial condition is set by $x_1(0) = 1.5$ and $x_2(0) = 0$. The desired tracking trajectory is $y_d = \sin(0.5t)$. The control parameters are selected as: $m = 3$ (i.e., $\beta = 0, 1, 2, 3$ and $\omega = 0.3, 0.4, 0.5, 0.6$), $T = 4$, $c_1 = 2$, $c_2 = 15$, $\sigma_1 = \sigma_2 = 0.01$, $r_1 = r_2 = 0.002$, $l = 0.6$, $\lambda = 2$, $\delta = 0.5$, $\rho_0 = 1$, $\rho_\infty = 0.4$, and $\varsigma = 1$.

(Case I) The simulation results of our control scheme are shown in Figs. 5.21 and 5.22. From Fig. 5.21, the initial tracking error $e_1(0)$ violates the initial performance constraints colored in brown, leading to the classical prescribed control methods being inapplicable. By our control scheme, we can regulate $e_1(0)$ into the allowable set within the given time T. With the aid of SPP, bounded yet customizable virtual performance boundaries are established during the initial interval as shown in blue color. Therefore, the severe overshoot or unconstrained behaviors over $[0, T)$ are effectively avoided. The control input signal is shown in Fig. 5.22.

(Case II) To further illustrate the effectiveness of our control scheme, we consider the introduction of external disturbance $d(t) = 3\cos 0.5t$ during 5-10s in the second row of (5.205). The comparative results under different threshold strategies are shown in Figs. 5.23 and 5.24. Our method can guarantee the tracking errors within the allowable set even the introduced external disturbances. The obtained results have demonstrated the robustness of our proposed control scheme against uncertain external disturbances.

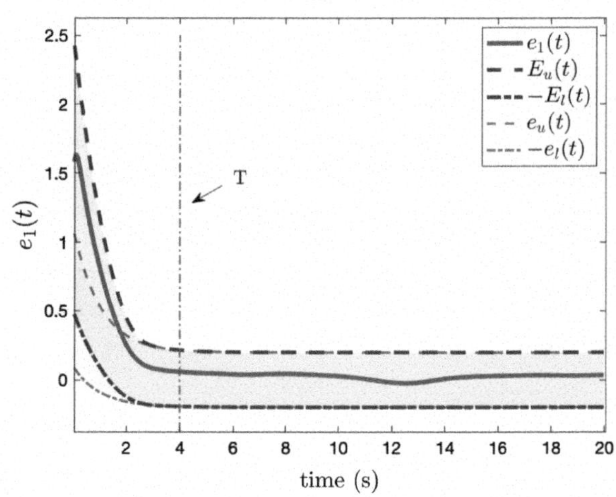

Fig. 5.21 The tracking error $e_1(t)$ for case I

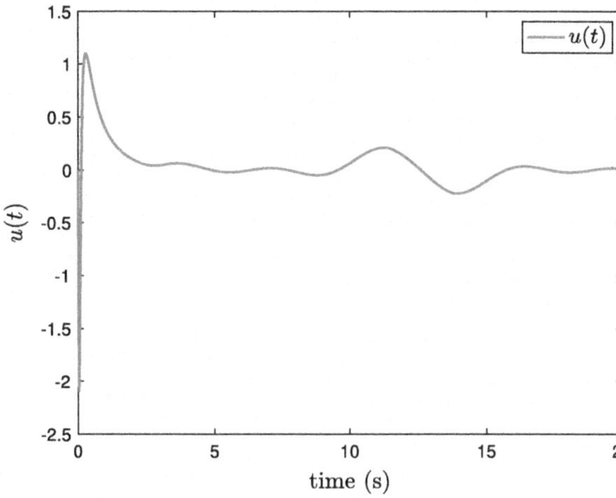

Fig. 5.22 The control input $u(t)$ for case I

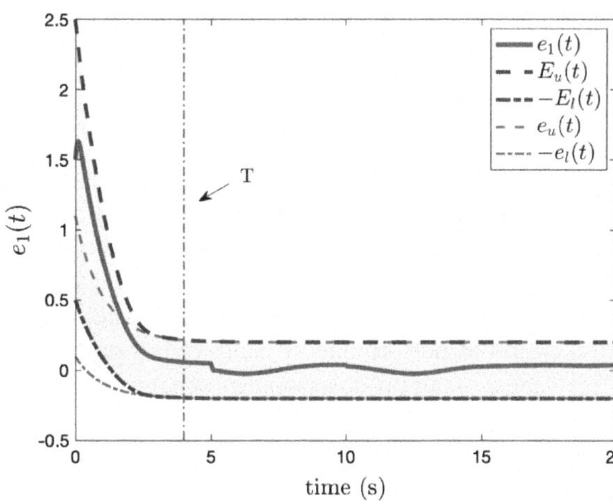

Fig. 5.23 The tracking error $e_1(t)$ for case II

5.4 Conclusions

In this chapter, we have addressed Entry Capture Problem for nonlinear systems with tunnel prescribed performance and unknown control directions. Based on the designed scaling function, two Error Self-tuning Function-based Tunnel Prescribed Control strategies have been proposed for nonlinear systems with known and unknown control directions, respectively. The proposed control schemes can ensure that the tracking errors are strictly constrained by the prescribed performance right after a user-given finite time, regardless of initial conditions. In this way, our proposed methods allow the initial conditions of nonlinear systems outside the allowable region defined by the tunnel prescribed performance. To handle unknown control

Fig. 5.24 The control input $u(t)$ for case II

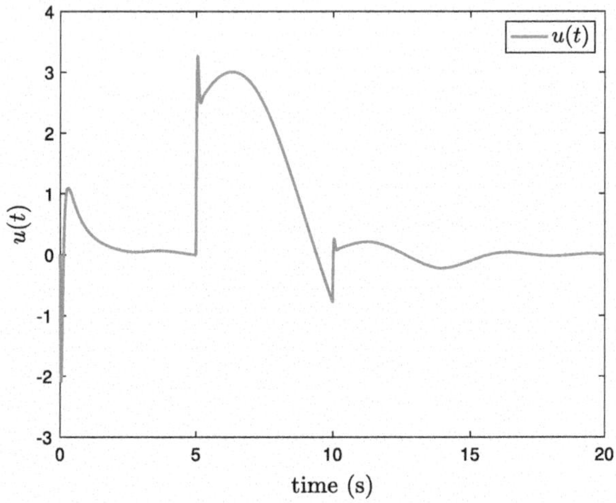

directions, an orientation function is utilized in the control design in lieu of the traditional Nussbaum technique, which reduces the control peaking problem and make the control design low complex. Moreover, a novel Self-adjustable Performance-based Tunnel Prescribed Control is proposed, which can handle the Entry Capture Problem for either symmetric or asymmetric performance constraints. With the aid of self-adjustable prescribed performance, the tracking error is regulated into an allowable set before a given time in a bounded yet customizable manner rather than no constraints there. In this way, the initial-condition constraints are removed, and better transient performance is achieved. Simulation results have demonstrated the effectiveness of our proposed control schemes in handling the Entry Capture Problem.

References

1. Bechlioulis, C.P., Rovithakis, G.A.: Robust adaptive control of feedback linearizable mimo nonlinear systems with prescribed performance. IEEE Transactions on Automatic Control **53**(9), 2090–2099 (2008)
2. Bechlioulis, C.P., Rovithakis, G.A.: Robust partial-state feedback prescribed performance control of cascade systems with unknown nonlinearities. IEEE Trans. Autom. Control **56**(9), 2224–2230 (2011)
3. Berger, T., Dennstädt, D., Ilchmann, A., Worthmann, K.: Funnel model predictive control for nonlinear systems with relative degree one. SIAM Journal on Control and Optimization **60**(6), 3358–3383 (2022)
4. Berger, T., Lê, H.H., Reis, T.: Funnel control for nonlinear systems with known strict relative degree. Automatica **87**, 345–357 (2018)
5. Berger, T., Puche, M., Schwenninger, F.L.: Funnel control for a moving water tank. Automatica **135**, 109999 (2022)
6. Berger, T., Rauert, A.L.: Funnel cruise control. Automatica **119**, 109061 (2020)
7. Berger, T., Reis, T.: Zero dynamics and funnel control for linear electrical circuits. Journal of the Franklin Institute **351**(11), 5099–5132 (2014)

8. Bu, X., He, G., Wei, D.: A new prescribed performance control approach for uncertain nonlinear dynamic systems via back-stepping. Journal of the Franklin Institute **355**(17), 8510–8536 (2018)
9. Bu, X., Qi, Q., Jiang, B.: A simplified finite-time fuzzy neural controller with prescribed performance applied to waverider aircraft. IEEE Trans. Fuzzy Syst. **30**(7), 2529–2537 (2022)
10. Cao, Y., Cao, J., Song, Y.: Practical prescribed time tracking control over infinite time interval involving mismatched uncertainties and non-vanishing disturbances. Automatica **136**, 110050 (2022)
11. Dimanidis, I.S., Bechlioulis, C.P., Rovithakis, G.A.: Output feedback approximation-free prescribed performance tracking control for uncertain mimo nonlinear systems. IEEE Trans. Autom. Control **65**(12), 5058–5069 (2020)
12. Gao, Z., Zhang, Y., Guo, G.: Prescribed-time control of vehicular platoons based on a disturbance observer. IEEE Transactions on Circuits and Systems II: Express Briefs **69**(9), 3789–3793 (2022)
13. Ge, S.S., Hong, F., Lee, T.H.: Robust adaptive control of nonlinear systems with unknown time delays. Automatica **41**(7), 1181–1190 (2005)
14. Ge, S.S., Wang, C.: Direct adaptive nn control of a class of nonlinear systems. IEEE Transactions on Neural Networks **13**(1), 214–221 (2002)
15. Ge, S.S., Zhang, J., Lee, T.H.: Adaptive mnn control for a class of non-affine narmax systems with disturbances. Systems & control letters **53**(1), 1–12 (2004)
16. Ha, M., Wang, D., Liu, D.: Discounted iterative adaptive critic designs with novel stability analysis for tracking control. IEEE/CAA Journal of Automatica Sinica **9**(7), 1262–1272 (2022)
17. Hua, C., Li, H., Li, K., Ding, W.: Low-computation tracking control of nonlinear systems with asymmetric full-state constraints and unknown control directions. IEEE Transactions on Systems, Man, and Cybernetics: Systems **53**(2), 1051–1059 (2023)
18. Ilchmann, A., Ryan, E.P., Sangwin, C.J.: Tracking with prescribed transient behaviour. ESAIM: Control, Optimisation and Calculus of Variations **7**, 471–493 (2002)
19. Ilchmann, A., Ryan, E.P., Townsend, P.: Tracking with prescribed transient behavior for nonlinear systems of known relative degree. SIAM Journal on Control and Optimization **46**(1), 210–230 (2007)
20. Ilchmann, A., Schuster, H.: Pi-funnel control for two mass systems. IEEE Transactions on Automatic Control **54**(4), 918–923 (2009)
21. Ji, R., Ge, S.S., Zhao, K., Li, H.: Event-triggered tracking control for nonlinear systems with prescribed performance. IEEE Transactions on Systems, Man, and Cybernetics: Systems (2024)
22. Ji, R., Li, D., Ge, S.S.: Saturation-tolerant prescribed control for nonlinear time-delay systems. IEEE Transactions on Fuzzy Systems (2022)
23. Ji, R., Li, D., Ge, S.S., Li, H.: Tunnel prescribed control of nonlinear systems with unknown control directions. IEEE Transactions on Neural Networks and Learning Systems (2023)
24. Jiang, Z.P., Mareels, I., Hill, D.J., Huang, J.: A unifying framework for global regulation via nonlinear output feedback: from iss to iiss. IEEE Transactions on Automatic Control **49**(4), 549–562 (2004)
25. Jin, X.: Adaptive fixed-time control for mimo nonlinear systems with asymmetric output constraints using universal barrier functions. IEEE Transactions on Automatic Control **64**(7), 3046–3053 (2018)
26. Kanakis, G.S., Rovithakis, G.A.: Guaranteeing global asymptotic stability and prescribed transient and steady-state attributes via uniting control. IEEE Trans. Autom. Control **65**(5), 1956–1968 (2019)
27. Krstic, M., Kokotovic, P.V.: Lean backstepping design for a jet engine compressor model. In: Proceedings Int. Conf. Control Appl., pp. 1047–1052. IEEE (1995)
28. Krstic, M., Kokotovic, P.V., Kanellakopoulos, I.: Nonlinear and adaptive control design. John Wiley & Sons, Inc. (1995)
29. Li, T., Li, Z., Wang, D., Chen, C.P.: Output-feedback adaptive neural control for stochastic nonlinear time-varying delay systems with unknown control directions. IEEE Transactions on Neural Networks and Learning Systems **26**(6), 1188–1201 (2014)

30. Li, Y., Tong, S.: Adaptive neural networks prescribed performance control design for switched interconnected uncertain nonlinear systems. IEEE transactions on neural networks and learning systems **29**(7), 3059–3068 (2017)
31. Lian, Y., Xia, J., Park, J.H., Sun, W., Shen, H.: Disturbance observer-based adaptive neural network output feedback control for uncertain nonlinear systems. IEEE Transactions on Neural Networks and Learning Systems (2022)
32. Liu, G., Park, J.H., Xu, H., Hua, C.: Reduced-order observer-based output-feedback tracking control for nonlinear time-delay systems with global prescribed performance. IEEE Transactions on Cybernetics (2022)
33. Lu, S., Chen, M., Liu, Y., Shao, S.: Adaptive nn tracking control for uncertain mimo nonlinear system with time-varying state constraints and disturbances. IEEE Transactions on Neural Networks and Learning Systems (2022)
34. Lv, M., De Schutter, B., Cao, J., Baldi, S.: Adaptive prescribed performance asymptotic tracking for high-order odd-rational-power nonlinear systems. IEEE Transactions on Automatic Control **68**(2), 1047–1053 (2023)
35. Ning, P., Hua, C., Li, K., Meng, R.: A blow-up function approach to global event-triggered prescribed tracking output feedback control of nonlinear systems. IEEE Transactions on Circuits and Systems I: Regular Papers **69**(10), 4225–4236 (2022)
36. Polycarpou, M.M., Ioannou, P.A.: A robust adaptive nonlinear control design. In: 1993 American control conference, pp. 1365–1369. IEEE (1993)
37. Song, Y., Huang, X., Wen, C.: Tracking control for a class of unknown nonsquare mimo nonaffine systems: A deep-rooted information based robust adaptive approach. IEEE Transactions on Automatic Control **61**(10), 3227–3233 (2015)
38. Song, Y., Wang, Y., Wen, C.: Adaptive fault-tolerant pi tracking control with guaranteed transient and steady-state performance. IEEE Transactions on Automatic Control **62**(1), 481–487 (2016)
39. Song, Y.D., Zhou, S.: Tracking control of uncertain nonlinear systems with deferred asymmetric time-varying full state constraints. Automatica **98**, 314–322 (2018)
40. Song, Y.D., Zhou, S.: Tracking control of uncertain nonlinear systems with deferred asymmetric time-varying full state constraints. Automatica **98**, 314–322 (2018)
41. Sui, S., Chen, C.P., Tong, S.: A novel adaptive nn prescribed performance control for stochastic nonlinear systems. IEEE Transactions on Neural Networks and Learning Systems **32**(7), 3196–3205 (2020)
42. Sui, S., Chen, C.P., Tong, S.: Finite-time adaptive fuzzy prescribed performance control for high-order stochastic nonlinear systems. IEEE Transactions on Fuzzy Systems (2021)
43. Swaroop, D., Hedrick, J.K., Yip, P.P., Gerdes, J.C.: Dynamic surface control for a class of nonlinear systems. IEEE transactions on automatic control **45**(10), 1893–1899 (2000)
44. Teel, A., Praly, L.: Tools for semiglobal stabilization by partial state and output feedback. SIAM Journal on Control and Optimization **33**(5), 1443–1488 (1995)
45. Wang, C., Wen, C., Lin, Y.: Adaptive actuator failure compensation for a class of nonlinear systems with unknown control direction. IEEE Transactions on Automatic Control **62**(1), 385–392 (2017)
46. Wang, J., Zhang, H., Ma, K., Liu, Z., Chen, C.P.: Neural adaptive self-triggered control for uncertain nonlinear systems with input hysteresis. IEEE transactions on neural networks and learning systems **33**(11), 6206–6214 (2021)
47. Yong, K., Chen, M., Shi, Y., Wu, Q.: Flexible performance-based robust control for a class of nonlinear systems with input saturation. Automatica **122**, 109268 (2020)
48. Zhang, J.X., Wang, Q.G., Ding, W.: Global output-feedback prescribed performance control of nonlinear systems with unknown virtual control coefficients. IEEE Transactions on Automatic Control (2021)
49. Zhang, T., Ge, S.S.: Adaptive neural network tracking control of mimo nonlinear systems with unknown dead zones and control directions. IEEE Transactions on Neural Networks **20**(3), 483–497 (2009)

50. Zhang, T., Xia, M., Yi, Y.: Adaptive neural dynamic surface control of strict-feedback nonlinear systems with full state constraints and unmodeled dynamics. Automatica **81**, 232–239 (2017)
51. Zhang, Z., Xie, X.J., Ge, S.S.: Adaptive tracking for uncertain mimo nonlinear systems with time-varying parameters and bounded disturbance. IEEE Transactions on Systems, Man, and Cybernetics: Systems **51**(7), 4479–4491 (2019)
52. Zhao, K., Lewis, F.L., Zhao, L.: Unifying performance specifications in tracking control of mimo nonlinear systems with actuation faults. Automatica **155**, 111102 (2023)
53. Zhao, K., Song, Y., Chen, C.P., Chen, L.: Adaptive asymptotic tracking with global performance for nonlinear systems with unknown control directions. IEEE Transactions on Automatic Control **67**(3), 1566–1573 (2021)
54. Zhao, K., Song, Y., Ma, T., He, L.: Prescribed performance control of uncertain euler–lagrange systems subject to full-state constraints. IEEE transactions on neural networks and learning systems **29**(8), 3478–3489 (2017)
55. Zhao, K., Song, Y., Ma, T., He, L.: Prescribed performance control of uncertain euler–lagrange systems subject to full-state constraints. IEEE Transactions on Neural Networks and Learning Systems **29**(8), 3478–3489 (2018)
56. Zhao, K., Wen, C., Song, Y., Lewis, F.L.: Adaptive uniform performance control of strict-feedback nonlinear systems with time-varying control gain. IEEE/CAA Journal of Automatica Sinica (2022)
57. Zhou, S., Song, Y., Wen, C.: Event-triggered practical prescribed time output feedback neuroadaptive tracking control under saturated actuation. IEEE Transactions on Neural Networks and Learning Systems (2021)

Chapter 6
Saturation-Tolerant Prescribed Control of MIMO Nonlinear Systems

6.1 Introduction

In practice, most systems are nonlinear and multivariable in character, which makes their control theories find immediate and increasing attention. Various control strategies in recent years have been proposed for these systems, such as robust control, backstepping control and sliding mode control. Compared with the single input and single output systems, the control design for MIMO nonlinear systems becomes much more complex and difficult since the couplings and uncertainties in systems. It is therefore essential to develop an appropriate control strategy for these MIMO nonlinear systems from both practical demand and theoretical challenge. Generally, almost all practical systems are facing various constraints, which can be classified into two categories, performance-related constraints (e.g., transient and steady state performance) and capability-related ones (e.g., input saturation). Recently, many studies have been devoted to various nonlinear systems simultaneously considering input and output constraints. Unfortunately, there exist three problems in the above studies. (a) The above literature complies with the strict hypothesis that the input and output constraints are assumed to be achieved simultaneously. However, the performance constraints are prespecified and keep fixed without flexible adaptive ability. If following the traditional PPC, the uncertainties and nonlinearities from actuator faults and input saturation may lead the prescribed performance constraints to be violated since the tracking ability is limited by the input saturation. (b) We cannot preselect such prescribed performance functions to always guarantee the tracking errors within the feasible region due to the complex and unknown nonlinearities. (c) The analysis of the above two constraints is independent without comprehensively considering their interconnections, where the over-relaxed constraints may lead to design conservatism. Motivated by the above discussion, how to accomplish a trade-off between input and output constraints remains open and deserves more in-depth study. Based on the above discussion, we, in this chapter, focus on how to effectively balance two contradict constraints, input saturation and prescribed

R. Ji et al., *Constrained Nonlinear Control: Performance and Robustness*,
https://doi.org/10.1007/978-981-96-2971-8_6

performance, especially for MIMO nonlinear systems with actuator faults and external disturbances. To this end, we propose two novel saturation-tolerant Prescribed Control strategies for MIMO nonlinear systems with known control directions and unknown control directions, respectively. It can guarantee the tracking errors always within the performance constraints, and effectively improve the robustness against these unexpected uncertainties.

6.2 Saturation-Tolerant Prescribed Control of MIMO Nonstrict Nonlinear Systems with Actuator Faults

6.2.1 Problem Formulation

Consider the following MIMO nonlinear non-strict feedback system

$$\begin{cases} \dot{x}_{j,i_j} = f_{j,i_j}(x_j) + x_{j,i_j+1} \\ \dot{x}_{j,\rho_j} = f_{j,\rho_j}(\bar{x}) + u_j(t) \\ y_j = x_{j,1} \end{cases} \tag{6.1}$$

where $j \in \mathbb{I}_{1:m}$, $x_j = [x_{j,1}, \ldots, x_{j,\rho_j}]^{\mathrm{T}}$, $\bar{x} = [x_{1,1}, \ldots, x_{1,\rho_1}, \ldots, x_{m,1}, \ldots, x_{m,\rho_m}]^{\mathrm{T}}$, y_j is the only measurable state, $f_{j,i_j}(\cdot)$ represents the unknown nonlinear smooth function, and $u_j(t)$ represents the control input suffering from actuator faults and saturations as,

$$u_j(t) = g_j(t, t_0)h_j(v_j) + h_{j,s}(t, t_1), \tag{6.2}$$

where $g_j(t, t_0)$ and $h_{j,s}(t, t_1)$ denote the actuator effectiveness and additive actuation faults where t_0 and t_1 being instants when these effects are stimulated, and $h_j(v_j)$ is

$$h_j(v_j) = \begin{cases} v_{j,\max}, & v_j > v_{j,\max} \\ v_j, & v_{j,\min} \le v_j \le v_{j,\max} \\ v_{j,\min}, & v_j < v_{j,\min} \end{cases} \tag{6.3}$$

where $v_{j,\max}$ and $v_{j,\min}$ are known constants, and v_j is the actual control input to be designed.

The control objective is to design a control scheme for the MIMO nonlinear systems such that (a) TPP is first designed in a concise form to simplify the control design, where smaller overshoot performance can be achieved in a positive manner. (b) To accomplish the trade-off between input and output constrains, SPP is then introduced to possess the ability of enlarging or recovering the performance boundaries when saturations occur or disappear. (c) SPC is proposed to guarantee the closed-loop system not only finite-time stability, but also the tracking errors evolve within the specified constraints although input saturations and actuator faults.

We give the following assumption.

Assumption 6.1 The actuator effectiveness $g_j(t, t_0)$ and the additive actuation faults $h_{j,s}(t, t_1)$ are bounded: $\underline{\rho}_j \leq g_j(t, t_0) \leq 1$ and $|h_{j,s}(t, t_1)| \leq \bar{h}_s$ with $\underline{\rho}_j$ and $\bar{h}_s$ being positive constants.

Remark 6.1 The discussed MIMO nonlinear systems (6.1) are very general and it is of certainty that their control design can find immediate and wide applications, e.g., unmanned aerial vehicles, servo motor, unmanned surface vehicle, etc. As most of the practical systems are composed of a series of interconnected subsystems with couplings in the forms of unknown nonlinearities, it imposes an unexpected control design burden for MIMO nonlinear systems compared with pure-feedback, strict-feedback, and SISO nonlinear systems in [23, 26, 42, 45, 47] since the stability analysis does not follow the standard backstepping design. To capture more general conditions, no linearly parameterization or growth conditions in [4, 12, 40, 46, 53, 56–58] are imposed on the unknown nonlinear functions in this section. Besides, it is worth noting that each subsystem can be different in order which can describe the interactions among multiple platforms.

Remark 6.2 Actuator faults and saturations are inevitable factors in almost all practical engineering, which may lead the systems instability or even catastrophic accident. In the previous literature [3, 5, 8, 14, 16, 23, 24, 28, 34, 36, 46, 54, 55, 61–65, 67, 69], either input saturations are taken into the control designs, or actuator faults. We in this section simultaneously consider both effects to obtain favorable tracking performances although both occur. The control design can pave the way for more general conditions.

6.2.2 *Saturation-Tolerant Prescribed Performance*

Before we present our Saturation-tolerant Prescribed Performance, we review the conventional Prescribed Performance Functions (PPFs) and Tunnel Prescribed Performance (TPP) defined in Chap. 4. Note that all the above PPFs and TPP are user-prespecified, where the boundaries are rigid without adaptive adjustment ability. However, almost all practical systems are subject to input saturations, actuator faults and external disturbance. The above uncertainties and nonlinearities may lead the TPP design infeasible or over-conservative. To illustrate this problem, consider $\ddot{x} = u$ system with u following (6.2) with $g_j(t, 0) = 1$ and $h_{j,s}(t, 10) = 0.75\tan(t - 10) - 0.75\tan(t - 11)$. Referring to (4.5), we select two boundaries, TPP1 and TPP2, where TPP2 has more relaxed constraints than TPP1. According to simulation results, control inputs are subject to saturation effects during 0s–2s and 10s–13s. And the tracking performance is effected by saturations nonlinearities as shown in Fig. 6.1. Especially when 10s–13s, there has obvious performance degradation. If we select TPP1 as user-specified performance constraints, the tracking errors approach or even cross the boundaries once saturations occur. Unfortunately, the

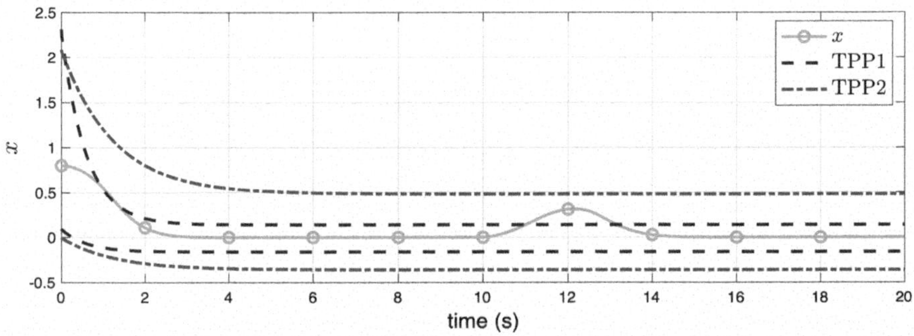

Fig. 6.1 System response and user-specified TPP

control design becomes singular when the tracking errors violate the specified constraints. Then we select other relaxed conditions like TPP2. Although the tracking errors can be guaranteed within the specified regions, there exist two problems: (i) almost all practical systems are subject to uncertainties and nonlinearities, which we cannot get full knowledge about in advance. Namely, we cannot select TPP2 such that tracking errors always satisfy the specified constraints. (ii) over-relaxed TPP2 not only makes the control design too conservative to some extend, but also leads to the superiority of TPP weak.

Motivated by the above discussion, we aim to develop a flexible performance boundaries related to the saturations. We thus introduce nonnegative modified signals into TPP as SPP, which can be formulated as

$$\begin{aligned} \bar{e}_{j,l}(t) &= e_{j,l}(t) + \eta_{1,l}(t), \\ \bar{e}_{j,u}(t) &= e_{j,u}(t) + \eta_{1,u}(t), \end{aligned} \tag{6.4}$$

where $\eta_{1,l}(t)$ and $\eta_{1,u}(t)$ are nonnegative modified signals. When input saturations occur, the positive modified signals can can enlarge the prescribed boundaries of SPP, and then fast converge to zero when saturations disappear. The proposed SPP have flexible ability to the uncertainties and nonlinearities in systems, and reduce the conservatism of the control design. The magnitudes of these signals are associated with input saturations errors, and their values can be obtained by the latter designed auxiliary system.

Remark 6.3 Compared with the traditional PPFs designed in [1, 9, 14, 16, 17, 30, 33, 34, 37, 49, 50, 64, 65], we propose TPP with concise form, which can not only simplify the control design, but also small overshoot performance can be achieved positively. We further extend TPP design mechanism to a more general concept, which can pave the way for more expressions of TPP. To compensate for saturations effects, SPP is proposed featuring the ability of enlarging or recovering the performance boundaries when saturations occur or disappear, which can guarantee the tracking errors within SPP constraints despite saturations and actuator faults. And these nonnegative modified signals construct the associations between input saturations and performance constraints.

6.2.3 Fuzzy State Observer

The only measurable state of the nonlinear system (6.1) is the output y_j. Utilizing FLSs, a fuzzy state observer is expected to be constructed to estimate the system state $x_{j,i_j}, i_j \in \mathbb{I}_{2:\rho_j}$ for the control design. For the unknown function f_{j,i_j}, we have

$$\hat{f}_{j,i_j}(\hat{x}_j|\hat{\theta}_{j,i_j}) = \hat{\theta}_{j,i_j}^{\mathrm{T}} \varphi_{j,i_j}(\hat{x}_j), \tag{6.5}$$

where $\hat{x}_j$ and $\hat{\theta}_{j,i_j}$ are the estimated vectors of x_j and θ_{j,i_j}. Then the nonlinear system (6.1) can be rewritten as

$$\begin{cases} \dot{x}_j = A_j x_j + K_j y_j + F_j(x_j) + B_{j,\rho_j} u_j \\ y_j = C_j x_j \end{cases} \tag{6.6}$$

where $F_j(x_j) = \sum_{k=1}^{\rho_j} B_{j,k} f_{j,k}(x_j)$, $B_{j,k} = [0, \dots 1, \dots 0]^{\mathrm{T}}$, $B_{j,\rho_j} = [0, \dots, 1]^{\mathrm{T}}$, and $C_j = [1, \dots, 0]$. A_j and K_j are with

$$A_j = \begin{bmatrix} -k_{j,1} & & \\ \vdots & I_{\rho_j - 1} & \\ -k_{j,\rho_j} & 0 \cdots & 0 \end{bmatrix}, \quad K_j = \begin{bmatrix} k_{j,1} \\ \vdots \\ k_{j,\rho_j} \end{bmatrix}, \tag{6.7}$$

where $k_{j,i_j} > 0$, $i_j \in \mathbb{I}_{1:\rho_j}$. And the state observer, hereby, can be constructed as

$$\begin{cases} \dot{\hat{x}}_j = A_j \hat{x}_j + K_j y_j + \hat{F}_j(\hat{x}_j) + B_{j,\rho_j} h_j(v_j) \\ y_j = C_j x_j \end{cases} \tag{6.8}$$

where $\hat{F}_j(\hat{x}_j) = \sum_{k=1}^{\rho_j} B_{j,k} \hat{f}_{j,k}(\hat{x}_j)$. For the sake of simplicity, the variables are left out in the brackets in the following design process. Combining (6.6) and (6.8), we can obtain

$$\dot{e}_{j,0} = A_j e_{j,0} + F_j - \hat{F}_j + B_{j,\rho_j}(u_j - h_j(v_j)), \tag{6.9}$$

where $e_{j,0} = x_j - \hat{x}_j$ with $e_{j,0} = [e_{j,0}^1, \dots, e_{j,0}^{\rho_j}]^{\mathrm{T}}$. We adopt the candidate Lyapunov function as $V_{j,0} = e_{j,0}^{\mathrm{T}} P_j e_{j,0}$ with P_j being the positive definite matrix. Considering Lemma 3.2 and substituting (6.2), (6.5) and (6.9) into the time derivative of $V_{j,0}$ yields

$$\dot{V}_{j,0} = 2e_j^{\mathrm{T}} P_j [\sum_{k=1}^{\rho_j} B_{j,k}(\varepsilon_{j,k} + \tilde{\theta}_{j,k}\varphi_{j,k}(\hat{x}_j)) + B_{j,\rho_j}(h_{j,s}$$
$$- h_j + g_j h_j)] + e_{j,0}^{\mathrm{T}}(A_j^{\mathrm{T}} P_j + P_j A_j) e_{j,0}, \tag{6.10}$$

where $\tilde{\theta}_{j,k} = \theta_{j,k}^* - \hat{\theta}_{j,k}$ with $\theta_{j,k}^*$ being the optimal parameter. From the Young's inequality, we have

$$2e_{j,0}^{\mathrm{T}} P_j B_{j,\rho_j}(h_{j,s} - (1 - g_j)h_j) \leq 2\|e_{j,0}\|^2 + \|P_j\|^2(\bar{h}_s^2 + (1 - g_j)^2 h_j^2) \tag{6.11}$$

$$2e_{j,0}^{\mathrm{T}} P_j \sum_{k=1}^{\rho_j} B_{j,k}(\tilde{\theta}_{j,k}\varphi_{j,k}(\hat{x}_j) + \varepsilon_{j,k})$$
$$\leq 2\rho_j \|e_{j,0}\|^2 + \|P_j\|^2 \sum_{k=1}^{\rho_j} (\tilde{\theta}_{j,k}^{\mathrm{T}} \tilde{\theta}_{j,k} + \varepsilon_{j,k}^2). \tag{6.12}$$

And we can design the parameter in z to make it be a strict Hurwitz matrix. Namely, there exists a positive definite matrix Q_j such that $A_j^{\mathrm{T}} P_j + P_j A_j = -Q_j$. Taking (6.11) and (6.12) into (6.10), we have

$$\dot{V}_{j,0} \leq - e_{j,0}^{\mathrm{T}} Q_j e_{j,0} + (2 + 2\rho_j)\|e_{j,0}\|^2 + \|P_j\|^2(\bar{h}_s^2$$
$$+ \sum_{k=1}^{\rho_j} \varepsilon_{j,k}^2 + (1 - g_j)^2 h_j^2) + \|P_j\|^2 \sum_{k=1}^{\rho_j} \tilde{\theta}_{j,k}^{\mathrm{T}} \tilde{\theta}_{j,k}$$
$$\leq - \lambda_{j,0}\|e_{j,0}\|^2 + \|P_j\|^2 \sum_{k=1}^{\rho_j} \tilde{\theta}_{j,k}^{\mathrm{T}} \tilde{\theta}_{j,k} + \epsilon_{j,0}, \tag{6.13}$$

where $\lambda_{j,0} = \lambda_{\min}(Q_j) - 2 - 2\rho_j$ and $\epsilon_{j,0} = \|P_j\|^2(\bar{h}_s^2 + \sum_{k=1}^{\rho_j} \varepsilon_{j,k}^2 + (1 - g_j)^2 h_j^2)$.

Remark 6.4 The fuzzy state observer is developed to approximate the unmeasurable system states, in which the actuator faults and saturations are synthesized. However, unlike the assumptions in [22, 39, 43, 44, 52], the knowledge of actuator faults is difficult to be obtained when the systems in operations. Therefore, the state observer are constructed without the requirement of actuator faults restrictions. From the practical point of view, the proposed state observer design can be adopted to a wide class of general conditions.

6.2.4 Auxiliary System Design

Auxiliary system is developed to provide nonnegative modified signals $\eta_{1,l}(t)$ and $\eta_{1,u}(t)$ in (6.4). To make SPP feature the ability of enlarging or recovering the prescribed performance constraints when saturations occur or disappear, developing the associations between input saturations errors (errors between the actual and desired control inputs) and the modified signals are significant.

Following ETF $T(\cdot)$, we can transform the constraint error e_j into an equivalent unconstrained variable $z_{j,1}$

$$e_j = \frac{\bar{e}_{j,u} + \bar{e}_{j,l}}{2} T(z_{j,1}) + \frac{\bar{e}_{j,u} - \bar{e}_{j,l}}{2}, \tag{6.14}$$

where $T(\cdot)$ follows the properties in [1] that (i) $-1 < T(z_{j,1}) <$ for $-\infty < z_{j,1} < \infty$, and (ii) $\lim_{z_{j,1}\to+\infty} T(z_{j,1}) = 1, \lim_{z_{j,1}\to-\infty} T(z_{j,1}) = -1$. Thus, the bounded equivalent variable $z_{j,1}$ can guarantee the tracking errors $e_j(t)$ can satisfy the inequality (4.5) for all time.

Motivated by the previous systems with input saturations in [5, 9, 13, 25, 29, 34, 48, 51, 54, 65], auxiliary system is a promising approach to tackle input saturations and performance constraints without neglecting their interaction and conflict. To establish the bridge between saturations errors and performance constraints, an auxiliary system is developed, where nonnegative signals $\eta_{i_j,u}$ and $\eta_{i_j,l}$ are provided for the modified signals $\eta_{1,u}$ and $\eta_{1,l}$. We first define $\Lambda = \partial z_{j,1}/\partial e_j$, $\Lambda_u = \partial z_{j,1}/\partial \bar{e}_{j,u}$, and $\Lambda_l = \partial z_{j,1}/\partial \bar{e}_{j,l}$. The auxiliary system can be designed as

$$\begin{cases} \dot{\eta}_{1,u} = -\Lambda_u^{\text{inv}} \Lambda(-\Gamma_1 \eta_{1,u} + \eta_{2,u}), \\ \dot{\eta}_{i_j,u} = -\Gamma_{i_j} \eta_{i_j,u} + \eta_{i_j+1,u}, \; i_j \in \mathbb{I}_{2:\rho_j}, \\ \dot{\eta}_{1,l} = \Lambda_l^{\text{inv}} \Lambda(-\Gamma_1 \eta_{1,l} + \eta_{2,l}), \\ \dot{\eta}_{i_j,l} = -\Gamma_{i_j} \eta_{i_j,l} + \eta_{i_j+1,l}, \; i_j \in \mathbb{I}_{2:\rho_j}, \end{cases} \tag{6.15}$$

where Γ_{i_j} represents the designed positive matrix, $\eta_{\rho_j+1,u} = G^-(v_j, v_{j,\min})$, $\eta_{\rho_j+1,l} = G^+(v_j, v_{j,\max})$, and the initial conditions for $\eta_{i_j,u}$ and $\eta_{i_j,l}$ are set to zeros. The auxiliary system has the following property.

Theorem 6.1 *Consider ETF (6.14) and SPP (6.4) with the modified signals updated by (6.15). If* $|v_j - u_j| \le \bar{h}$ *and* $-\bar{e}_{j,l}(t) \le e_j \le \bar{e}_{j,u}(t)$ *hold for* $\forall t \ge 0$, *then* $0 \le \eta_{i_j,u}, \eta_{i_j,l} \le k_{ul}\bar{h}$, $i_j \in \mathbb{I}_{1:\rho_j}$ *with* k_{ul} *being positive constant.*

Proof: Following the Lemma 2.6 and some simple derivations, we can easily obtain the above theorem. The proof of this theorem is thus omitted for clarity and conciseness.

Actually, the auxiliary system (6.15) establish a mechanism between saturations errors and performance boundaries. According to the definition of $\eta_{\rho_j+1,u}$ and $\eta_{\rho_j+1,l}$, the absolute values of saturations errors are tactfully extracted, where the unreachable increasing or decreasing additional signals $\eta_{i_j,u}$ or $\eta_{i_j,l}$ can be obtained for x_{j,i_j}.

Referring to (6.15), the modified signals $\eta_{1,u}$ and $\eta_{1,l}$ are obtained by gradually integrating, which will be further provided for SPP (6.4). And these signals are utilized to enlarge the performance boundaries once the saturations are stimulated. Moreover, the positive matrix Γ_{i_j} are significant to guarantee the stability of the auxiliary system, which can make the modified signals fast converge to zero when saturations disappear. Namely, SPP will recover to the user-specified performance boundaries.

Remark 6.5 It is remarkable that the auxiliary system establishes a mechanism between saturations errors and modified signals, which first excavate the interconnections between performance functions and saturations errors. The modified signals $\eta_{\rho_j+1,u}$ and $\eta_{\rho_j+1,l}$ represent the unreachable increasing or decreasing rates for x_{j,i_j}. And $\eta_{1,u}$ and $\eta_{1,l}$ are utilized as additional signals for SPP, which can make SPP have the ability to elastically enlarge performance boundaries according to the auxiliary system when saturations, and then recover the user-specified performance when saturations disappear to degrade the user-specified tracking performance to achieve a balance among various constraints in a positive manner. Additionally, SPP can not only guarantee the tracking errors not to violate the performance constraints, but also accomplish a trade-off between input saturations and performance related constraints.

6.2.5 *Saturation-Tolerant Prescribed Control*

In this section, we propose a novel SPC for the MIMO nonlinear system simultaneously considering actuator faults, saturations, and unmeasurable states. We first define the following auxiliary variables as

$$e_j = \frac{\bar{e}_{j,u} + \bar{e}_{j,l}}{2} T(z_{j,1}) + \frac{\bar{e}_{j,u} - \bar{e}_{j,l}}{2} \tag{6.16}$$

$$z_{j,i_j} = \hat{x}_{j,i_j} + \eta_{i_j,l} - \eta_{i_j,u} - \alpha^f_{j,i_j}, \quad i_j \in \mathbb{I}_{2:\rho_j}, \tag{6.17}$$

where α^f_{j,i_j} is the first-order filter output by directly passing through α_{j,i_j} to the filter. The virtual and actual control inputs can be developed as

$$\begin{aligned}\alpha_{j,1} = & -\hat{\theta}^{\mathrm{T}}_{j,1}\varphi(\hat{x}_{j,1}) + \dot{y}_{j,d} - \Lambda^{\mathrm{inv}}(\Lambda_u \dot{e}_{j,u} + \Lambda_l \dot{e}_{j,l}) \\ & - \Gamma_1(\eta_{1,u} - \eta_{1,l}) - \Lambda^{\mathrm{inv}}(c_{j,1}z_{j,1} + b_{j,1}z^{\beta}_{j,1}),\end{aligned} \tag{6.18}$$

$$\begin{aligned}\alpha_{j,2} = & -k_{j,2}e^1_{j,0} - \Gamma_2(\eta_{2,u} - \eta_{2,l}) - \frac{\omega_{j,2}}{\delta_{j,2}} - \Lambda z_{j,1} \\ & - \hat{\theta}^{\mathrm{T}}_{j,2}\varphi_{j,2}(\hat{\bar{x}}_{j,2}) - c_{j,2}z_{j,2} - b_{j,2}z^{\beta}_{j,2},\end{aligned} \tag{6.19}$$

$$\alpha_{j,i_j} = -k_{j,i_j}e_{j,0}^1 - \Gamma_{i_j}(\eta_{i_j,u} - \eta_{i_j,l}) - \frac{\omega_{j,i_j}}{\delta_{j,i_j}} \\ - z_{j,i_j-1} - \hat{\theta}_{j,i_j}^{\mathrm{T}}\varphi_{j,i_j}(\hat{\bar{x}}_{j,i_j}) - c_{j,i_j}z_{j,i_j} - b_{j,i_j}z_{j,i_j}^{\beta},\ i \in \mathbb{I}_{3:\rho_j-1}, \tag{6.20}$$

$$v_j = \Gamma_{\rho_j}(\eta_{\rho_j,l} - \eta_{\rho_j,u}) - z_{j,\rho_j-1} - \frac{\omega_{j,\rho_j}}{\delta_{j,\rho_j}} \\ - c_{j,\rho_j}z_{j,\rho_j} - b_{j,\rho_j}z_{j,\rho_j}^{\beta} - k_{j,\rho_j}e_{j,0}^1, \tag{6.21}$$

with

$$\dot{\hat{\theta}}_{j,1} = \gamma_{j,1}(z_{j,1}\Lambda\varphi_{j,1}(\hat{x}_{j,1}) - \sigma_{j,1}\hat{\theta}_{j,1}), \tag{6.22}$$

$$\dot{\hat{\theta}}_{j,2} = \gamma_{j,2}(z_{j,2}\varphi_{j,2}(\hat{\bar{x}}_{j,2}) - \sigma_{j,2}\hat{\theta}_{j,2}), \tag{6.23}$$

$$\dot{\hat{\theta}}_{j,i_j} = \gamma_{j,i_j}(z_{j,i_j}\varphi_{j,i_j}(\hat{\bar{x}}_{j,i_j}) - \sigma_{j,i_j}\hat{\theta}_{j,i_j}),\ i_j \in \mathbb{I}_{3:\rho_j-1}, \tag{6.24}$$

$$\dot{\hat{\theta}}_{j,\rho_j} = -\gamma_{j,\rho_j}(z_{j,\rho_j}\varphi_{j,\rho_j}(\hat{x}_{j,\rho_j}) + \sigma_{j,\rho_j}\hat{\theta}_{j,\rho_j}), \tag{6.25}$$

where $c_{j,i_j}, b_{j,i_j}, \delta_{j,i_j}, \gamma_{j,i_j}$ and σ_{j,i_j} are positive constants, $\omega_{j,i_j+1} = \alpha_{j,i_j+1}^f - \alpha_{j,i_j}$, $\hat{\bar{x}}_{j,i_j} = [\hat{x}_{j,1}, \ldots, \hat{x}_{j,i_j}]^{\mathrm{T}}$, and $0 < \beta < 1$.

According to the Lemma 2.1, we have the following main result of this section:

Theorem 6.2 *Consider the closed-loop system consisting of MIMO nonlinear system (6.1) suffering from actuator faults (6.2) and saturations (6.3), the state observer (6.8), virtual control inputs (6.18)–(6.20), actuate control input (6.21), adaptive laws (6.22)–(6.25), and SPP (6.4) associated with the auxiliary system (6.15). Then all the system states can be regulated to a small neighborhood around the desired trajectories within finite time, and the tracking errors are guaranteed within SPP even saturations and faults occur.*

Proof *Step 1:* From (6.16), the time derivative of $z_{j,1}$ is,

$$\dot{z}_{j,1} = \Lambda(x_{j,2} + \eta_{2,l} - \eta_{2,u} + f_{j,1}(x_j) - \dot{y}_{j,d}) \\ + \Lambda_u\dot{e}_{j,u} + \Lambda_l\dot{e}_{j,l} + \Lambda\Gamma_1(\eta_{1,u} - \eta_{1,l}). \tag{6.26}$$

Let $z_{j,2} = \hat{x}_{j,2} + \eta_{2,l} - \eta_{2,u} - \alpha_{j,2}^f$ and $\omega_{j,2} = \alpha_{j,2}^f - \alpha_{j,1}$, where $\alpha_{j,2}^f$ is the first-order filter output with the designed virtual control $\alpha_{j,1}$ as input. Due to $e_{j,0}^2 = x_{j,2} - \hat{x}_{j,2}$, we have,

$$\dot{z}_{j,1} = \Lambda(e_{j,0}^2 + z_{j,2} + \omega_{j,2} + \alpha_{j,1} + f_{j,1}(x_j)) \\ + \Lambda_u\dot{e}_{j,u} + \Lambda_l\dot{e}_{j,l} + \Lambda\Gamma_1(\eta_{1,u} - \eta_{1,l}) - \Lambda\dot{y}_{j,d}, \tag{6.27}$$

Consider the candidate Lyapunov function as,

$$V_{j,1} = V_{j,0} + \frac{1}{2}z_{j,1}^2 + \frac{1}{2\gamma_{j,1}}\tilde{\theta}_{j,1}^{\mathrm{T}}\tilde{\theta}_{j,1}. \tag{6.28}$$

Substituting (6.18), (6.22) and (6.27) into the time derivative of $V_{j,1}$ yields,

$$\begin{aligned}\dot{V}_{j,1} =& \dot{V}_{j,0} + z_{j,1}\dot{z}_{j,1} - \frac{1}{\gamma_{j,1}}\tilde{\theta}_{j,1}^{\mathrm{T}}\dot{\hat{\theta}}_{j,1} \\ =& \dot{V}_{j,0} + z_{j,1}\Lambda(\theta_{j,1}^{\mathrm{T}}\varphi_{j,1}(\hat{x}_{j,1}) - \hat{\theta}_{j,1}^{\mathrm{T}}\varphi_{j,1}(\hat{x}_{j,1}) + \omega_{j,2} + \theta_{j,1}^{\mathrm{T}}\varphi_{j,1}(\hat{x}_j) - \theta_{j,1}^{\mathrm{T}}\varphi_{j,1}(\hat{x}_{j,1})) \\ &- \tilde{\theta}_{j,1}^{\mathrm{T}}(z_{j,1}\Lambda\varphi_{j,1}(\hat{x}_{j,1}) - \sigma_{j,1}\hat{\theta}_{j,1}) - c_{j,1}z_{j,1}^2 \\ &+ z_{j,1}\Lambda(e_{j,0}^2 + z_{j,2} + \varepsilon_{j,1}) - b_{j,1}|z_{j,1}|^{\beta+1}. \end{aligned} \tag{6.29}$$

According to the Young's inequality, we can obtain,

$$z_{j,1}\Lambda(\theta_{j,1}^{*\mathrm{T}}\varphi_{j,1}(\hat{x}_j) - \theta_{j,1}^{*\mathrm{T}}\varphi_{j,1}(\hat{x}_{j,1})) \le z_{j,1}^2 + \Lambda^2\|\theta_{j,1}^*\|^2 \tag{6.30}$$

$$z_{j,1}\Lambda(e_{j,0}^2 + \varepsilon_{j,1}) \le z_{j,1}^2 + \frac{1}{2}\Lambda^2\|e_{j,0}\|^2 + \frac{1}{2}\Lambda^2\varepsilon_{j,1}^2. \tag{6.31}$$

Due to $\tilde{\theta}_{j,1} = \theta_{j,1}^* - \hat{\theta}_{j,1}$, we further have,

$$\begin{aligned}\sigma_{j,1}\tilde{\theta}_{j,1}^{\mathrm{T}}\hat{\theta}_{j,1} =& \sigma_{j,1}\tilde{\theta}_{j,1}^{\mathrm{T}}(\theta_{j,1}^* - \tilde{\theta}_{j,1}) \\ \le& -\frac{\sigma_{j,1}}{2}\|\tilde{\theta}_{j,1}\|^2 + \frac{\sigma_{j,1}}{2}\|\theta_{j,1}^*\|^2. \end{aligned} \tag{6.32}$$

Substituting (6.30)–(6.32) into (6.29) yields, we can obtain,

$$\begin{aligned}\dot{V}_{j,1} \le& -\lambda_{j,1}\|e_{j,0}\|^2 + \|P_j\|^2\sum_{k=1}^{\rho_j}\|\tilde{\theta}_{j,k}\|^2 - \bar{c}_{j,1}z_{j,1}^2 \\ &- b_{j,1}|z_{j,1}|^{\beta+1} + z_{j,1}\Lambda z_{j,2} + z_{j,1}\Lambda\omega_{j,2} \\ &- \frac{\sigma_{j,1}}{2}\|\tilde{\theta}_{j,1}\|^2 + \epsilon_{j,1}, \end{aligned} \tag{6.33}$$

where $\lambda_{j,1} = \lambda_{j,0} - \frac{1}{2}\Lambda^2$, $\bar{c}_{j,1} = c_{j,1} - 2$ and $\epsilon_{j,1} = \frac{\sigma_{j,1}}{2}\|\theta_{j,1}^*\|^2 + \epsilon_{j,0} + \Lambda^2\|\theta_{j,1}^*\|^2 + \frac{1}{2}\Lambda^2\varepsilon_{j,1}^2$. A first-order filter is introduced to avoid the prohibitively complicated computation in the backstepping process,

$$\delta_{j,2}\dot{\alpha}_{j,2}^f + \alpha_{j,2}^f = \alpha_{j,1},\ \alpha_{j,2}^f(0) = \alpha_{j,1}(0), \tag{6.34}$$

Therefore, we have,

$$\dot{\omega}_{j,2} = -\frac{\omega_{j,2}}{\delta_{j,2}} - \dot{\alpha}_{j,1}. \tag{6.35}$$

Step 2: According to (6.17), the time derivative of $z_{j,2}$ is,

$$\begin{aligned}\dot{z}_{j,2} =& \dot{\hat{x}}_{j,2} + \dot{\eta}_{2,l} - \dot{\eta}_{2,u} - \dot{\alpha}^f_{j,2} \\ =& k_{j,2}e^1_{j,0} + \hat{x}_{j,3} + \hat{\theta}^{\mathrm{T}}_{j,2}\varphi_{j,2}(\hat{x}_j) + \Gamma_2(\eta_{2,u} \\ &- \eta_{2,l}) + \eta_{3,l} - \eta_{3,u} + \frac{\omega_{j,2}}{\delta_{j,2}}.\end{aligned} \tag{6.36}$$

Let $z_{j,3} = \hat{x}_{j,3} + \eta_{3,l} - \eta_{3,u} - \alpha^f_{j,3}$ be the tracking error and $\omega_{j,3} = \alpha^f_{j,3} - \alpha_{j,2}$, where $\alpha^f_{j,3}$ is obtained by through the virtual control $\alpha_{j,2}$ into the filter. We then have,

$$\begin{aligned}\dot{z}_{j,2} =& z_{j,3} + \omega_{j,3} + \alpha_{j,2} + k_{j,2}e^1_{j,0} + \Gamma_2(\eta_{2,u} \\ &- \eta_{2,l}) + \frac{\omega_{j,2}}{\delta_{j,2}} + \hat{\theta}^{\mathrm{T}}_{j,2}\varphi_{j,2}(\hat{x}_j).\end{aligned} \tag{6.37}$$

The virtual control $\alpha_{j,2}$ and the adaptive law for $\hat{\theta}_{j,2}$ are designed as (6.19) and (6.23). Construct the Lyapunov function,

$$V_{j,2} = V_{j,1} + \frac{1}{2}z^2_{j,2} + \frac{1}{2\gamma_{j,2}}\tilde{\theta}^{\mathrm{T}}_{j,2}\tilde{\theta}_{j,2} + \frac{1}{2}\omega^2_{j,2}. \tag{6.38}$$

Substituting (6.19), (6.23), and (6.35)–(6.37) into the time derivative of $V_{j,2}$, we have,

$$\begin{aligned}\dot{V}_{j,2} =& \dot{V}_{j,1} + z_{j,2}(z_{j,3} + \omega_{j,3} - \Lambda z_{j,1} - c_{j,2}z_{j,2} \\ &+ \theta^{*\mathrm{T}}_{j,2}\varphi_{j,2}(\hat{\bar{x}}_{j,2}) - \hat{\theta}^{\mathrm{T}}_{j,2}\varphi_{j,2}(\hat{\bar{x}}_{j,2}) \\ &+ \theta^{*\mathrm{T}}_{j,2}\varphi_{j,2}(\hat{x}_j) - \theta^{*\mathrm{T}}_{j,2}\varphi_{j,2}(\hat{\bar{x}}_{j,2}) \\ &- \tilde{\theta}^{\mathrm{T}}_{j,2}\varphi_{j,2}(\hat{x}_j) - b_{j,2}z^{\beta}_{j,2}) - \omega_{j,2}\bigg(\frac{\omega_{j,2}}{\delta_{j,2}} \\ &+ \dot{\alpha}_{j,1}\bigg) - \tilde{\theta}^{\mathrm{T}}_{j,2}(z_{j,2}\varphi_{j,2}(\hat{\bar{x}}_{j,2}) - \sigma_{j,2}\hat{\theta}_{j,2}).\end{aligned} \tag{6.39}$$

According to the Young's inequality and $\tilde{\theta}_{j,2} = \theta^*_{j,2} - \hat{\theta}_{j,2}$, we have,

$$z_{j,2}(\theta^{*\mathrm{T}}_{j,2}\varphi_{j,2}(\hat{x}_j) - \theta^{*\mathrm{T}}_{j,2}\varphi_{j,2}(\hat{\bar{x}}_{j,2})) \le z^2_{j,2} + \|\theta^*_{j,2}\|^2, \tag{6.40}$$

$$- z_{j,2}\tilde{\theta}^{\mathrm{T}}_{j,2}\varphi_{j,2}(\hat{x}_j) \le \frac{1}{2}z^2_{j,2} + \frac{1}{2}\|\tilde{\theta}_{j,2}\|^2 \tag{6.41}$$

$$\begin{aligned}&\sigma_{j,2}\tilde{\theta}^{\mathrm{T}}_{j,2}\hat{\theta}_{j,2} = \sigma_{j,2}\tilde{\theta}^{\mathrm{T}}_{j,2}(\theta^*_{j,2} - \tilde{\theta}_{j,2}) \\ &\le -\frac{\sigma_{j,2}}{2}\|\tilde{\theta}_{j,2}\|^2 + \frac{\sigma_{j,2}}{2}\|\theta^*_{j,2}\|^2.\end{aligned} \tag{6.42}$$

Taking (6.33) and (6.40)–(6.42) into (6.39), we can obtain,

$$
\begin{aligned}
\dot{V}_{j,2} \leq & -\lambda_{j,1}\|e_{j,0}\|^2 - \bar{c}_{j,1}z_{j,1}^2 - b_{j,1}|z_{j,1}|^{\beta+1} \\
& -(c_{j,2}-\frac{3}{2})z_{j,2}^2 - b_{j,2}|z_{j,2}|^{\beta+1} + z_{j,1}\Lambda\omega_{j,2} \\
& -\frac{\sigma_{j,1}}{2}\|\tilde{\theta}_{j,1}\|^2 - \frac{\sigma_{j,2}-1}{2}\|\tilde{\theta}_{j,2}\|^2 + \epsilon_{j,1} \\
& +\|P_j\|^2\sum_{k=1}^{\rho_j}\|\tilde{\theta}_{j,k}\|^2 + z_{j,2}(z_{j,3}+\omega_{j,3}) \\
& -\omega_{j,2}\left(\frac{\omega_{j,2}}{\delta_{j,2}}+\dot{\alpha}_{j,1}\right) + \frac{\sigma_{j,2}+2}{2}\|\theta_{j,2}^*\|^2 \\
\leq & -\lambda_{j,1}\|e_{j,0}\|^2 - \sum_{k=1}^{2}\bar{c}_{j,k}z_{j,k}^2 - \sum_{k=1}^{2}b_{j,k}|z_{j,k}|^{\beta+1} \\
& -\sum_{k=1}^{2}\bar{\sigma}_{j,k}\|\tilde{\theta}_{j,k}\|^2 + \|P_j\|^2\sum_{k=1}^{\rho_j}\|\tilde{\theta}_{j,k}\|^2 \\
& + z_{j,2}(z_{j,3}+\omega_{j,3}) + z_{j,1}\Lambda\omega_{j,2} + \epsilon_{j,2} \\
& -\omega_{j,2}\left(\frac{\omega_{j,2}}{\delta_{j,2}}+\dot{\alpha}_{j,1}\right).
\end{aligned} \tag{6.43}
$$

where $\bar{c}_{j,2} = c_{j,2} - \frac{3}{2}$, $\bar{\sigma}_{j,1} = \frac{\sigma_{j,1}}{2}$, $\bar{\sigma}_{j,2} = \frac{\sigma_{j,2}-1}{2}$, and $\epsilon_{j,2} = \epsilon_{j,1} + \frac{\sigma_{j,2}+2}{2}\|\theta_{j,2}^*\|^2$. Let $\alpha_{j,2}$ pass through the following first-order filter,

$$
\delta_{j,3}\dot{\alpha}_{j,3}^f + \alpha_{j,3}^f = \alpha_{j,2},\ \alpha_{j,3}^f(0) = \alpha_{j,3}(0), \tag{6.44}
$$

The time derivative of $\omega_{j,3}$ can be described as,

$$
\dot{\omega}_{j,3} = -\frac{\omega_{j,3}}{\delta_{j,3}} - \dot{\alpha}_{j,2}. \tag{6.45}
$$

Step i_j ($3 \leq i_j \leq \rho_j - 1$): We define the tracking error z_{j,i_j} as $z_{j,i_j} = \hat{x}_{j,i_j} + \eta_{i_j,l} - \eta_{i_j,u} - \alpha_{j,i_j}^f$. Then the time derivative of z_{j,i_j} can be formulated as,

$$
\begin{aligned}
\dot{z}_{j,i_j} =& \dot{\hat{x}}_{j,i_j} + \dot{\eta}_{i_j,l} - \dot{\eta}_{i_j,u} - \dot{\alpha}_{j,i_j}^f \\
=& k_{j,i_j}e_{j,0}^1 + \hat{x}_{j,i_j+1} + \hat{\theta}_{j,i_j}^{\mathrm{T}}\varphi_{j,i_j}(\hat{x}_j) + \Gamma_{i_j}(\eta_{i_j,u} - \eta_{i_j,l}) \\
& + \eta_{i_j+1,l} - \eta_{i_j+1,u} + \frac{\omega_{j,i_j}}{\delta_{j,i_j}}.
\end{aligned} \tag{6.46}
$$

Let $z_{j,i_j+1} = \hat{x}_{j,i_j+1} + \eta_{i_j+1,l} - \eta_{i_j+1,u} - \alpha_{j,i_j+1}^f$ and $\omega_{j,i_j+1} = \alpha_{j,i_j+1}^f - \alpha_{j,i_j}$. The tracking error z_{j,i_j} can be rewritten as,

$$\dot{z}_{j,i_j} = z_{j,i_j+1} + \omega_{j,i_j+1} + \alpha_{j,i_j} + k_{j,i_j} e^1_{j,0} + \Gamma_{i_j}(\eta_{i_j,u} - \eta_{i_j,l}) + \frac{\omega_{j,i_j}}{\delta_{j,i_j}} + \hat{\theta}^{\mathrm{T}}_{j,i_j} \varphi_{j,i_j}(\hat{x}_j). \tag{6.47}$$

The virtual control input α_{j,i_j} and the adaptive law $\hat{\theta}_{j,i_j}$ can be designed as (6.20) and (6.24). Then, the Lyapunov function is constructed as,

$$V_{j,i_j} = V_{j,i_j-1} + \frac{1}{2} z^2_{j,i_j} + \frac{1}{2\gamma_{j,i_j}} \tilde{\theta}^{\mathrm{T}}_{j,i_j} \tilde{\theta}_{j,i_j} + \frac{1}{2} \omega^2_{j,i_j}. \tag{6.48}$$

Taking (6.20), (6.24), (6.45), and (6.47) into the time derivative of V_{j,i_j} yields,

$$\begin{aligned} \dot{V}_{j,i_j} = & \dot{V}_{j,i_j-1} + z_{j,i_j}(z_{j,i_j+1} + \omega_{j,i_j+1} - z_{j,i_j-1} + \theta^{*\mathrm{T}}_{j,i_j} \varphi_{j,i_j}(\hat{\bar{x}}_{j,i_j}) - \hat{\theta}^{\mathrm{T}}_{j,i_j} \varphi_{j,i_j}(\hat{\bar{x}}_{j,i_j}) \\ & - b_{j,i_j} z^{\beta}_{j,i_j} + \theta^{*\mathrm{T}}_{j,i_j} \varphi_{j,i_j}(\hat{x}_j) - \theta^{*\mathrm{T}}_{j,i_j} \varphi_{j,i_j}(\hat{\bar{x}}_{j,i_j}) - c_{j,i_j} z_{j,i_j} \\ & - \tilde{\theta}^{\mathrm{T}}_{j,i_j} \varphi_{j,i_j}(\hat{x}_j)) - \tilde{\theta}^{\mathrm{T}}_{j,i_j}(z_{j,i_j} \varphi_{j,i_j}(\hat{\bar{x}}_{j,i_j}) \\ & - \sigma_{j,i_j} \hat{\theta}_{j,i_j}) - \omega_{j,i_j} \left(\frac{\omega_{j,i_j}}{\delta_{j,i_j}} + \dot{\alpha}_{j,i_j-1} \right). \end{aligned} \tag{6.49}$$

According to the Young's inequality and the definition $\tilde{\theta}_{j,i_j} = \theta^*_{j,i_j} - \hat{\theta}_{j,i_j}$, we can have,

$$z_{j,i_j}(\theta^{*\mathrm{T}}_{j,i_j} \varphi_{j,i_j}(\hat{x}_j) - \theta^{*\mathrm{T}}_{j,i_j} \varphi_{j,i_j}(\hat{\bar{x}}_{j,i_j})) \le z^2_{j,i_j} + \|\theta^*_{j,i_j}\|^2, \tag{6.50}$$

$$- z_{j,i_j} \tilde{\theta}^{\mathrm{T}}_{j,i_j} \varphi_{j,i_j}(\hat{x}_j) \le \frac{1}{2} z^2_{j,i_j} + \frac{1}{2} \|\tilde{\theta}_{j,i_j}\|^2 \tag{6.51}$$

$$\sigma_{j,i} \tilde{\theta}^{\mathrm{T}}_{j,i_j} \hat{\theta}_{j,i_j} = \sigma_{j,i_j} \tilde{\theta}^{\mathrm{T}}_{j,i_j}(\theta^*_{j,i_j} - \tilde{\theta}_{j,i_j}) \le - \frac{\sigma_{j,i_j}}{2} \|\tilde{\theta}_{j,i_j}\|^2 + \frac{\sigma_{j,i_j}}{2} \|\theta^*_{j,i_j}\|^2. \tag{6.52}$$

Taking (6.50)–(6.52) into (6.49), we have,

$$\begin{aligned} \dot{V}_{j,i_j} \le & \dot{V}_{j,i_j-1} + z_{j,i_j}(z_{j,i_j+1} + \omega_{j,i_j+1} - z_{j,i_j-1} + \tilde{\theta}^{\mathrm{T}}_{j,i_j} \varphi_{j,i_j}(\hat{\bar{x}}_{j,i_j}) - b_{j,i_j} z^{\beta}_{j,i_j} - c_{j,i_j} z_{j,i_j}) \\ & + \frac{1}{2} z^2_{j,i_j} + \frac{1}{2} \|\tilde{\theta}_{j,i_j}\|^2 + z^2_{j,i_j} + \|\theta^*_{j,i_j}\|^2 - \tilde{\theta}^{\mathrm{T}}_{j,i_j} z_{j,i_j} \varphi_{j,i_j}(\hat{\bar{x}}_{j,i_j}) - \frac{\sigma_{j,i_j}}{2} \|\tilde{\theta}_{j,i_j}\|^2 \\ & + \frac{\sigma_{j,i_j}}{2} \|\theta^*_{j,i_j}\|^2 - \omega_{j,i_j} \left(\frac{\omega_{j,i_j}}{\delta_{j,i_j}} + \dot{\alpha}_{j,i_j-1} \right) \\ \le & - \lambda_{j,1} \|e_{j,0}\|^2 - \sum_{k=1}^{i_j} \bar{c}_{j,k} z^2_{j,k} - \sum_{k=1}^{i_j} b_{j,k} |z_{j,k}|^{\beta+1} - \sum_{k=1}^{i_j} \bar{\sigma}_{j,k} \|\tilde{\theta}_{j,k}\|^2 \\ & + \|P_j\|^2 \sum_{k=1}^{\rho_j} \|\tilde{\theta}_{j,k}\|^2 + \sum_{k=2}^{i_j} z_{j,k} \omega_{j,k+1} + z_{j,1} \Lambda \omega_{j,2} + z_{j,i_j} z_{j,i_j+1} \\ & - \sum_{k=2}^{i_j} \omega_{j,k} \left(\frac{\omega_{j,k}}{\delta_{j,k}} + \dot{\alpha}_{j,k-1} \right) + \epsilon_{j,i_j}, \end{aligned} \tag{6.53}$$

where $\bar{c}_{j,i_j} = c_{j,i_j} - \frac{3}{2}$, $\epsilon_{j,i_j} = \epsilon_{j,i_j-1} + \frac{\sigma_{j,i_j}+2}{2}\|\theta^*_{j,i_j}\|^2$, and $\bar{\sigma}_{j,i_j} = \frac{\sigma_{j,i_j}-1}{2}$. Let a_{j,i_j} through the first-order filter,

$$\delta_{j,i_j+1}\dot{\alpha}^f_{j,i_j+1} + \alpha^f_{j,i_j+1} = \alpha_{j,i_j},\ \alpha^f_{j,i_j+1}(0) = \alpha_{j,i_j}(0), \tag{6.54}$$

where δ_{j,i_j+1} is a positive constant. Therefore, we have,

$$\dot{\omega}_{j,i_j+1} = -\frac{\omega_{j,i_j+1}}{\delta_{j,i_j+1}} - \dot{\alpha}_{j,i_j}. \tag{6.55}$$

Step ρ_j: Let $z_{j,\rho_j} = \hat{x}_{j,\rho_j} + \eta_{\rho_j,l} - \eta_{\rho_j,u} - \alpha^f_{j,\rho_j}$. And according to the definition $\omega_{j,\rho_j} = \alpha^f_{j,\rho_j} - \alpha_{j,\rho_j-1}$, we obtain the time derivative of z_{j,ρ_j} as,

$$\begin{aligned}\dot{z}_{j,\rho_j} =& \dot{\hat{x}}_{j,\rho_j} + \dot{\eta}_{\rho_j,l} - \dot{\eta}_{\rho_j,u} - \dot{\alpha}^f_{j,\rho_j}\\ =& h_j(v_j) + \hat{\theta}^{\mathrm{T}}_{j,\rho_j}\varphi_{j,\rho_j}(\hat{x}_j) + \eta_{\rho_j+1,l} - \eta_{\rho_j+1,u}\\ &+ \Gamma_{\rho_j}(\eta_{\rho_j,u} - \eta_{\rho_j,l}) + \frac{\omega_{j,\rho_j}}{\delta_{j,\rho_j}} + k_{j,\rho_j}e^1_{j,0}.\end{aligned} \tag{6.56}$$

Considering the auxiliary system designed in (6.15), the term $h_j(v_j) + \eta_{\rho_j+1,l} - \eta_{\rho_j+1,u} = v_j$ holds. The actual control input v_j and the adaptive law $\hat{\theta}_{j,\rho_j}$ can be described as (6.21) and (6.25). The Lyapunov function is designed as,

$$V_{j,\rho_j} = V_{j,\rho_j-1} + \frac{1}{2}z^2_{j,\rho_j} + \frac{\tilde{\theta}^{\mathrm{T}}_{j,\rho_j}\tilde{\theta}_{j,\rho_j}}{2\gamma_{j,\rho_j}} + \frac{1}{2}\omega^2_{j,\rho_j}. \tag{6.57}$$

According to (6.21), (6.25), and (6.56), we have the time derivative of V_{j,ρ_j} as,

$$\begin{aligned}\dot{V}_{j,\rho_j} =& \dot{V}_{j,\rho_j} + z_{j,\rho_j}(-z_{j,\rho_j-1} - c_{j,\rho_j}z_{j,\rho_j} + \theta^{*\mathrm{T}}_{j,\rho_j}\varphi_{j,\rho_j}(\hat{x}_j)\\ &- \tilde{\theta}^{\mathrm{T}}_{j,\rho_j}\varphi_{j,\rho_j}(\hat{x}_j) - b_{j,\rho_j}z^{\beta}_{j,\rho_j}) + \tilde{\theta}^{\mathrm{T}}_{j,\rho_j}(\sigma_{j,\rho_j}\hat{\theta}_{j,\rho_j}\\ &+ z_{j,\rho_j}\varphi_{j,\rho_j}(\hat{x}_{j,\rho_j})) + \omega_{j,\rho_j}\dot{\omega}_{j,\rho_j}.\end{aligned} \tag{6.58}$$

Due to the Young's inequality and $\tilde{\theta}_{j,\rho_j} = \theta^*_{j,\rho_j} - \hat{\theta}_{j,\rho_j}$, we have,

$$z_{j,\rho_j}\theta^{*\mathrm{T}}_{j,\rho_j}\varphi_{j,\rho_j}(\hat{x}_j) \le \frac{1}{2}z^2_{j,\rho_j} + \frac{1}{2}\|\theta^*_{j,\rho_j}\|^2 \tag{6.59}$$

$$\sigma_{j,\rho_j}\tilde{\theta}^{\mathrm{T}}_{j,\rho_j}\hat{\theta}_{j,\rho_j} = \sigma_{j,\rho_j}\tilde{\theta}^{\mathrm{T}}_{j,\rho_j}(\theta^*_{j,\rho_j} - \tilde{\theta}_{j,\rho_j}) \le -\frac{\sigma_{j,\rho_j}}{2}\|\tilde{\theta}_{j,\rho_j}\|^2 + \frac{\sigma_{j,\rho_j}}{2}\|\theta^*_{j,\rho_j}\|^2. \tag{6.60}$$

Substituting (6.53), (6.55), (6.59), and (6.60) into (6.58), we have,

$$\begin{aligned}\dot{V}_{j,\rho_j} \leq & -\lambda_{j,1}\|e_{j,0}\|^2 - \sum_{k=1}^{\rho_j} \bar{c}_{j,k} z_{j,k}^2 + z_{j,1}\Lambda\omega_{j,2} - \sum_{k=1}^{\rho_j} b_{j,k}|z_{j,k}|^{\beta+1} - \sum_{k=1}^{\rho_j} \bar{\sigma}_{j,k}\|\tilde{\theta}_{j,k}\|^2 \\ & + \|P_j\|^2 \sum_{k=1}^{\rho_j} \|\tilde{\theta}_{j,k}\|^2 + \sum_{k=2}^{\rho_j-1} z_{j,k}\omega_{j,k+1} - \sum_{k=2}^{\rho_j} \omega_{j,k}\left(\frac{\omega_{j,k}}{\delta_{j,k}} + \dot{\alpha}_{j,k-1}\right) + \epsilon_{j,\rho_j},\end{aligned} \tag{6.61}$$

where $\bar{c}_{j,\rho_j} = c_{j,\rho_j} - \frac{1}{2}$, $\bar{\sigma}_{j,\rho_j} = \frac{\sigma_{j,\rho_j}}{2}$, and $\epsilon_{j,\rho_j} = \epsilon_{j,\rho_j-1} + \frac{\sigma_{j,\rho_j}+1}{2}\|\theta^*_{j,\rho_j}\|^2$. From Yong's inequality, we have,

$$z_{j,1}\Lambda\omega_{j,2} \leq \frac{1}{2}z_{j,1}^2 + \frac{1}{2}\Lambda^2\omega_{j,2}^2 \tag{6.62}$$

$$\sum_{k=2}^{\rho_j-1} z_{j,k}\omega_{j,k+1} \leq \sum_{k=2}^{\rho_j-1} \frac{1}{2}z_{j,k}^2 + \sum_{k=3}^{\rho_j} \frac{1}{2}\omega_{j,k}^2 \tag{6.63}$$

$$-\sum_{k=2}^{\rho_j} \omega_{j,k}\dot{\alpha}_{j,k-1} \leq \sum_{k=2}^{\rho_j} \frac{1}{2}\omega_{j,k}^2 + \sum_{k=2}^{\rho_j} \frac{1}{2}\dot{\alpha}_{j,k-1}^2. \tag{6.64}$$

Taking (6.62)–(6.64) into (6.61) yields,

$$\begin{aligned}\dot{V}_{j,\rho_j} \leq & -\lambda_{j,1}\|e_{j,0}\|^2 - \sum_{k=1}^{\rho_j} l_{j,k} z_{j,k}^2 - \sum_{k=1}^{\rho_j} b_{j,k}|z_{j,k}|^{\beta+1} \\ & - \sum_{k=2}^{\rho_j} \bar{\delta}_{j,k}\omega_{j,k}^2 - \sum_{k=1}^{\rho_j} p_{j,k}\|\tilde{\theta}_{j,k}\|^2 + \bar{\epsilon}_{j,\rho_j},\end{aligned} \tag{6.65}$$

where $\bar{\epsilon}_{j,\rho_j} = \sum_{k=2}^{\rho_j} \frac{1}{2}\dot{\alpha}_{j,k-1}^2 + \epsilon_{j,\rho_j}$, $p_{j,k} = \bar{\sigma}_{j,k} - \|P_j\|^2$, $l_{j,k} = \bar{c}_{j,k} - \frac{1}{2}$ for $k \in \mathbb{I}_{1:\rho_j}$, $\bar{\delta}_{j,2} = \frac{1}{\delta_{j,2}} - \frac{1+\Lambda^2}{2}$, and $\bar{\delta}_{j,k} = \frac{1}{\delta_{j,k}} - 1$ for $k \in \mathbb{I}_{3:\rho_j}$. From Lemma 2.2, we can obtain,

$$\begin{aligned}-\|\tilde{\theta}_{j,k}\|^2 = & -\|\tilde{\theta}_{j,k}\|^2 + \|\tilde{\theta}_{j,k}\|^{\beta+1} - \|\tilde{\theta}_{j,k}\|^{\beta+1} \\ \leq & \frac{1-\beta}{2} - \frac{1-\beta}{2}\|\tilde{\theta}_{j,k}\|^2 - \|\tilde{\theta}_{j,k}\|^{\beta+1}\end{aligned} \tag{6.66}$$

The terms $-\lambda_{j,1}\|e_{j,0}\|^2$ and $-\sum_{k=2}^{\rho_j} \bar{\delta}_{j,k}\omega_{j,k}^2$ follow the same process. (6.65) can be,

$$
\begin{aligned}
\dot{V}_{j,\rho_j} \leq & -\frac{1-\beta}{2}\lambda_{j,1}\|e_{j,0}\|^2 - \lambda_{j,1}\|e_{j,0}\|^{\beta+1} + \bar{\epsilon} - \frac{1-\beta}{2}\sum_{k=1}^{\rho_j} p_{j,k}\|\tilde{\theta}_{j,k}\|^2 \\
& - \sum_{k=1}^{\rho_j} p_{j,k}\|\tilde{\theta}_{j,k}\|^{\beta+1} - \frac{1-\beta}{2}\sum_{k=1}^{\rho_j} \bar{\delta}_{j,k}\omega_{j,k}^2 - \sum_{k=1}^{\rho_j} \bar{\delta}_{j,k}\omega_{j,k}^{\beta+1} \\
& - \sum_{k=1}^{\rho_j} l_{j,k}z_{j,k}^2 - \sum_{k=1}^{\rho_j} b_{j,k}|z_{j,k}|^{\beta+1}
\end{aligned} \tag{6.67}
$$

where $\bar{\epsilon} = \bar{\epsilon}_{j,\rho_j} + \frac{\lambda_{j,1}(1-\beta)}{2} + \sum_{k=1}^{\rho_j} \frac{\bar{\delta}_{j,k}(1-\beta)}{2} + \sum_{k=1}^{\rho_j} \frac{p_{j,k}(1-\beta)}{2}$. According to Lemma 2.4, (6.67) can be formulated as,

$$
\begin{aligned}
\dot{V}_{j,\rho_j} \leq & - c_1\left(\frac{1}{2}\|e_{j,0}\|^2 + \frac{1}{2}\sum_{k=1}^{\rho_j} z_{j,k}^2 + \sum_{k=2}^{\rho_j} \frac{1}{2}\omega_{j,k}^2 + \sum_{k=1}^{\rho_j} \frac{\tilde{\theta}_{j,k}^{\mathrm{T}}\tilde{\theta}_{j,k}}{2\gamma_{j,k}}\right) \\
& - c_2\left(\frac{1}{2}\|e_{j,0}\|^2 + \frac{1}{2}\sum_{k=1}^{\rho_j} z_{j,k}^2 + \sum_{k=2}^{\rho_j} \frac{1}{2}\omega_{j,k}^2 + \sum_{k=1}^{\rho_j} \frac{\tilde{\theta}_{j,k}^{\mathrm{T}}\tilde{\theta}_{j,k}}{2\gamma_{j,k}}\right)^{\bar{\beta}} + \bar{\epsilon} \\
\leq & -c_1 V_{j,\rho_j} - c_2(V_{j,\rho_j})^{\bar{\beta}} + \bar{\epsilon},
\end{aligned} \tag{6.68}
$$

where $c_1 = \min((1-\beta)\lambda_{j,1}, 2l_{j,k}, (1-\beta)\bar{\delta}_{j,k}, (1-\beta)p_{j,k}\gamma_{j,k})$, and $c_2 = \min(2^{\bar{\beta}}\lambda_{j,1}, 2^{\bar{\beta}}b_{j,k}, 2^{\bar{\beta}}\bar{\delta}_{j,k}, 2^{\bar{\beta}}p_{j,k})$ with $\bar{\beta} = \frac{1+\beta}{2}$.

From (6.68) and Lemma 2.1, the following inequality holds

$$
\dot{V}_{j,\rho_j} \leq -c_1 V_{j,\rho_j} - c_2 c_3 V_{j,\rho_j}^{\bar{\beta}} + \bar{\epsilon} - c_2(1-c_3)V_{j,\rho_j}^{\bar{\beta}}. \tag{6.69}
$$

where c_3 is a positive constant. We define $\Omega_V = \{\xi | V_{j,\rho_j}^{\bar{\beta}} \leq \frac{\bar{\epsilon}}{c_2(1-c_3)}\}$ and $\bar{\Omega}_V = \{\xi | V_{j,\rho_j}^{\bar{\beta}} > \frac{\bar{\epsilon}}{c_2(1-c_3)}\}$.

Case 1: If $\xi \in \bar{\Omega}_V$, the following inequality holds

$$
\dot{V}_{j,\rho_j} \leq -c_1 V_{j,\rho_j} - c_2 c_3 V_{j,\rho_j}^{\bar{\beta}}. \tag{6.70}
$$

Integrating the above inequality over $[t_0, T]$, we have

$$
\int_{t_0}^{T} \frac{V_{j,\rho_j}^{-\bar{\beta}}}{c_2 c_3 + c_1 V_{j,\rho_j}^{1-\bar{\beta}}} \mathrm{dV} \leq -\int_{t_0}^{T} \mathrm{dt}. \tag{6.71}
$$

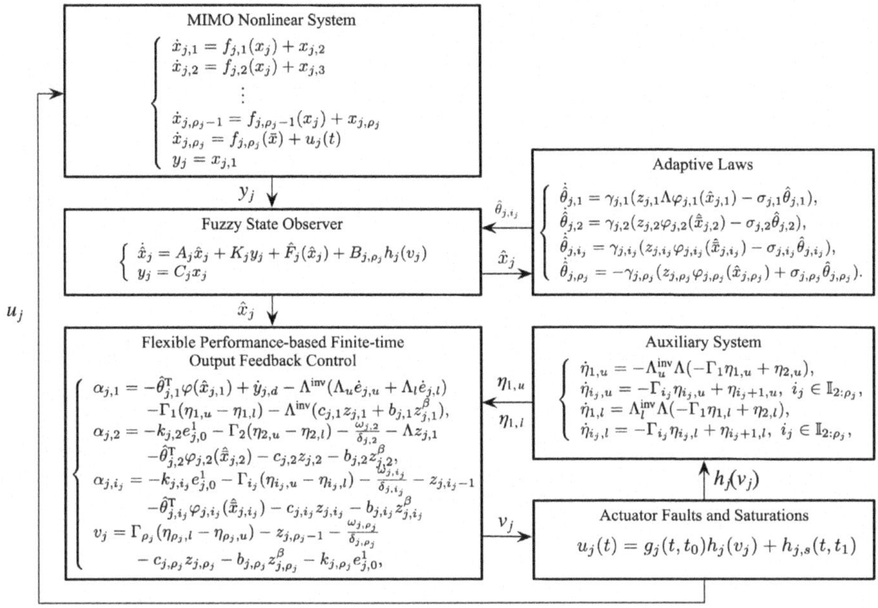

Fig. 6.2 Block diagram of the control scheme

Then the settling time can be described as

$$T = t_0 + \frac{1}{c_1(1-\bar{\beta})} \ln \left(\frac{c_2 c_3 + c_1 V_{j,\rho_j}^{1-\bar{\beta}}(t_0)}{c_2 c_3} \right). \tag{6.72}$$

Namely, $\xi \in \Omega_V$ for $\forall t \geq T$.

Case 2: Considering the system state ξ in the region Ω_V, from the case 1 above, ξ cannot escape the set Ω_V. This completes the proof. Additionally, the proposed control design is shown in Fig. 6.2 to make the designed approach in a clear manner.

□

Remark 6.6 Compared with the existing control designs in [9, 14, 16, 34, 64, 65], SPC establishes the bridge between input saturations and performance constraints instead of assuming both constraints being satisfied simultaneously. Compared with [1, 9, 14, 16, 17, 30, 33, 34, 37, 49, 50, 64, 65], the proposed SPC can not only guarantee finite-time stability, but also guarantee the tracking errors to evolve within performance constraints despite the existence of saturations and actuator faults. Namely, SPC can flexibly enlarge or recover the performance boundaries according to saturation signals which can dramatically decrease the control conservatism and achieve favorable tracking performance although various constraints.

6.2.6 Numerical Simulations

In this section, simulation examples are conducted to further demonstrate the feasibility and effectiveness of the proposed control scheme. We first consider the dynamics of the tilting quadcopter in [20]

$$\begin{cases} \dot{x}_{j,1} = f_{j,1}(x_j) + x_{j,2} \\ \dot{x}_{j,2} = f_{j,2}(\bar{x}) + u_j(t) \\ y_j = x_{j,1} \end{cases} \tag{6.73}$$

where $j \in \mathbb{I}_{1:3}$, $x_j = [x_{j,1}, x_{j,2}]^{\mathrm{T}}$, $\bar{x} = [x_1^{\mathrm{T}}, x_2^{\mathrm{T}}, x_3^{\mathrm{T}}]^{\mathrm{T}}$, $f_{j,1}(x_j) = 0$, $f_{1,2}(\bar{x})$, $f_{2,2}(\bar{x})$ and $f_{3,2}(\bar{x})$ are the components of the vector $-M_0(\bar{x})^{-1}(C_0(\bar{x})\chi + D_0(\bar{x})\chi)$ with $\chi = [x_{1,2}, x_{2,2}, x_{3,2}]^{\mathrm{T}}$. For the terms $M_0(\bar{x})$, $C_0(\bar{x})$ and $D_0(\bar{x})$, please refer to [20] for details and we omit here for brevity. Following the proposed control scheme, the control inputs and adaptive law can be designed as

$$\begin{aligned} \alpha_{j,1} = {} & \dot{y}_{j,d} - \Lambda^{\mathrm{inv}}(\Lambda_u \dot{e}_{j,u} + \Lambda_l \dot{e}_{j,l}) - \Gamma_1(\eta_{1,u} \\ & - \eta_{1,l}) - \Lambda^{\mathrm{inv}}(c_{j,1} z_{j,1} + b_{j,1} z_{j,1}^{\beta}), \end{aligned} \tag{6.74}$$

$$v_j = \Gamma_2(\eta_{2,l} - \eta_{2,u}) - \Lambda z_{j,1} - \frac{\omega_{j,2}}{\delta_{j,2}} - c_{j,2} z_{j,2} - b_{j,2} z_{j,2}^{\beta} - k_{j,2} e_{j,0}^{1}, \tag{6.75}$$

$$\dot{\theta}_{j,2} = -\gamma_{j,2}(z_{j,2}\varphi_{j,2}(\hat{x}_{j,2}) + \sigma_{j,2}\hat{\theta}_{j,}), \tag{6.76}$$

and the auxiliary system is

$$\begin{cases} \dot{\eta}_{1,u} = -\Lambda_u^{\mathrm{inv}}\Lambda(-\Gamma_1\eta_{1,u} + \eta_{2,u}), \\ \dot{\eta}_{2,u} = -\Gamma_2\eta_{2,u} + \eta_{3,u}, \\ \dot{\eta}_{1,l} = \Lambda_l^{\mathrm{inv}}\Lambda(-\Gamma_1\eta_{1,l} + \eta_{2,l}), \\ \dot{\eta}_{2,l} = -\Gamma_2\eta_{2,l} + \eta_{3,l}, \end{cases} \tag{6.77}$$

where $\Gamma_1 = \Gamma_2 = 2$ and all the other parameters for simulations can be found in the following Table 6.1.

To showcase the effectiveness and flexibility of the proposed control scheme, comparative simulations are carried out to explore the merit of SPC. We design the comparative control without the auxiliary system, where the performance constraints are always TPP and the simulation results are denoted as Tunnel Prescribed Control (TPC). Simulation results can be found in Figs. 6.3, 6.4, 6.5 and 6.6. The tracking errors trajectories of SPC and TPC are exhibited in Fig. 6.3. The tracking errors of SPC can be guaranteed to converge to a small neighborhood around the equilibrium within finite time and the tracking errors are evolved within the specified performance constraints of SPP despite the existence of input saturations and actuator faults. Specifically, we here take the tracking error $e_{1,1}$ of SPC as an example. According to the control signals in Fig. 6.4, the control input h_1 of SPC stimulates the saturation effect during 0s–5s, 6s–11s and 18s–24s caused by nonlinearities and

Table 6.1 Parameters for simulations

Parameters	Values	Parameters	Values
$c_{j,1}, c_{j,2}$	6	$b_{j,1}, b_{j,2}$	1
Γ_1, Γ_2	2	$v_{j,\max}$	5
$v_{j,\min}$	−5	$k_{j,1}$	5
$k_{j,2}$	80	$\sigma_{j,2}$	2
$\delta_{j,2},$	0.1	$\gamma_{j,2}$	1.2
β	0.75	$\varrho_1(t)$	$1.1e^{-t}+0.1$
$\varrho_2(t)$	$0.7e^{-t}+0.1$	$\varrho_3(t)$	$1.5e^{-t}+0.1$
$g_1(t,2)$	$1-0.9\lvert\sin(0.55t)\rvert$	$g_2(t,2)$	0.6
$g_3(t,2)$	$1-0.9\lvert\cos(0.43t)\rvert$	$h_{1,s}(t,2)$	$0.4\sin(0.9t)$
$h_{2,s}(t,2)$	$\sin(0.3t)$	$h_{3,s}(t,2)$	0.35
$y_{1,d}$	$0.9\cos(0.25t-0.3)$	$y_{2,d}$	$0.6\cos(0.4t)$
$y_{3,d}$	$0.8\cos(0.3t-0.4)+0.3$	$\hat{\theta}_{j,2}(0)$	$[0.2,\ldots,0.2]^{\mathrm{T}}$
$e_{j,l}(t)$	$(0.2-\mathrm{sign}(e_j(0)))\varrho_j(t)+0.1\mathrm{sign}(e_j(0))$		
$e_{j,u}(t)$	$(0.7+\mathrm{sign}(e_j(0)))\varrho_j(t)-0.1\mathrm{sign}(e_j(0))$		

uncertainties in systems. Therefore, the tracking error $e_{1,1}$ of SPC shows obvious hills or volatility respectively during the saturations periods. To compensate for these tracking degradation, the auxiliary system is introduced to generate modified signals as the red dotted line in Fig. 6.6. The signals $\eta_{2,1,u}$ and $\eta_{2,1,l}$ represents unreachable increasing or decreasing rates for $x_{1,2}$, and $\eta_{1,1,u}$ and $\eta_{1,1,l}$ are utilized into SPP for adjusting performance constraints. With the help of these signals, SPC has the ability of enlarging SPP constraints to guarantee favorable performance, where the tracking error $e_{1,1}$ does not violate SPP boundaries as shown in Fig. 6.3 although under input saturations and actuator faults conditions. As input saturations disappear, these modified signals $\eta_{2,1,u}$, $\eta_{2,1,l}$, $\eta_{1,1,u}$ and $\eta_{1,1,l}$ shown in Fig. 6.6 fast converge to zero, which can further lead the enlarged SPP to fast recover the specified performance as shown in Fig. 6.3. Namely, this mechanism accomplish a trade-off between the system abilities related constraints (e.g., input saturations and unmeasurable system states) and the performance related constraints (e.g., specified convergence rate and overshoot), and avoids conservative control design. Moreover, due to the tight space shaped by the SPP, SPP can achieve smaller overshoot performance in a positive manner and the tracking errors trajectories in Fig. 6.3 can verify its superiority significantly. The other tracking errors analysis of SPC are omitted here for brevity. However, TPC cannot guarantee the favorable tracking performance under the same conditions. From Figs. 6.3 and 6.4, it can be seen that the tracking error $e_{3,1}$ of TPC approaches the boundary and violates the performance after 6.71 s resulting in the singularity of the control design. It means we cannot assume that input saturations and performance constraints are satisfied simultaneously and handle these constraints with neglecting their interconnections and conflict. And the user-specified performance constraints of TPC may be not achieved once saturations and actuator faults,

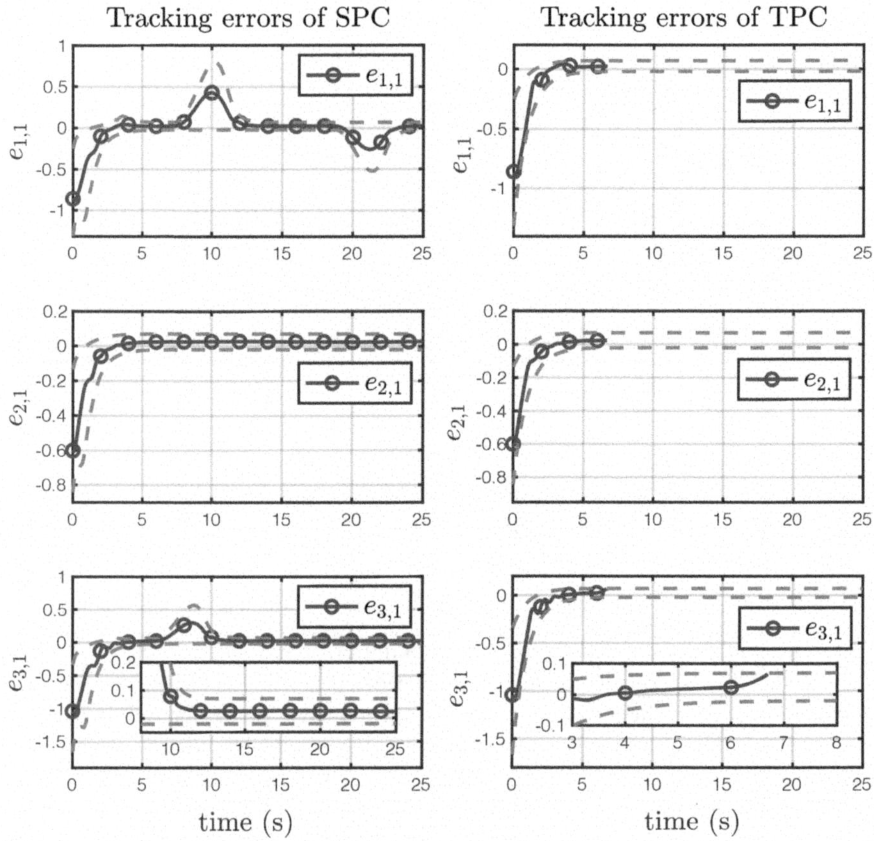

Fig. 6.3 The tracking error trajectories of $e_{1,1}$, $e_{2,1}$, and $e_{3,1}$

and the closed-loop systems may be severely degraded or even instability. According to the simulation results, the effectiveness and superiority of SPC has been verified. In addition, Fig. 6.5 describes the estimated states of fuzzy state observer of SPC, and the observer errors can achieve finite-time stability. And Fig. 6.6 is the auxiliary system states of SPC, which bridge the associations between input saturations and performance constraints. When input saturations are stimulated, the auxiliary system can generate modified signals for enlarging SPP boundaries. When input saturations disappear, the modified signals fast converge to zero to recover the enlarged SPP to the user-specified performance.

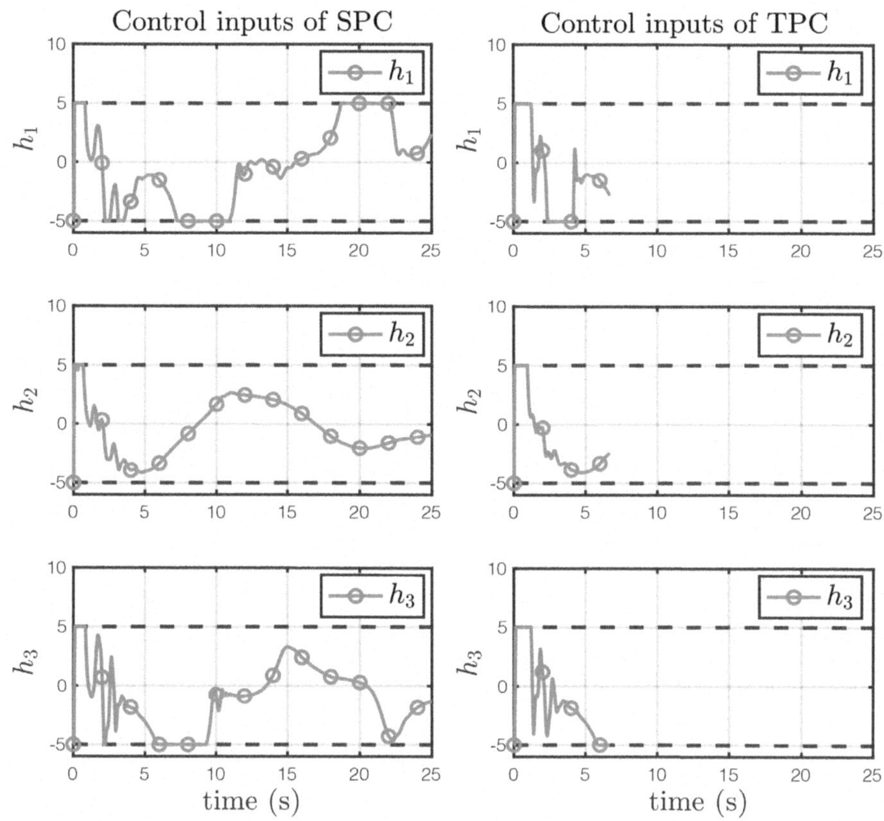

Fig. 6.4 The control inputs of SPC and TPC

6.3 Saturation-Tolerant Prescribed Control for MIMO Nonlinear Systems with Unknown Control Directions and External Disturbances

6.3.1 Problem Formulation

Consider the following MIMO nonlinear systems:

$$\begin{cases} \dot{x}_{i,j_i} = f_{i,j_i}(\bar{x}_{i,j_i}) + g_{i,j_i}x_{i,j_i+1} + d_{i,j_i}(t) \\ \dot{x}_{i,n_i} = f_{i,n_i}(\bar{x}) + g_{i,n_i}u_i + d_{i,n_i}(t) \\ y_i = x_{i,1} \end{cases} \tag{6.78}$$

where $i \in \mathbb{I}_{1:m}$, $\bar{x}_{i,j_i} = [x_{i,1}, \ldots, x_{i,j_i}]^T \in \mathbb{R}^{j_i}$ for $j_i \in \mathbb{I}_{1:n_i}$, $\bar{x} = \mathrm{col}(\bar{x}_{1,n_1}, \ldots, \bar{x}_{m,n_m})$, $f_{i,j_i}(\cdot)$ and g_{i,j_i}, $j_i \in \mathbb{I}_{1:n_i}$ denote known smooth function and unknown

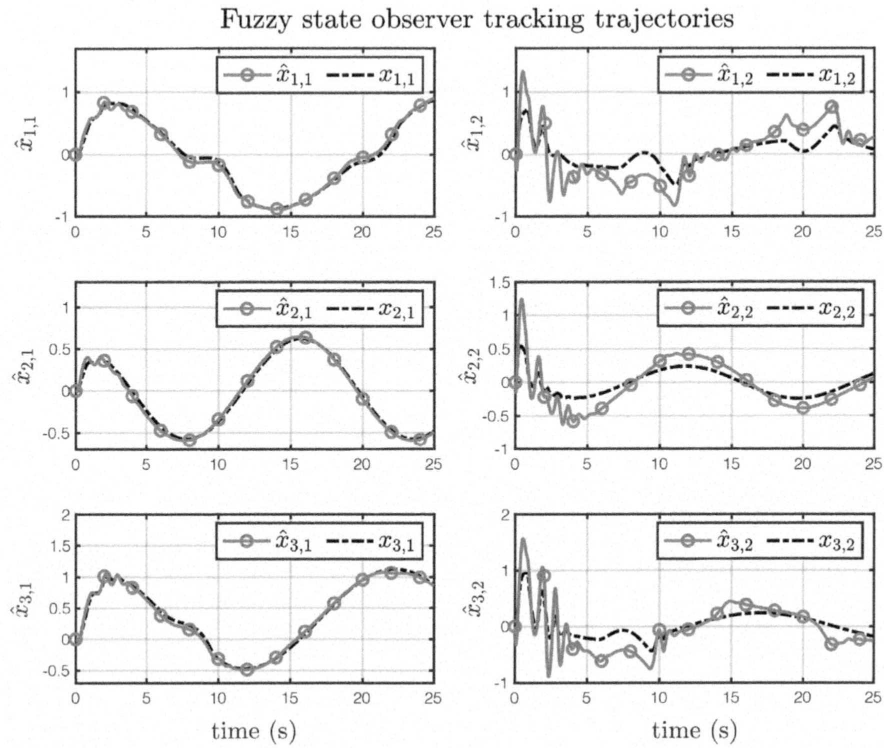

Fig. 6.5 Fuzzy state observer tracking trajectories

control coefficient, d_{i,j_i} denotes an unknown external disturbance, y_i represents the system output and u_i stands for the actual control input with asymmetric saturation nonlinearities as

$$u_i = \begin{cases} u_{i,\max}, & v_i > u_{i,\max} \\ v_i, & u_{i,\min} \leq v_i \leq u_{i,\max} \\ u_{i,\min}, & v_i < u_{i,\min} \end{cases} \tag{6.79}$$

where v_i denotes the desired control input, $u_{i,\max}$ and $u_{i,\min}$ are constraints on u_i.

Control objective: The control problem in this work is how to design a control strategy for the above nonlinear systems with external disturbances and unknown control directions such that: (i) The tracking error $e_{i,1} = y_i - y_{i,d}$ is ultimately bounded when $u_i \neq v_i$, where $y_{i,d}$ is the desired trajectory. (ii) When $u_i = v_i$, the tracking error $e_{i,1}$ can achieve finite-time convergence into the specified FTPP as $\{e_{i,1} : -e_{i,l} < e_{i,1} < e_{i,u}\}$ where $e_{i,u}$ and $-e_{i,l}$ represent the performance boundaries of FTPP, respectively. (iii) Explore an auxiliary system to bridge a connection between input and performance requirements for less conservative control design.

To proceed, the following assumptions are first given.

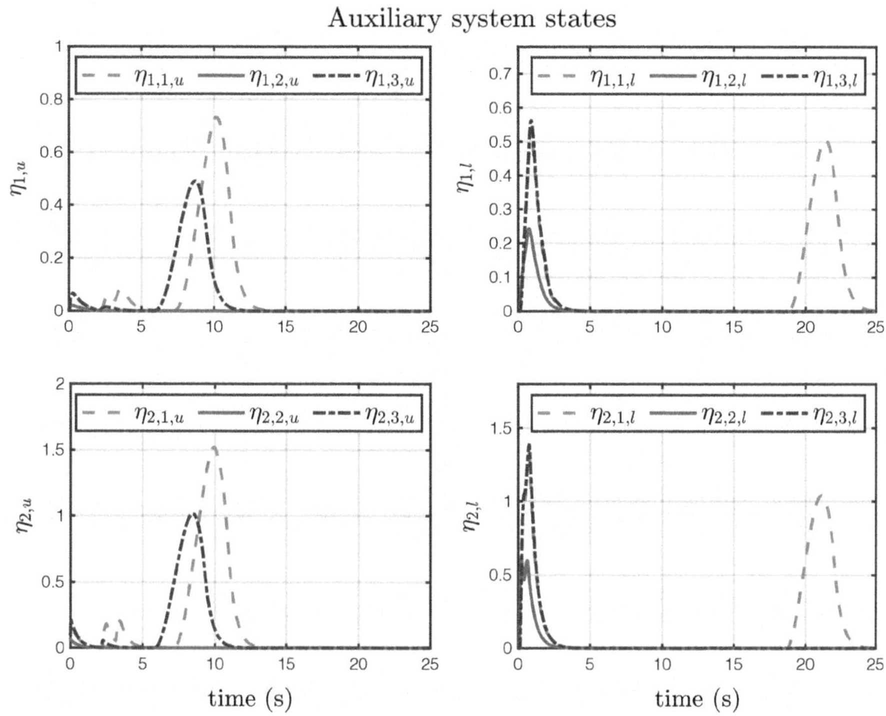

Fig. 6.6 The auxiliary system states

Assumption 6.2 The sign of g_{i,j_i}, $j_i \in \mathbb{I}_{1:n_i}$ is unknown, which satisfies $0 < \underline{g}_{i,j_i} \leq |g_{i,j_i}| \leq \bar{g}_{i,j_i}$ with $\underline{g}_{i,j_i}$ and $\bar{g}_{i,j_i}$ being unknown constants.

Assumption 6.3 The external disturbance and its derivative are bounded, that is, $|d_{i,j_i}| \leq \bar{D}_{i,j_i}$ and $|\dot{d}_{i,j_i}| \leq \bar{\dot{D}}_{i,j_i}$, where $\bar{D}_{i,j_i}$ and $\bar{\dot{D}}_{i,j_i}$ are unknown positive constants.

Assumption 6.4 Considering the nonlinear system in (6.78) with input constraint and external disturbance, there should exist a feasible control strategy such that the control objective holds.

Remark 6.7 As for Assumption 6.2, the boundedness of control coefficient is guaranteed and it is necessary to require that g_{i,j_i} be away from zero for the nonlinear systems to be controllable [31, 61]. Assumption 6.3 indicates the boundedness of external disturbance and its time derivative since many disturbances satisfy such conditions as presented in [6, 34]. For the explanation of Assumption 6.4, please refer to the following remark for more details.

Remark 6.8 It can be known that not all the nonlinear systems in (6.78) can be stabilized due to the input saturation effect. Specifically, consider the following simple case of system (6.78) as $\dot{x}_1 = x_1 + u$. Intuitively, if $x_1 > u_{\max}$ or $x_1 < u_{\min}$, no feasible control strategy exists to stabilize such system. However, the practical systems always operate within a feasible region for safety and mission accomplishment, i.e., flight systems and manipulator systems. Hence, there usually exist feasible control strategies to complete the control objective even under input saturation [7, 51, 60]. It indicates that Assumption 6.4 is reasonable and serves as the basis of the subsequent control development.

Remark 6.9 Compared with SISO, pure feedback and strict feedback nonlinear systems in [23, 26, 42, 47], the discussed MIMO nonlinear systems (6.78) are much more complicated due to the couplings and nonlinearities among subsystems. Moreover, both input saturation and performance constraints are simultaneously considered. Although some efforts have been made in [6, 15, 32, 34, 38, 59, 68] under the same conditions, these results are based on known control directions. How to handle MOMO nonlinear systems subject to unknown control directions under the above conditions is more difficult yet significant.

6.3.2 *Finite-Time Tunnel Prescribed Performance*

Before the FTPP, we report the classical Prescribed Performance Functions (PPFs) which ensure the tracking error $e_{i,1}$ evolving within the predefined boundaries

$$\begin{cases} -\delta_i \rho_i(t) < e_{i,1} < \rho_i(t), \text{ if } e_{i,1}(0) \geq 0 \\ -\rho_i(t) < e_{i,1} < \delta_i \rho_i(t), \text{ if } e_{i,1}(0) < 0 \end{cases} \tag{6.80}$$

where $0 < \delta_i \leq 1$ and $\rho_i(t) = (\rho_{i,0} - \rho_{i,\infty})e^{-\ell_i t} + \rho_{i,\infty}$ with $\ell_i > 0$ and $\rho_{i,0} > \rho_{i,\infty} > 0$.

The tracking trajectory of $e_{i,1}$ and PPFs are plotted in Fig. 6.7. We will evaluate the classical PPFs from the following aspects. (i) PPFs in (6.80) are divided into two cases in the light of different initial conditions, leading to the control development more complex. (ii) As shown in Fig. 6.7, the performance boundaries of PPFs are distributed on different sides, where only a funnel-shape allowable set is established. However, such loose PPFs may lead to obvious overshoot, which is unexpected in practical application. (iii) The convergence time to the steady-state performance tends to infinite, (i.e., $\rho(t) \to \rho_\infty$ if $t \to +\infty$). In practical applications, it is desirable that the tracking error can be regulated into the prescribed set in finite time.

Motivated by the above discussion, we therefore propose Finite-time Tunnel Prescribed Performance (FTPP) as

$$\rho_i(t) = \begin{cases} (\rho_{i,0}^{\ell_i} - \ell_i \psi t)^{\frac{1}{\ell_i}} + \rho_{i,t_f}, & 0 \leq t \leq t_f \\ \rho_{i,t_f}, & t > t_f \end{cases} \tag{6.81}$$

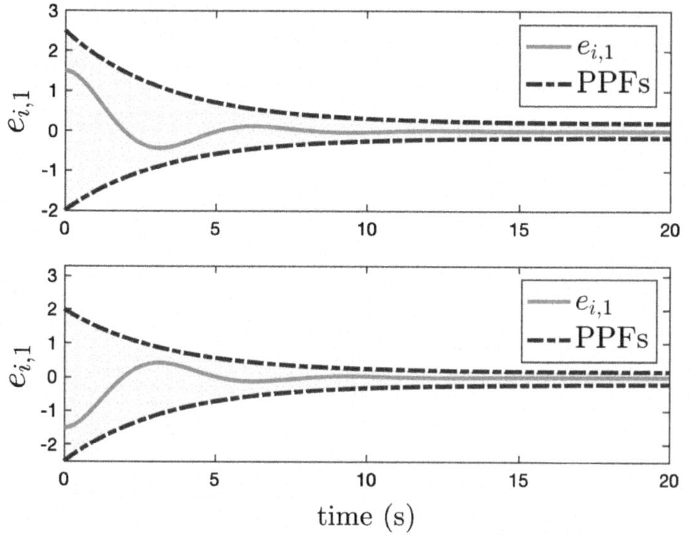

Fig. 6.7 The performance of PPFs in the case of $e_{i,1}(0) \geq 0$ and $e_{i,1}(0) < 0$

where $\rho_{i,0}$, ρ_{i,t_f} and ψ are positive constants to be suitably selected, $\rho_i(0) = \rho_{i,0} + \rho_{i,t_f}$, $\ell_i\psi = \rho_{i,0}^{\ell_i}/t_f$, $\ell_i = q_1/q_2 \in (0, 1)$ where q_1 and q_2 are positive odd and even integers. The FTPP possesses the properties that $\lim_{t\to t_f} \rho_i(t) = \rho_{i,t_f}$ and $\rho_i(t) = \rho_{i,t_f}$ for all $t \geq t_f$. Moreover, $\rho_i(t)$ is continuous and differentiable, and its time derivative is continuous at $t = t_f$.

Then, the tracking error $e_{i,1}$ can be formulated as

$$-e_{i,l} < e_{i,1}(t) < e_{i,u}, \tag{6.82}$$

with

$$\begin{cases} e_{i,l} = (\delta_{i,l} - \text{sign}(e_{i,1}(0)))\rho_i(t) + \rho_{i,t_f}\text{sign}(e_{i,1}(0)) \\ e_{i,u} = (\delta_{i,u} + \text{sign}(e_{i,1}(0)))\rho_i(t) - \rho_{i,t_f}\text{sign}(e_{i,1}(0)) \end{cases} \tag{6.83}$$

where $0 < \delta_{i,l}, \delta_{i,u} \leq 1$.

To better illustrate the typical features of FTPP, the tracking error signal $e_{i,1}$ and FTPP are depicted in Fig. 6.8. (i) By using sign function and scale parameters ($\delta_{i,l}$ and $\delta_{i,u}$) in (6.83), the FTPP can adaptive adjust its orientation according to different initial tracking error conditions as shown in Fig. 6.8, which facilitates the control implementation. (ii) Different from the above funnel-shape allowable set, FTPP establishes a tunnel-shape performance constraints, which guarantees stricter transient performance, i.e., smaller overshoot. (iii) The settling time of FTPP can be user specified in advance (i.e., $\rho_i(t) \to \rho_{i,t_f}$ when $t > t_f$). Therefore, the tracking

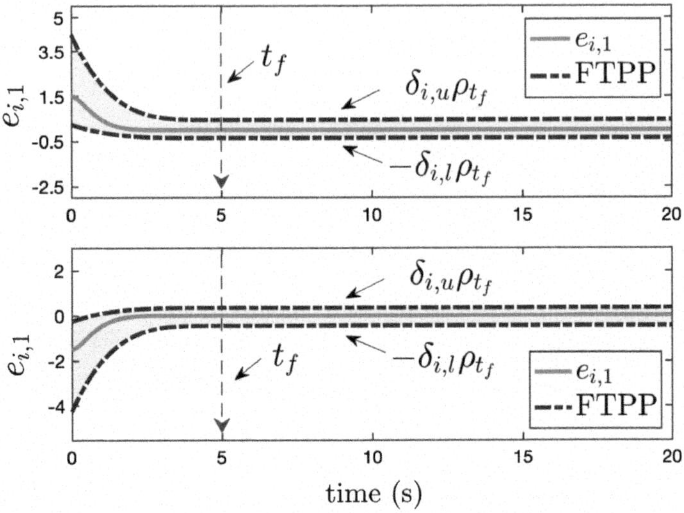

Fig. 6.8 The performance of FTPP in the case of $e_{i,1}(0) \geq 0$ and $e_{i,1}(0) < 0$

error is regulated into the residual set in finite time, instead of relying on the initial conditions or control parameters.

To characterize a general family of FTPP, we propose the following definition for future research.

Definition 6.1 A differential and smooth function $\rho(t)$ can be called Finite-time Prescribed Performance, if it has the following properties: (i) $\rho_i(t) > 0$, (ii) $\dot{\rho}_i(t) \leq 0$, (iii) $\lim_{t \to t_f} \rho_i(t) = \rho_{i,t_f}$ with $\rho_{i,tf} > 0$ and t_f a set time, (iv) $\rho_i(t) = \rho_{i,t_f}$ for all $t > t_f$. Further, if constraint boundaries are on the same side similar to a tunnel shape allowable set, we can call it Finite-time Tunnel Prescribed Performance.

Remark 6.10 We pause here to reveal the motivation of such tunnel-shape FTPP. With the increasing of tracking demands, the practical systems are always required to track and even interact with target, especially for flight systems. Therefore, there exists stricter tracking allowable set for the tracking process to avoid target losing due to the limited perception region. In comparison with PPFs in [2, 11, 27, 41, 66], such FTPP provides a better solution for such problem by redeveloping performance functions to establish tighter performance constraints. Moreover, user predefined settling time is also desirable in most practical applications instead of relying on initial conditions or control parameters in [10, 18, 35, 37]. By abstracting its properties, we present a general definition of FTPP to promote other FTPP expressions not limited to (6.83).

6.3.3 Finite-Time Saturation-Tolerant Prescribed Performance

Note that either the proposed FTPP or the PPFs are user predefined, whose expressions are fixed from the beginning of system operation. In this way, the performance constraints could be violated as the tracking capability is constrained by the input saturation. Although we can select more conservative performance functions, it is difficult to cover all conditions by using the redeveloped ones since the external disturbances are always time varying and unknown.

To clearly illustrate the aforesaid problem, we consider a system $\ddot{x}_1 = u_1 + d_1$ with input saturation (6.79), where $u_{1,\max} = 0.5$, $u_{1,\min} = -0.5$ and $d_1 = -0.8\tanh(t-5) + 0.85\tanh(t-4)$. According to (6.83), for FTPP1, let $\delta_{i,l} = 0.6$, $\delta_{i,u} = 0.5$, $\rho_{i,0} = 1$, $\rho_{i,t_f} = 0.8$, $\ell_i = 0.5$ and $t_f = 2$; for FTPP2, let $\delta_{i,l} = 0.5$, $\delta_{i,u} = 0.9$, $\rho_{i,0} = 1$, $\rho_{i,t_f} = 0.3$, $\ell_i = 0.5$ and $t_f = 2$. The tracking results are depicted in Fig. 6.9. From results, the control input encounters saturation nonlinearities during 0–2s and 5–8s due to disturbances, where the loss of tracking performance in gray shadow is obvious. If FTPP1 with narrow feasible space is selected, the tracking trajectory approaches or even violates the performance boundaries, which leads to control singularity problems. Additionally, if the boundaries are chosen in a conservative manner like FTPP2, there exist two problems. First, we cannot preassign such conservative boundaries to adapt to any initial conditions and unknown disturbances. Second, this manner is not in line with high precision application requirements and degrade the superiority of the FTPP.

We then naturally focus on constructing more flexible performance boundaries to balance input saturation and prescribed performance. By introducing nonnegative signals into FTPP (6.82), a novel finite-time Saturation-tolerant Prescribed Performance (SPP) can be formulated as

$$-E_{i,l} < e_{i,1}(t) < E_{i,u}, \tag{6.84}$$

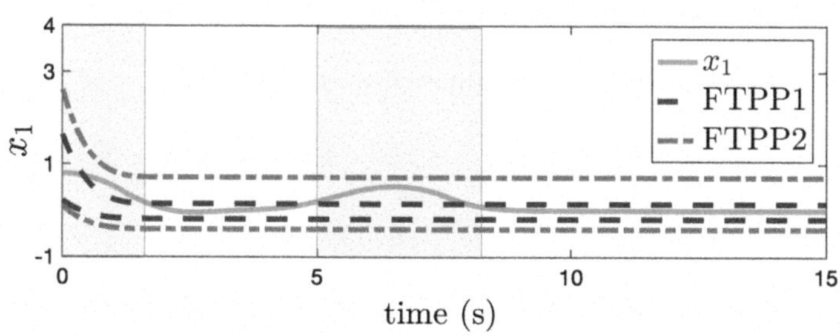

Fig. 6.9 System response and user-specified FTPP

where

$$E_{i,l}(t) = e_{i,l} + \eta_{1,l},\ E_{i,u}(t) = e_{i,u} + \eta_{1,u}, \tag{6.85}$$

where $\eta_{1,l}$ and $\eta_{1,u}$ are nonnegative variables generated by the auxiliary system (6.95), that is, finite-time SPP possesses the ability of enlarging the boundaries to reduce the performance requirements when necessary. And the magnitude of degradation should be related to the input saturation error. For more details about these generated signals $\eta_{1,l}$ and $\eta_{1,u}$, please refer to the auxiliary system design in the next section. In what follows, we employ the following error transformation function (ETF), by using which $e_{i,1}$ can be transformed to an equivalent unconstrained one $z_{i,1}$

$$z_{i,1} = \ln\left(\frac{E_{i,l} + e_{i,1}}{E_{i,u} - e_{i,1}}\right). \tag{6.86}$$

The bounded variable $z_{i,1}$ can guarantee the constraints in (6.84) to be satisfied.

Remark 6.11 Different from most PPFs with fixed expressions, finite-time SPP employs flexible performance boundaries to deal with the conflict between input and performance constraints. It is well known that there exists no additional space for control input due to physical limits, which motivates finite-time SPP to enlarge its boundaries to guarantee the above two constraints. In contrast to [6, 15, 32, 34, 38, 59, 68], finite-time SPP removes the assumption in these studies that the above two constraints are always required to be satisfied. As discussed above, these existing control strategies suffer from singularity problem once the tracking error violates the performance constraints since the classical PPFs are not defined there. In this study, finite-time SPP provides an effective solution, by enlarging or recovering the performance boundaries in the light of different input saturation conditions, to guarantee both two constraints even under unknown external disturbances.

6.3.4 *State Observer and Nonlinear Disturbance Observer*

We introduce the coordinate transformation as $\xi_{i,j_i} = x_{i,j_i}\prod_{k=j_i}^{n_i} g_{i,k}^{-1}$ and rewrite nonlinear systems (6.78) as

$$\begin{cases} \dot{\xi}_{i,j_i} = \bar{f}_{i,j_i}(\bar{\xi}_{i,j_i}) + \xi_{i,j_i+1} + \bar{d}_{i,j_i} \\ \dot{\xi}_{i,n_i} = \bar{f}_{i,n_i}(\bar{\xi}) + u_i(t) + \bar{d}_{i,n_i} \end{cases} \tag{6.87}$$

where $\bar{\xi}_{i,j_i} = [\xi_{i,1}, \ldots, \xi_{i,j_i}]^T$, $\bar{\xi} = \mathrm{col}(\xi_{1,n_1}, \ldots, \xi_{m,n_m})$, $\bar{f}_{i,j_i}(\cdot) = f_{i,j_i}(\cdot)\prod_{k=j_i}^{n_j} g_{i,k}^{-1}$ and $\bar{d}_{i,j_i} = d_{i,j_i}\prod_{k=j_i}^{n_j} g_{i,k}^{-1}$ with $j_i \in \mathbb{I}_{1:n_i}$. Based on Lemma 3.2, we have $\bar{f}_{i,j_i}(\cdot) = \theta_{i,j_i}\varphi_{i,j_i}(\bar{\xi}_{i,j_i}) + \varepsilon_{i,j_i}$. Then, establish the fuzzy state observer to approximate the unavailable transformed state ξ_{i,j_i}

$$\begin{cases} \dot{\hat{\xi}}_{i,j_i} = -k_{i,j_i}\hat{\xi}_{i,1} + \hat{\bar{f}}_{i,j_i}(\hat{\bar{\xi}}_{i,j_i}) + \hat{\xi}_{i,j_i+1} + \hat{\bar{d}}_{i,j_i} \\ \dot{\hat{\xi}}_{i,n_i} = -k_{i,n_i}\hat{\xi}_{i,1} + \hat{\bar{f}}_{i,n_i}(\hat{\bar{\xi}}) + u_i + \hat{\bar{d}}_{i,n_i} \end{cases} \tag{6.88}$$

where $\hat{\xi}_{i,j_i}$ and $\hat{\bar{d}}_{i,j_i}$ denote the estimated state and disturbance, respectively, $\hat{\bar{f}}_{i,j_i}(\hat{\bar{\xi}}_{i,j_i}) = \hat{\theta}_{i,j_i}\varphi_{i,j_i}(\hat{\bar{\xi}}_{i,j_i})$ with $\hat{\theta}_{i,j_i}$ being the estimated weight. Let $\xi_i = \mathrm{col}(\xi_{i,1},\ldots,\xi_{i,n_i})$, $\hat{\xi}_i = \mathrm{col}(\hat{\xi}_{i,1},\ldots,\hat{\xi}_{i,n_i})$, $\bar{d}_i = \mathrm{col}(\bar{d}_{i,1},\ldots,\bar{d}_{i,n_i})$ and $\hat{\bar{d}}_i = \mathrm{col}(\hat{\bar{d}}_{i,1},\ldots,\hat{\bar{d}}_{i,n_i})$. The dynamics of estimation error $\tilde{\xi}_i = \xi_i - \hat{\xi}_i$ is

$$\dot{\tilde{\xi}}_i = A_i\tilde{\xi}_i + K_i\xi_{i,1} + f_i - \hat{f}_i + \tilde{\bar{d}}_i, \tag{6.89}$$

where $L_{i,k} = [0,\ldots 1,\ldots 0]^T$, $L_{i,n_i} = [0,\ldots,1]^T$, $f_i = \sum_{k=1}^{n_i} L_{i,k}\bar{f}_{i,k}(\bar{\xi}_{i,k})$, $\hat{f}_i = \sum_{k=1}^{n_i} L_{i,k}\hat{\bar{f}}_{i,k}(\hat{\bar{\xi}}_{i,k})$, and $\tilde{\bar{d}}_i = \bar{d}_i - \hat{\bar{d}}_i$. A_i and K_i are

$$A_i = \begin{bmatrix} -k_{i,1} & & & \\ \vdots & & I_{n_i-1} & \\ -k_{i,n_i} & 0 & \cdots & 0 \end{bmatrix},\quad K_i = \begin{bmatrix} k_{i,1} \\ \vdots \\ k_{i,n_i} \end{bmatrix}, \tag{6.90}$$

where k_{i,j_i} is chosen such that the polynomial $p(s) = s^{n_i} + k_{i,1}s^{n_i-1} + \cdots + k_{i,n_i-1}s + k_{i,n_i}$ to be Hurwitz. And for any positive matrix $P = P^T$, there exists a positive definite matrix $Q = Q^T$ such that $A^TP + PA = -Q$.

To estimate the external disturbance $\bar{d}_{i,j_i}$, $j_i \in \mathbb{I}_{1:n_i}$, we define an auxiliary state $h_{i,j_i} = \bar{d}_{i,j_i} - k_{d,j_i}\xi_{i,j_i}$ with k_{d,j_i} a positive constant. Considering (6.87), the time derivative of h_{i,j_i}, $i_i \in \mathbb{I}_{1:n_i-1}$ is $\dot{h}_{i,j_i} = \dot{\bar{d}}_{i,j_i} - k_{d,j_i}(\bar{f}_{i,j_i}(\bar{\xi}_{i,j_i}) + \xi_{i,j_i+1} + h_{i,j_i} + k_{d,j_i}\xi_{i,j_i})$. The estimation of h_{i,j_i} is given by $\dot{\hat{h}}_{i,j_i} = -k_{d,j_i}(\hat{\bar{f}}_{i,j_i}(\hat{\bar{\xi}}_{i,j_i}) + \hat{\xi}_{i,j_i+1} + \hat{h}_{i,j_i} + k_{d,j_i}\hat{\xi}_{i,j_i})$. Define $\tilde{h}_{i,j_i} = h_{i,j_i} - \hat{h}_{i,j_i}$, we have

$$\begin{aligned} \dot{\tilde{h}}_{i,j_i} = & -k_{d,j_i}\tilde{h}_{i,j_i} + \dot{\bar{d}}_{i,j_i} - k_{d,j_i}(\bar{f}_{i,j_i} - \hat{\bar{f}}_{i,j_i}) \\ & - k_{d,j_i}(\tilde{\xi}_{i,j_i+1} + k_{d,j_i}\tilde{\xi}_{i,j_i}). \end{aligned} \tag{6.91}$$

For h_{i,n_i}, we have $\dot{h}_{i,n_i} = \dot{\bar{d}}_{i,n_i} - k_{d,n_i}(\bar{f}_{i,n_i}(\bar{\xi}) + u_i + h_{i,n_i} + k_{d,j_i}\xi_{i,n_i})$. The estimation of h_{i,n_i} is $\dot{\hat{h}}_{i,n_i} = -k_{d,n_i}(\hat{\bar{f}}_{i,n_i}(\hat{\bar{\xi}}) + u_i + \hat{h}_{i,n_i} + k_{d,j_i}\hat{\xi}_{i,n_i})$. The time derivative of $\tilde{h}_{i,n_i} = h_{i,n_i} - \hat{h}_{i,n_i}$ is given

$$\dot{\tilde{h}}_{i,n_i} = \dot{\bar{d}}_{i,n_i} - k_{d,n_i}(\tilde{h}_{i,n_i} + \bar{f}_{i,n_i} - \hat{\bar{f}}_{i,n_i} + k_{d,n_i}\tilde{\xi}_{i,n_i}). \tag{6.92}$$

Define the Lyapunov candidate function

$$V_0 = \tilde{\xi}_i^T P\tilde{\xi}_i + \frac{1}{2}\sum_{k=1}^{n_i}\tilde{h}_{i,k}^2. \tag{6.93}$$

According to Assumption 6.3, its time derivative is

$$\dot{V}_0 \leq -\lambda_0\|\tilde{\xi}_i\|^2 - \sum_{k=1}^{n_i}\lambda_1\tilde{h}_{i,k}^2 + \sum_{k=1}^{n_i}\lambda_2\|\tilde{\theta}_{i,k}\|^2 + \xi_{i,1}^2 + \epsilon_0 \tag{6.94}$$

where $\lambda_0 = \lambda_{\min}(Q) - \|PK_i\|^2 - 2\|P\|^2 - l_i\|P\|^2 - \frac{5}{2}\bar{k}_d^2 - \frac{1}{2}\bar{k}_d^4$, $\lambda_1 = k_{d,k} - 4 - \frac{l_i}{2}$, $\lambda_2 = 1 + \frac{k_{d,k}^2}{2}$, $\epsilon_0 = \frac{1}{2}\sum_{k=1}^{n_i} k_{d,k}^2\varepsilon_{i,k}^2 + \sum_{k=1}^{n_i}\varepsilon_{i,k}^2 + \frac{1}{2}\sum_{k=1}^{n_i}\bar{\dot{D}}_{i,k}^2$, l_i denotes the node number of FLSs and $\bar{k}_d = \max(k_{d,j_i})$ for $j_i \in \mathbb{I}_{1:n_i}$.

Remark 6.12 Time-varying external disturbances widely exists in most practical systems. To effectively improve the robustness of the nonlinear systems against unknown disturbances, nonlinear disturbance observer is a straightforward method to tackle this problem. From the nonlinear disturbance observer design, it indicates that if $\tilde{h}_{i,j_i}$ and $\tilde{\xi}_{i,j_i}$ are guaranteed to be bounded, so does $\tilde{\tilde{d}}_{i,j_i}$ which guarantees the system robustness. Moreover, such disturbance observer can be also extended to various conditions since many other uncertainties can be added in the disturbance item. Therefore, it is significant from both theoretical and practical viewpoints.

6.3.5 Auxiliary System Design

Before the control development, we construct an auxiliary system to generate the expected signals $\eta_{1,l}$ and $\eta_{1,u}$ in (6.86). Under the constraint (6.84) and the transformation (6.86), we define the following variables $\Lambda = \partial z_{i,1}/\partial e_{i,1}$, $\Lambda_l = \partial z_{i,1}/\partial E_{i,l}$ and $\Lambda_u = \partial z_{i,1}/\partial E_{i,u}$. As a result, we build the auxiliary system as

$$\begin{cases} \dot{\eta}_{1,u} = -\Lambda_u^{\text{inv}}\Lambda(-\Gamma_{1,u}\eta_{1,u} + \eta_{2,u}), \\ \dot{\eta}_{j_i,u} = -\Gamma_{j_i,u}\eta_{j_i,u} + \eta_{j_i+1,u}, \ j_i \in \mathbb{I}_{2:n_i}, \\ \dot{\eta}_{1,l} = \Lambda_l^{\text{inv}}\Lambda(-\Gamma_{1,l}\eta_{1,l} + \eta_{2,l}), \\ \dot{\eta}_{j_i,l} = -\Gamma_{j_i,l}\eta_{j_i,l} + \eta_{j_i+1,l}, \ j_i \in \mathbb{I}_{2:n_i}, \end{cases} \tag{6.95}$$

where $\eta_{n_i+1,u} = (\frac{3}{2} - \frac{1}{2}\text{sign}(v_i - u_i))|v_i - u_i|$ and $\eta_{n_i+1,l} = (\frac{3}{2} + \frac{1}{2}\text{sign}(v_i - u_i))$ $|v_i - u_i|$ are associated to the input saturation error, $\Gamma_{j_i,u}$ and $\Gamma_{j_i,l}$ as positive definite matrices. We set the initial system states as $\eta_{j_i,u}(0) = 0$ and $\eta_{j_i,l}(0) = 0$.

According to the auxiliary system design, it establishes a link between input and performance constraints. The modified signals are related to the input saturation error, where $\eta_{1,u}$ and $\eta_{1,l}$ are finally added into the performance boundaries. Then, a flexible performance called Finite-time SPP can be obtained even under strict input constraint. This novel feedback mechanism is analyzed as follows. Due to the unknown control directions of the MIMO nonlinear systems, both $\eta_{n_i+1,l}$ and $\eta_{n_i+1,u}$ are updated based on the condition of the input saturation error, which provide the unreachable increasing or decreasing rates for x_{i,n_i}. Similarly, the generated signals $\eta_{j_i+1,u}$ and $\eta_{j_i+1,l}$ are used to compensate for the saturation effect in x_{i,j_i} with

$j_i \in \mathbb{I}_{2:n_i}$. Subsequently, the nonnegative signals $\eta_{1,u}$ and $\eta_{1,l}$ are obtained to offer unattainable performance boundaries for the tracking error. Therefore, the auxiliary system establishes the connection between the performance boundaries and the input saturation, naturally. In the case of input saturation, these nonnegative signals are updated for enlarging the performance constraints, namely, to relax the performance constraints temporarily. When no saturation, $\eta_{n_i+1,u}$ and $\eta_{n_i+1,l}$ are equal to zero, and then $\eta_{1,u}$ and $\eta_{1,l}$ are fast driven to zero, which means that the enlarged boundaries recover back to the predefined FTPP. Overall, the proposed auxiliary system has the following property.

Theorem 6.3 *Consider the nonlinear systems* (6.78) *under the Finite-time SPP* (6.86) *and the ETF* (6.86). *The nonnegative signals in Finite-time SPP are provided by the proposed system* (6.95). *If the Finite-time SPP constraints* (6.84) *hold and there is a positive constant $\bar{u}$ having $|u_i - v_i| \le \bar{u}$, we can have $0 \le \eta_{j_i,u}, \eta_{j_i,l} \le k_{j_i,ul}\bar{u}$ with $k_{j_i,ul} = \max(\bar{u}\prod_{j_i}^{n_i}\Gamma_{j_i,u}^{inv}, \bar{u}\prod_{j_i}^{n_i}\Gamma_{j_i,l}^{inv})$.*

Proof We set the initial auxiliary system states as $\eta_{j_i,u}(0) = \eta_{j_i,l}(0) = 0$, $j_i \in \mathbb{I}_{1:n_i}$. $\Gamma_{j_i,u}$ and $\Gamma_{j_i,l}$ are positive matrices. According to the definitions of $\eta_{n_i+1,u}$ and $\eta_{n_i+1,l}$ in (6.95), we always have $\eta_{n_i+1,u}, \eta_{n_i+1,l} \ge 0$. Based on Lemma 2.6, one has $\eta_{n_i,u}, \eta_{n_i,l} \ge 0$. Upon the inequality $|u_i - v_i| \le \bar{u}$, if $\eta_{n_i,u} > \Gamma_{n_i,u}^{\text{inv}}\bar{u}$, we get $\dot{\eta}_{n_i,u} < 0$, which indicates $0 \le \eta_{n_i,u} \le \Gamma_{n_i,u}^{\text{inv}}\bar{u}$. Similarly, we have $0 \le \eta_{n_i,l} \le \Gamma_{n_i,l}^{\text{inv}}\bar{u}$. It can be inductively concluded that $0 \le \eta_{j_i,u} \le \bar{u}\prod_{j_i}^{n_i}\Gamma_{j_i,u}^{\text{inv}}$ and $0 \le \eta_{j_i,l} \le \bar{u}\prod_{j_i}^{n_i}\Gamma_{j_i,l}^{\text{inv}}$ for $j_i \in \mathbb{I}_{2:n_i-1}$.

Therefore, $\eta_{2,u}$ and $\eta_{2,l}$ satisfy $0 \le \eta_{2,u} \le \bar{u}\prod_{j_i=2}^{n_i}\Gamma_{j_i,u}^{\text{inv}}$ and $0 \le \eta_{2,l} \le \bar{u}\prod_{j_i=2}^{n_i}\Gamma_{j_i,l}^{\text{inv}}$. As the Finite-time SPP constraints (6.84) hold, based on the ETF (6.86) and Lemma 2.6, $\Lambda_u^{\text{inv}}\Lambda\Gamma_{1,u}$ and $-\Lambda_l^{\text{inv}}\Lambda\Gamma_{1,l}$ are Metzler matrices such that $\eta_{1,u}, \eta_{1,l} \ge 0$. Likewise, we have $\eta_{1,u} \le \bar{u}\prod_{j_i=1}^{n_i}\Gamma_{j_i,u}^{\text{inv}}$ and $\eta_{1,l} \le \bar{u}\prod_{j_i=1}^{n_i}\Gamma_{j_i,l}^{\text{inv}}$. By further considering no saturation scenario, namely, $v_i = u_i$, the auxiliary system states $\eta_{j_i,u}$ and $\eta_{j_i,l}$, $j_i \in \mathbb{I}_{1:n_i}$ will fast converge to zero and we omit it here for conciseness. □

Remark 6.13 Based on the absolute value of input saturation error, the auxiliary system establishes a connection between input saturation and performance constraint. The sizes of its generated signals $\eta_{1,u}$ and $\eta_{1,l}$ determines the level to which the performance boundaries are enlarged. In this way, it can not only improve the robustness of control strategy against external disturbances, but also achieves high tracking accuracy with less the control conservatism. Although some great studies have been reported for nonlinear systems under known control directions in [21, 60], how to tackle MIMO nonlinear system with unknown control directions is still under exploration and makes the control design even more challenging.

Remark 6.14 It is worthy noting that not only external disturbances can degrade the tracking performance, but also actuator faults, time delay, input quantization, etc. In this way, the proposed auxiliary system as a baseline technique can be easily extended to many existing control schemes, to handle the nonlinear systems simultaneously having input saturation and performance constraints. Therefore, this strategy may find immediate development in various practical applications and theories.

6.3.6 Nussbaum-Based Saturation-Tolerant Prescribed Control

To showcase the control diagram, the overall scheme is presented in Fig. 6.10. Define the tracking error variables

$$z_{i,1} = \ln\left(\frac{E_{i,l} + e_{i,1}}{E_{i,u} - e_{i,1}}\right), \tag{6.96}$$

$$z_{i,j_i} = \eta_{j_i,l} - \eta_{j_i,u} - \alpha^f_{i,j_i} + \hat{\xi}_{i,j_i}, \ j_i \in \mathbb{I}_{2:n_i} \tag{6.97}$$

$$\omega_{i,j_i} = \alpha^f_{i,j_i} - \alpha_{i,j_i-1}, \ j_i \in \mathbb{I}_{2:n_i}, \tag{6.98}$$

where α_{i,j_i} is the input and α^f_{i,j_i} denotes the output of the first-order filter to be defined later. The design process is developed.

Step 1: Considering the definitions of $z_{i,1}$, $z_{i,2}$, $\omega_{i,2}$ and $\tilde{\xi}_{i,2}$, and differentiating $z_{i,1}$ yields,

$$\dot{z}_{i,1} = \Lambda[\bar{g}_i(\alpha_{i,1} + \tilde{\xi}_{i,2} + z_{i,2} + \omega_{i,2}) + d_{i,1}] + F_1 + \Lambda(1 - \bar{g}_i)(\eta_{2,l} - \eta_{2,u}) \tag{6.99}$$

where $F_1 = \Lambda(f_{i,1}(\bar{x}_{i,1}) - \dot{y}_{i,d}) + \Lambda_u \dot{e}_{i,u} + \Lambda_l \dot{e}_{i,l} + \Lambda(\Gamma_{1,u}\eta_{1,u} - \Gamma_{1,l}\eta_{1,l})$, $\bar{g}_i = \prod_{k=1}^{n_i} g_{i,k}$. We give the virtual control input α_1 and adaptive law for $\hat{\theta}_{i,1}$ as

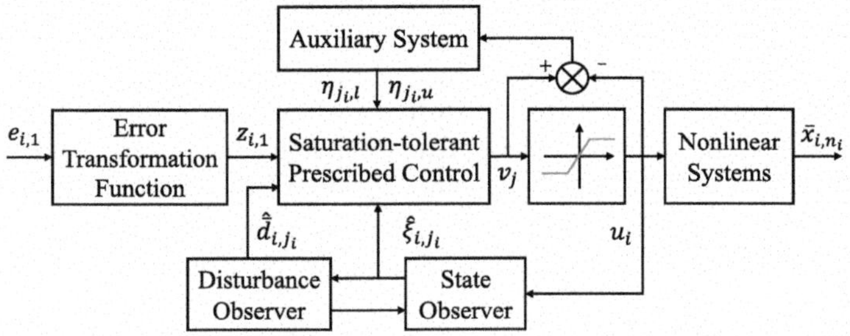

Fig. 6.10 The block diagram of Finite-time SPC

$$\alpha_{i,1} = N(\zeta)\bar{\alpha}_{i,1},\ \dot{\zeta} = \Lambda z_{i,1}\bar{\alpha}_{i,1},$$
$$\dot{\hat{\theta}}_{i,1} = \gamma_{i,1}(z_{i,1}\Lambda\varphi_{i,1}(\hat{\xi}_{i,1}) - \sigma_{i,1}\hat{\theta}_{i,1}), \tag{6.100}$$

where $\bar{\alpha}_{i,1} = \Lambda^{\mathrm{inv}}(c_{i,1}z_{i,1} + b_{i,1}z_{i,1}^{\beta} + F_1) + \dot{\hat{d}}_{i,1}$, $N(\zeta)$ denotes a Nussbaum-type function, and $\gamma_{i,1}, \sigma_{i,1} > 0$.

Establish the following Lyapunov candidate function

$$V_1 = V_0 + \frac{1}{2}z_{i,1}^2 + \frac{1}{2\gamma_{i,1}}\tilde{\theta}_{i,1}^T\tilde{\theta}_{i,1}. \tag{6.101}$$

Along (6.99) and (6.100), we can get

$$\begin{aligned}\dot{V}_1 \leq &\dot{V}_0 - c_{i,1}z_{i,1}^2 - b_{i,1}z_{i,1}^{\beta+1} - z_{i,1}\Lambda\dot{\hat{d}}_{i,1} + (\bar{g}_iN(\zeta) + 1)\dot{\zeta}\\ &+ z_{i,1}\Lambda\bar{g}_i(\tilde{\xi}_{i,2} + z_{i,2} + \omega_{i,2}) + z_{i,1}\Lambda d_{i,1} + \tilde{\theta}_{i,1}^T(\sigma_{i,1}\hat{\theta}_{i,1}\\ &- z_{i,1}\Lambda\varphi_{i,1}(\hat{\xi}_{i,1})) + z_{i,1}\Lambda(1 - \bar{g}_i)(\eta_{2,l} - \eta_{2,u})\end{aligned} \tag{6.102}$$

From the nonlinear disturbance observer and Assumption 6.3, we get $\tilde{h}_{i,1} = \bar{d}_{i,1} - \hat{\bar{d}}_{i,1} - k_{d,1}\tilde{\xi}_{i,1}$, which yields

$$\begin{aligned}z_{i,1}\Lambda d_{i,1} - z_{i,1}\Lambda\dot{\hat{d}}_{i,1} &= (\bar{g}_i - 1)\Lambda z_{i,1}\bar{d}_{i,1} + z_{i,1}\Lambda(\tilde{h}_{i,1} + k_{d,1}\tilde{\xi}_{i,1})\\ &\leq \frac{3}{2}\Lambda^2z_{i,1}^2 + \frac{(\bar{g}_i - 1)^2}{2\bar{g}_i^2}\bar{D}_{i,1}^2 + \frac{1}{2}\tilde{h}_{i,1}^2 + \frac{1}{2}k_{d,1}^2\|\tilde{\xi}_i\|^2.\end{aligned} \tag{6.103}$$

Based on Young's inequality, it yields

$$\Lambda z_{i,1}\bar{g}_i\tilde{\xi}_{i,2} \leq \frac{1}{2}\Lambda^2\bar{g}_i^2z_{i,1}^2 + \frac{1}{2}\|\tilde{\xi}_i\|^2 \tag{6.104}$$
$$\Lambda z_{i,1}\bar{g}_iz_{i,2} \leq \frac{1}{2}\Lambda^2\bar{g}_i^2z_{i,1}^2 + \frac{1}{2}z_{i,2}^2 \tag{6.105}$$
$$\Lambda z_{i,1}\bar{g}_i\omega_{i,2} \leq \frac{1}{2}\Lambda^2\bar{g}_i^2z_{i,1}^2 + \frac{1}{2}\omega_{i,2}^2 \tag{6.106}$$

According to (6.87) and $e_{i,1} = x_{i,1} - y_{i,d}$, we get $\xi_{i,1}^2 = \frac{2}{\bar{g}_i^2}(e_{i,1}^2 + y_{i,d}^2)$, and based on Assumption 6.4 and (6.86), there is $\epsilon_2 > 0$ such that $e_{i,1}^2 \leq \epsilon_2$.

Substituting (6.103)–(6.106) into (6.102), we can further obtain

$$\begin{aligned}\dot{V}_1 \leq &- \bar{c}_{i,1}z_{i,1}^2 - b_{i,1}z_{i,1}^{\beta+1} - \bar{\lambda}_0\|\tilde{\xi}_i\|^2 - \sum_{k=1}^{n_i}\bar{\lambda}_1\tilde{h}_{i,k}^2 - \bar{\sigma}_{i,1}\|\tilde{\theta}_{i,1}\|^2\\ &+ \sum_{k=1}^{n_i}\lambda_2\|\tilde{\theta}_{i,k}\|^2 + \frac{1}{2}z_{i,2}^2 + \frac{1}{2}\omega_{i,2}^2 + (\bar{g}_iN(\zeta) + 1)\dot{\zeta} + \epsilon_{i,1}\end{aligned} \tag{6.107}$$

where $\bar{c}_{i,1} = c_{i,1} - 2\Lambda^2 - \frac{3}{2}\Lambda^2\bar{g}_i^2 - \frac{l_i}{2}\Lambda^2$, $\bar{\lambda}_0 = \lambda_0 - \frac{1}{2} - \frac{1}{2}k_{d,1}^2$, $\bar{\lambda}_1 = \lambda_1 - \frac{1}{2}$, $\bar{\sigma}_{i,1} = \frac{\sigma_{i,1}-1}{2}$ and $\epsilon_{i,1} = \epsilon_0 + \frac{(\bar{g}_i-1)^2}{2\bar{g}_i^2}\bar{D}_{i,1}^2 + \frac{2}{\bar{g}_i^2}(\epsilon_2 + y_{i,d}^2) + \frac{1}{2}(1-\bar{g}_i)^2(\eta_{2,l} - \eta_{2,u})^2 + \frac{\sigma_{i,1}}{2}\|\theta_{i,1}\|^2$.

To alleviate the computation burden in the backstepping procedure, let $\alpha_{i,1}$ be an input of a first-order filter

$$\delta_{i,2}\dot{\alpha}_{i,2}^f + \alpha_{i,2}^f = \alpha_{i,1},\ \alpha_{i,2}^f(0) = \alpha_{i,1}(0), \tag{6.108}$$

where $\delta_{i,2}$ is a positive constant.

Step 2: From (6.95), (6.97), (6.98) and (6.108), taking the time derivative of $z_{i,2}$, we have

$$\dot{z}_{i,2} = F_2 + z_{i,3} + \omega_{i,3} + \alpha_{i,2} \tag{6.109}$$

where $F_2 = -k_{i,2}\hat{\xi}_{i,1} + \Gamma_{2,u}\eta_{2,u} - \Gamma_{2,l}\eta_{2,l} + \frac{\omega_{i,2}}{\delta_{i,2}} + \hat{\bar{f}}_{i,2}(\hat{\bar{\xi}}_{i,2}) + \hat{\bar{d}}_{i,2}$. We define the virtual control input $\alpha_{i,2}$ and the adaptive law $\theta_{i,2}$

$$\begin{aligned}\alpha_{i,2} &= -c_{i,2}z_{i,2} - b_{i,2}z_{i,2}^{\beta} - F_2,\\ \dot{\hat{\theta}}_{i,2} &= \gamma_{i,2}(z_{i,2}\varphi_{i,2}(\hat{\bar{\xi}}_{i,2}) - \sigma_{i,2}\hat{\theta}_{i,2}),\end{aligned} \tag{6.110}$$

where $c_{i,2}, b_{i,2}, \gamma_{i,2}$ and $\sigma_{i,2}$ are positive constants. Referring to (6.98) and (6.108), we have $\dot{\omega}_{i,2} = -\frac{1}{\delta_{i,2}}\omega_{i,2} - \dot{\alpha}_{i,1}$. And there is $\epsilon_{\alpha,1} > 0$ such that $\|\dot{\alpha}_{i,1}\|^2 \le \epsilon_{\alpha,1}$. Consider the Lyapunov function V_2

$$V_2 = V_{i,1} + \frac{1}{2}z_{i,2}^2 + \frac{1}{2\gamma_{i,2}}\tilde{\theta}_{i,2}^T\tilde{\theta}_{i,2} + \frac{1}{2}\omega_{i,2}^2. \tag{6.111}$$

Substituting (6.110) into (6.111), we have

$$\begin{aligned}\dot{V}_2 \le &-\sum_{k=1}^{2}\bar{c}_{i,k}z_{i,k}^2 - \sum_{k=1}^{2}b_{i,k}|z_{i,k}|^{\beta+1} - \bar{\lambda}_0\|\tilde{\xi}_i\|^2 + \frac{1}{2}\omega_{i,3}^2\\ &-\sum_{k=1}^{n_i}\bar{\lambda}_1\tilde{h}_{i,k}^2 - \bar{\delta}_{i,2}\omega_{i,2}^2 - \sum_{k=1}^{2}\bar{\sigma}_{i,k}\|\tilde{\theta}_{i,k}\|^2 + \frac{1}{2}z_{i,3}^2\\ &+\sum_{k=1}^{n_i}\lambda_2\|\tilde{\theta}_{i,k}\|^2 + (\bar{g}_iN(\zeta)+1)\dot{\zeta} + \epsilon_{i,2},\end{aligned} \tag{6.112}$$

where $\bar{c}_{i,2} = c_{i,2} - \frac{3}{2} - \frac{l_i}{2}$, $\bar{\sigma}_{i,2} = \frac{\sigma_{i,2}-1}{2}$, $\bar{\delta}_{i,2} = \frac{1}{\delta_{i,2}} - 1$, and $\epsilon_{i,2} = \frac{\sigma_{i,2}}{2}\|\theta_{i,2}\|^2 + \frac{1}{2}\epsilon_{\alpha,1} + \epsilon_{i,1}$. Let $\alpha_{j,2}$ through the following filter

$$\delta_{i,3}\dot{\alpha}_{i,3}^{f}+\alpha_{i,3}^{f}=\alpha_{i,2},\ \alpha_{i,3}^{f}(0)=\alpha_{i,3}(0) \tag{6.113}$$

with $\delta_{j,3}$ being a positive constant.

Step j_i $(3\leq j_i\leq n_i-1)$**:** Recalling (6.97) and (6.98), taking the time derivative of z_{i,j_i} yields

$$\dot{z}_{i,j_i}=F_{j_i}+\omega_{i,j_i+1}+z_{i,j_i+1}+\alpha_{i,j_i} \tag{6.114}$$

where $F_{j_i}=-k_{i,j_i}\hat{\xi}_{i,1}+\Gamma_{j_i,u}\eta_{j_i,u}-\Gamma_{j_i,l}\eta_{j_i,l}+\frac{\omega_{i,j_i}}{\delta_{i,j_i}}+\hat{f}_{i,j_i}(\hat{\bar{\xi}}_{i,j_i})+\hat{\bar{d}}_{i,j_i}$. Let the virtual control input and the adaptive law as

$$\begin{aligned}\alpha_{i,j_i}&=-c_{i,j_i}z_{i,j_i}-b_{i,j_i}z_{i,j_i}^{\beta}-F_{j_i},\\ \dot{\hat{\theta}}_{i,j_i}&=\gamma_{i,j_i}(z_{i,j_i}\varphi_{i,j_i}(\hat{\bar{\xi}}_{i,j_i})-\sigma_{i,j_i}\hat{\theta}_{i,j_i}),\end{aligned} \tag{6.115}$$

where c_{i,j_i}, b_{i,j_i}, γ_{i,j_i} and σ_{i,j_i} are positive constants. Likewise, we have $\dot{\omega}_{i,j_i}=-\frac{1}{\delta_{i,j_i}}\omega_{i,j_i}-\dot{\alpha}_{i,j_i-1}$ and there exists $\epsilon_{\alpha,j_i}>0$ such that$\|\dot{\alpha}_{i,j_i-1}\|^2\leq\epsilon_{\alpha,j_i-1}$. The Lyapunov function is defined

$$V_{j_i}=V_{i,j_i-1}+\frac{1}{2}z_{i,j_i}^2+\frac{1}{2\gamma_{i,j_i}}\tilde{\theta}_{i,j_i}^T\tilde{\theta}_{i,j_i}+\frac{1}{2}\omega_{i,j_i}^2. \tag{6.116}$$

In the light of (6.114) and (6.115), the time derivative of V_{j_i} is

$$\begin{aligned}\dot{V}_{j_i}\leq&\frac{1}{2}\omega_{i,j_i+1}^2-\sum_{k=1}^{j_i}b_{i,k}|z_{i,k}|^{\beta+1}+(\bar{g}_iN(\zeta)+1)\dot{\zeta}\\ &-\sum_{k=1}^{j_i}\bar{c}_{i,k}z_{i,k}^2-\sum_{k=1}^{j_i}\bar{\sigma}_{i,k}\|\tilde{\theta}_{i,k}\|^2-\sum_{k=2}^{j_i}\bar{\delta}_{i,k}\omega_{i,k}^2-\bar{\lambda}_0\|\tilde{\xi}_i\|^2\\ &+\frac{1}{2}z_{i,j_i+1}^2+\sum_{k=1}^{n_i}\lambda_2\|\tilde{\theta}_{i,k}\|^2-\sum_{k=1}^{n_i}\bar{\lambda}_1\tilde{h}_{i,k}^2+\epsilon_{i,j_i},\end{aligned} \tag{6.117}$$

where $\bar{c}_{i,j_i}=c_{i,j_i}-\frac{3}{2}-\frac{l_i}{2}$, $\bar{\sigma}_{i,j_i}=\frac{\sigma_{i,j_i}-1}{2}$, $\bar{\delta}_{i,j_i}=\frac{1}{\delta_{i,j_i}}-1$, and $\epsilon_{i,j_i}=\frac{\sigma_{i,j_i}}{2}\|\theta_{i,j_i}\|^2+\frac{1}{2}\epsilon_{\alpha,j_i-1}+\epsilon_{i,j_i}$. Let α_{j,j_i} through the following filter

$$\delta_{i,j_i+1}\dot{\alpha}_{i,j_i+1}^{f}+\alpha_{i,j_i+1}^{f}=\alpha_{i,j_i}, \tag{6.118}$$

where $\delta_{i,j_i+1}>0$ and the initial condition is $\alpha_{i,j_i+1}^{f}(0)=\alpha_{i,j_i+1}(0)$.

Step n_i**:** According to $\eta_{n_i+1,l}$ and $\eta_{n_i+1,u}$ defined in (6.95), we analyze how such auxiliary system compensates for the input saturation nonlinearities from the following three conditions. (i) If no input saturation occurs, $\eta_{n_i+1,l}=\eta_{n_i+1,u}=0$, we therefore have $u_i+\eta_{n_i+1,l}-\eta_{n_i+1,l}=v_i$. (ii) If the control input signal v_i is larger than the maximum signal, namely, $v_i \geq u_{i,\max} \geq u_i$. Then, we have $\eta_{n_i+1,u} = |v_i - u_i|$ and $\eta_{n_i+1,l} = 2|v_i - u_i|$. It yields $u_i + \eta_{n_i+1,l} - \eta_{n_i+1,u} = u_i + 2|v_i - u_i| - |v_i - u_i| = v_i$. (iii) If the control input signal v_i is smaller than the minimum signal, namely, $v_i \leq u_{i,\min} \leq u_i$. Then, we have $\eta_{n_i+1,u} = 2|v_i - u_i|$ and $\eta_{n_i+1,l} = |v_i - u_i|$. It yields $u_i + \eta_{n_i+1,l} - \eta_{n_i+1,u} = u_i + |v_i - u_i| - 2|v_i - u_i| = v_i$. Consequently, as discussed above, we can always have $v_i = u_i + \eta_{n_i+1,l} - \eta_{n_i+1,u}$. From the definition of z_{i,n_i} in (6.97), its time derivative can be therefore obtained as

$$\dot{z}_{i,n_i} = v_i + \Gamma_{n_i,u}\eta_{n_i,u} - \Gamma_{n_i,l}\eta_{n_i,l} + \frac{\omega_{i,n_i}}{\delta_{i,n_i}} - k_{i,n_i}\hat{\xi}_{i,1} + \hat{\bar{f}}_{i,n_i}(\hat{\bar{\xi}}) + \hat{\bar{d}}_{i,n_i}, \tag{6.119}$$

Develop the control input v_i and the adaptive law θ_{n_i}:

$$\begin{aligned} v_i &= -c_{i,n_i}z_{i,n_i} - b_{i,n_i}z_{i,n_i}^{\beta} - \Gamma_{n_i,u}\eta_{n_i,u} + \Gamma_{n_i,l}\eta_{n_i,l} \\ &\quad - \frac{\omega_{i,n_i}}{\delta_{i,n_i}} + k_{i,n_i}\hat{\xi}_{i,1} - \hat{\bar{f}}_{i,n_i}(\hat{\bar{\xi}}) - \hat{\bar{d}}_{i,n_i}, \\ \dot{\hat{\theta}}_{i,n_i} &= \gamma_{i,n_i}(z_{i,n_i}\varphi_{i,n_i}(\hat{\bar{\xi}}_{i,n_i}) - \sigma_{i,n_i}\hat{\theta}_{i,n_i}), \end{aligned} \tag{6.120}$$

where c_{i,n_i}, b_{i,n_i}, γ_{i,n_i} and σ_{i,n_i} are positive constants. For Lyapunov function V_{n_i} is constructed as

$$V_{i,n_i} = V_{i,n_i-1} + \frac{1}{2}z_{i,n_i}^2 + \frac{1}{2}\omega_{i,n_i}^2 + \frac{1}{2\gamma_{i,n_i}}\tilde{\theta}_{i,n_i}^T\tilde{\theta}_{i,n_i}. \tag{6.121}$$

Substituting (6.119) and (6.120) into V_{n_i}, we can get

$$\begin{aligned} \dot{V}_{i,n_i} \leq &-\sum_{k=1}^{n_i}\bar{c}_{i,k}z_{i,k}^2 - \sum_{k=1}^{n_i}b_{i,k}|z_{i,k}|^{\beta+1} - \sum_{k=1}^{n_i}\bar{\lambda}_1\tilde{h}_{i,k}^2 \\ &- \sum_{k=1}^{n_i}\ell_{i,k}\|\tilde{\theta}_{i,k}\|^2 - \sum_{k=2}^{n_i}\bar{\delta}_{i,k}\omega_{i,k}^2 - \bar{\lambda}_0\|\tilde{\xi}_i\|^2 + (\bar{g}_iN(\zeta)+1)\dot{\zeta} + \epsilon_{i,n_i}, \end{aligned} \tag{6.122}$$

where $\bar{c}_{i,n_i} = c_{i,n_i} - \frac{l_i}{2} - \frac{1}{2}$, $\bar{\sigma}_{i,n_i} = \frac{\sigma_{i,n_i}-1}{2}$, $\bar{\delta}_{i,n_i} = \frac{1}{\delta_{i,n_i}} - 1$, $\ell_{i,j_i} = \bar{\sigma}_{i,j_i} - \lambda_2$, $j_i \in \mathbb{I}_{1:n_i}$, and $\epsilon_{i,n_i} = \epsilon_{i,n_i-1} + \frac{1}{2}\dot{\alpha}_{i,n_i-1}^2 + \frac{1}{2}\|\theta_{i,n_i}\|^2 + \frac{1}{2}\bar{u}^2$. According to Lemma 2.2, one obtains

$$-\|\tilde{h}_{i,k}\|^2 \leq \frac{1-\beta}{2} - \frac{1-\beta}{2}\|\tilde{h}_{i,k}\|^2 - \|\tilde{h}_{i,k}\|^{\beta+1}, \tag{6.123}$$

Similarly, other terms like $\tilde{\theta}_{i,k}$ and $\omega_{i,k}$ follow the same procedure. According to Lemma 2.4, we rewrite (6.122) as

$$\dot{V}_{i,n_i} \leq -\lambda_3 V_{i,n_i} - \lambda_4 V_{i,n_i}^{\bar{\beta}} + (\bar{g}_i N(\zeta) + 1)\dot{\zeta} + \epsilon \tag{6.124}$$

where $\lambda_3 = \min(2\bar{c}_{i,k}, \bar{\lambda}_1(1-\beta), \ell_{i,k}(1-\beta), \bar{\delta}_{i,k}(1-\beta), \bar{\lambda}_0(1-\beta))$, $\bar{\beta} = \frac{1+\beta}{2}$, $\lambda_4 = \min(2^{\bar{\beta}} b_{i,k}, 2^{\bar{\beta}}\bar{\lambda}_1, 2^{\bar{\beta}}\ell_{i,k}, 2^{\bar{\beta}}\bar{\delta}_{i,k}, 2^{\bar{\beta}}\bar{\lambda}_0)$, and $\epsilon = \epsilon_{i,n_i} + \frac{1-\beta}{2}(\bar{\lambda}_1 + \sum_{k=1}^{n_i} \ell_{i,k} + \sum_{k=1}^{n_i} \bar{\delta}_{i,k} + \bar{\lambda}_0)$.

The following theorem is given for the SPC of the nonlinear systems (6.78) subject to unknown control directions and disturbances.

Theorem 6.4 *Consider the nonlinear systems* (6.78) *under Assumption* 6.4*, fuzzy state observer* (6.88)*, and NDO* (6.91) *and* (6.92)*. The ETF is given by* (6.86)*, and Finite-time SPP is specified as* (6.84) *with modification signals generated by the auxiliary system* (6.95)*. Under the development of SPC* (6.100), (6.110), (6.115), (6.120)*, the closed-loop system satisfies:*

(i) The solution of the closed-loop system is SGPFS;
(ii) The tracking error $e_{i,1}(t)$ *into a prescribed-set* (6.82) *within finite time;*
(iii) The Finite-time SPP constraints (6.84) *are always guaranteed despite external disturbances and input saturation.*

Proof From the inequality (6.124), it can be known that there is $\epsilon_\zeta > 0$ such that $|\bar{g}_i N(\zeta) + 1| \leq \epsilon_\zeta$ according to Lemma 2.5. Let $\bar{\epsilon} = \epsilon + \epsilon_\zeta$, and the inequality (6.124) can be rewritten as

$$\begin{aligned} \dot{V}_{i,n_i} \leq & -\lambda_3 V_{i,n_i} - \lambda_4 V_{i,n_i}^{\bar{\beta}} + \bar{\epsilon} \\ \leq & -\lambda_3 V_{i,n_i} - \lambda_4 \lambda_5 V_{i,n_i}^{\bar{\beta}} - \lambda_4(1-\lambda_5) V_{i,n_i}^{\bar{\beta}} + \bar{\epsilon} \end{aligned} \tag{6.125}$$

where $0 < \lambda_5 < 1$. Define $\Omega = \{z_{i,j_i}, \omega_{i,j_i}, \tilde{\theta}_{i,j_i} | V_{i,n_j}^{\bar{\beta}} \leq \frac{\bar{\epsilon}}{\lambda_4(1-\lambda_5)}\}$ and $\bar{\Omega} = \{z_{i,j_i}, \omega_{i,j_i}, \tilde{\theta}_{i,j_i} | V_{i,n_i}^{\bar{\beta}} > \frac{\bar{\epsilon}}{\lambda_4(1-\lambda_5)}\}$.

Case 1: If $z_{i,j_i}, \omega_{i,j_i}, \tilde{\theta}_{i,j_i} \in \bar{\Omega}$, we have $-\lambda_4(1-\lambda_5)V_{i,n_i}^{\bar{\beta}} + \bar{\epsilon} < 0$. Therefore, the inequality (6.125) can be given by

$$\dot{V}_{i,n_i} \leq -\lambda_3 V_{i,n_i} - \lambda_4 \lambda_5 V_{i,n_i}^{\bar{\beta}}. \tag{6.126}$$

Dividing both side by $V_{n_i}^{\bar{\beta}}$ yields

$$\int_{t_0}^{t} \frac{V_{i,n_i}^{-\bar{\beta}}}{\lambda_4\lambda_5 + \lambda_3 V_{i,n_i}^{1-\bar{\beta}}} \mathrm{dV} \leq -\int_{t_0}^{t} \mathrm{dt}, \tag{6.127}$$

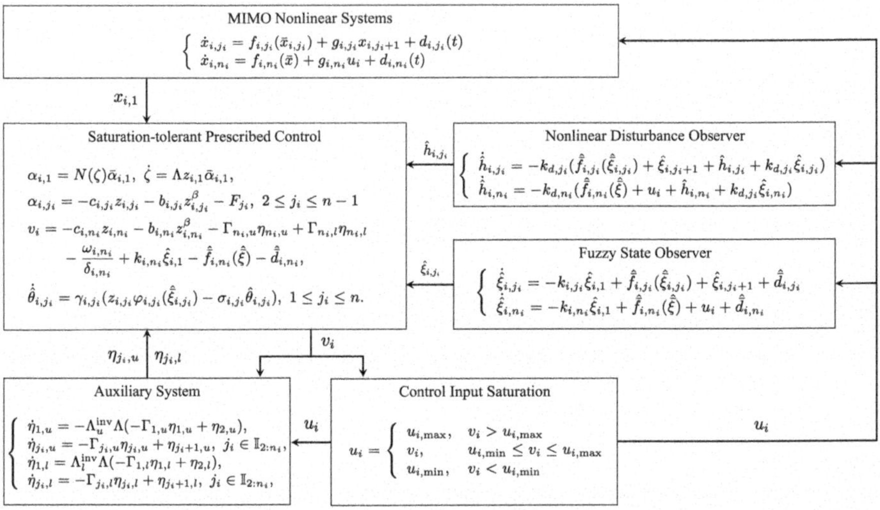

Fig. 6.11 The control procedure and signals

where t_0 as the initial time. Based on Lemma 2.1, the settling time is expressed by

$$T = t_0 + \frac{1}{\lambda_3(1-\bar{\beta})} \ln \left(\frac{\lambda_4\lambda_5 + \lambda_3 V_{i,n_i}^{1-\bar{\beta}}(t_0)}{\lambda_4\lambda_5} \right). \tag{6.128}$$

The above analysis indicates that the system states $z_{i,j_i}, \omega_{i,j_i}, \tilde{\theta}_{i,j_i} \in \Omega$ for $\forall t \geq T$.

Case 2: If $z_{i,j_i}, \omega_{i,j_i}, \tilde{\theta}_{i,j_i} \in \Omega$, z_{i,j_i} will remain within Ω.

Consequently, from Lemma 2.1, the SGPFS is guaranteed, with all the tracking errors to a residual set within finite time. Based on the Finite-time Tunnel Prescribed Performance (6.81) and the ETF (6.86), note that $e_{i,1}$ is guaranteed to a prescribed set (6.82) in t_f. Moreover, the bounded closed-loop system indicates that the Saturation-tolerant Prescribed Performance (6.84) is always satisfied. To clearly illustrate the control strategy, the details of the control procedure and signals are shown in Fig. 6.11. □

Remark 6.15 A guidance on selecting control parameters is provided. We divide these parameters into three categories: First, k_{i,j_i} and k_{d,j_i} in the state and disturbance observers influence the estimation performance. If we select large values, better estimations will be obtained. Note that their values will not lead to input saturation when estimations are acceptable. Second, the control parameters c_{i,j_i} and b_{i,j_i} determine the convergence rate and the tracking accuracy. Although larger values lead to faster convergence rate, it will increase the input saturation error which may degrade the tracking performance. Therefore, certain trade-off between convergence rate and tracking performance is normally necessary in practice. Moreover, larger values of

γ_{i,j_i} and σ_{i,j_i} in adaptive laws lead to faster estimation convergence but poor tracking accuracy. Third, for $\Gamma_{j_i,u}$ and $\Gamma_{j_i,l}$ in the auxiliary system, too larger values will generate smaller nonnegative signals. In this condition, Finite-time SPP cannot provide a large enough allowable set to guarantee both input and performance constraints. On the contrary, too small values will increase the control design conservatism. Therefore, their values belong to a balance product.

Remark 6.16 By introducing an auxiliary system and FTPP, the proposed control strategy provides a solution for nonlinear systems under unknown control directions and external disturbances to accomplish a balance between input saturation and performance requirements. Note that such trade-off strategy by introducing auxiliary system can be also extended to a more general nonlinear systems with time-varying or state-dependent control coefficients. The control design and stability analysis follow the same steps. However, we mainly focus on how to develop the auxiliary system for the underlying problem. Therefore, the control coefficients in this work are considered as unknown constants.

6.3.7 Numerical Simulations

Comparative simulations are carried out to illustrate the robustness of SPC. We take into account the tilting quadcopter dynamics

$$\begin{cases} \dot{x}_{i,1} = f_{i,1}(x_1) + g_{i,1}x_{i,2} + d_{i,1} \\ \dot{x}_{i,2} = f_{i,2}(\bar{x}) + g_{i,2}u_i + d_{i,2} \\ y_i = x_{i,1} \end{cases} \tag{6.129}$$

with $i \in \mathbb{I}_{1:3}$, $g_{i,1} = g_{i,2} = 1$, $d_{1,1} = 0.8\cos 05t$, $d_{2,1} = 0.3$, $d_{3,1} = 0.5\sin t$, $d_{1,2} = 15$, $d_{2,2} = 0.2\sin 1.5t$, $d_{3,2} = 5$. For the expressions of $f_{i,1}$ and $f_{i,2}$, please refer to [19]. Based on the above control development, we have

$$\begin{aligned} \alpha_{i,1} &= N(\zeta)\bar{\alpha}_{i,1},\ \dot{\zeta} = \Lambda z_{i,1}\bar{\alpha}_{i,1}, \\ v_i &= -c_{i,2}z_{i,2} - b_{i,2}z_{i,2}^{\beta} - \Gamma_{2,u}\eta_{2,u} + \Gamma_{2,l}\eta_{2,l} - \frac{\omega_{i,2}}{\delta_{i,2}} \\ &\quad + k_{i,2}\hat{\xi}_{i,1} - \hat{f}_{i,2}(\hat{\xi}) - \hat{d}_{i,2}, \\ \dot{\hat{\theta}}_{i,2} &= \gamma_{i,2}(z_{i,2}\varphi_{i,2}(\hat{\xi}_{i,2}) - \sigma_{i,2}\hat{\theta}_{i,2}), \end{aligned} \tag{6.130}$$

where $N(\cdot) = \zeta^2\cos\zeta$, $F_1 = \Lambda(f_{i,1}(\bar{x}_{i,1}) - \dot{y}_{i,d}) + \Lambda_u\dot{e}_{i,u} + \Lambda_l\dot{e}_{i,l} + \Lambda(\Gamma_{1,u}\eta_{1,u} - \Gamma_{1,l}\eta_{1,l})$, and $\bar{\alpha}_{i,1} = \Lambda^{\text{inv}}(c_{i,1}z_{i,1} + b_{i,1}z_{i,1}^{\beta} + F_1) + \hat{d}_{i,1}$. Considering the FTPP (6.82), SPP (6.84), state observer (6.88), NDO, auxiliary (6.95) and SPC (6.130), all parameters utilized in simulations are summarized as the following Table 6.2. Besides, all the initial conditions of systems are set to zero, the desired trajectories are $y_{1,d} = 0.9\cos(0.25t - 0.33)$, $y_{2,d} = 0.6\cos(0.39t)$ and $y_{3,d} = 0.8\cos(0.3t -$

Table 6.2 Parameters in simulations

Parameter	Value	Parameter	Value	Parameter	Value	Parameter	Value	Parameter	Value
$c_{i,1}$	5.0	$c_{i,2}$	4.0	$b_{i,1}$	0.5	$t_{f,2}$	3.0	$t_{f,3}$	5.0
$b_{i,2}$	0.5	β	0.75	$\gamma_{i,2}$	2.4	$t_{f,1}$	3.0	ρ_{1,t_f}	0.5
$\delta_{i,2}$	0.01	$\sigma_{i,2}$	2.0	$\Gamma_{1,u}$	2.0	ℓ_i	0.5	$\rho_{3,0}$	1.8
$\Gamma_{2,u}$	5.0	$\Gamma_{1,l}$	3.0	$\Gamma_{2,l}$	3.0	ρ_{3,t_f}	0.5	ρ_{2,t_f}	0.7
$k_{i,1}$	5.0	$k_{i,2}$	50.0	$k_{d,1}$	12.0	$k_{d,2}$	15	$\delta_{1,l}$	0.2
$\delta_{i,u}$	0.8	$\delta_{2,l}$	0.1	$\delta_{3,l}$	0.1	$\rho_{1,0}$	2.0	$\rho_{2,0}$	3.0

$0.42) + 0.3$, and the control boundaries are $u_{1,\max} = u_{3,\max} = 20$, $u_{2,\max} = 40$, $u_{1,\min} = u_{3,\min} = -20$ and $u_{2,\min} = -40$. To better show the advantages of the SPC, we employ the classical PPC (6.80) without the auxiliary system as a comparative one. The tracking results are shown in Figs. 6.12, 6.13, 6.15 and 6.16.

The comparative simulation results of the tracking error $e_{i,1}$, $i \in \mathbb{I}_{1:3}$ under SPC and PPC are shown in Fig. 6.12. We first analyze the tracking performance under the proposed SPC. From Fig. 6.12, all the tracking errors $e_{i,1}$, $i \in \mathbb{I}_{1:3}$ of SPC are bounded and always evolve within the allowable set. With the aid of tunnel-shape SPP, better transient performance (i.e., smaller overshoot) can be obtained, and the settling time for the tracking errors to the steady state boundary can be flexibly defined by users in advance instead of relying on the initial tracking conditions or the control parameters. Moreover, the simulation results in Fig. 6.12 also illustrate the robustness of SPC against external disturbances. The salient feature associated with SPC is that the SPP possess flexible performance boundaries in the light of different input saturation conditions. Therefore, the tracking error $e_{i,1}$ is always guaranteed within the allowable set even under input saturation and external disturbances.

Let us take the tracking error $e_{1,1}$ of SPC to show the mechanism behind this work. The others follow the same analysis process. Considering the existence of external disturbances, the tracking performance of $e_{1,1}$ is often seriously deteriorated (i.e., 0–2s and 11–14s), leading $e_{1,1}$ closed to the performance boundaries. Correspondingly, the control input is increasing in this condition but is constrained within a reasonable region which limits the tracking ability. As shown in Fig. 6.13, input saturation occurs during 0–2s and 11–14s. To guarantee both input and performance constraints, the auxiliary system is invoked during the aforementioned period to generate additional nonnegative signals based on the input saturation error as shown in Fig. 6.14. By introducing $\eta_{1,u,1}$ and $\eta_{1,l,1}$ into the performance functions, SPP possesses the ability to temporarily enlarge the performance boundaries to avoid the constraint violation as shown in Fig. 6.12. When there is no input saturation, these additional signals $\eta_{1,u,1}$ and $\eta_{1,l,1}$ are regulated to zero and SPP therefore recovers back to the original performance boundaries. In this way, the auxiliary system establishes a relationship between input and performance constraints to accomplish a balance strategy. Such control strategy not only improves the robustness against the external disturbance, but also achieves high tracking precision in less conservatism. The estimation errors

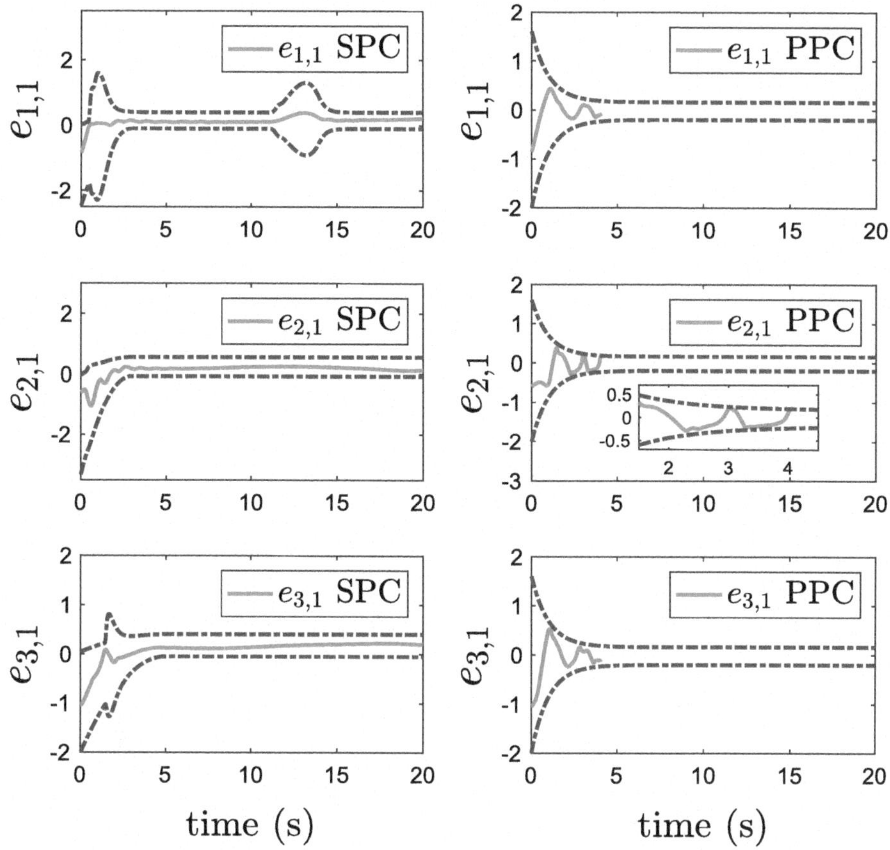

Fig. 6.12 Constraints on tracking errors $e_{1,1}$, $e_{2,1}$ and $e_{3,1}$

of fuzzy state and disturbance observer are plotted in Figs. 6.15 and 6.16, where all estimation errors are finite-time stability.

We then consider the tracking performance of the classical PPC. As shown in Fig. 6.12, its funnel-shape performance cannot impose strict constraints on the tracking trajectory since the loose allowable set of PPFs, where the unexpected overshoot and oscillation often exist. Specifically, we take $e_{2,1}$ of the PPC as an example. Due to the unexpected uncertainties, $e_{2,1}$ shows poor transient tracking performance in Fig. 6.12, which is always closed to the performance boundaries. In this condition, it can easily reach the input saturation condition as shown in Fig. 6.13, leading to limited tracking ability. Therefore, $e_{2,1}$ as shown in Fig. 6.12 violates the performance constraints after 4.11 s, resulting in control singularity and system instability problem since the classical PPC is undefined there. It indicates that the classical PPC cannot exhibit strong robustness against external disturbances, especially under input saturation.

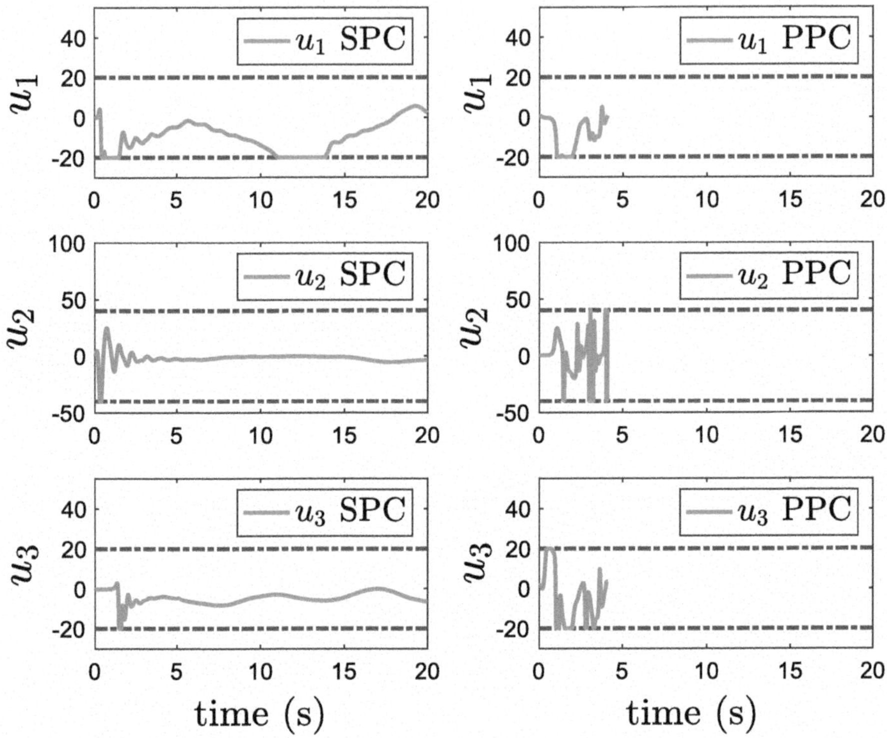

Fig. 6.13 The control inputs of u_1, u_2 and u_3

6.4 Conclusion

In this chapter, we address the tracking control problem for a more general MIMO nonlinear systems simultaneously considering input saturation and prescribed performance constraints. To tackle the trade-off between input and output constraints, two novel auxiliary systems have been established based on the input saturation error for nonlinear systems with known control directions and unknown control directions, respectively. By introducing the nonnegative signals generated by auxiliary systems, the resulted Saturation-tolerant Prescribed Performance has the ability of relaxing or recovering the performance boundaries according to the input saturation conditions. Specifically, when input saturation occurs, the Saturation-tolerant Prescribed Performance relaxes the boundaries to avoid the constraint violation; when the saturation disappears, the relaxed performance boundaries fast recover back to the original ones. It has been proven that the tracking errors are always confined in the Saturation-tolerant Prescribed Performance despite external disturbances, actuator faults and input saturation. Moreover, the tracking errors are regulated to the prescribed region within finite time under our proposed control schemes. In this way, we guarantee better tracking behavior with less control conservatism.

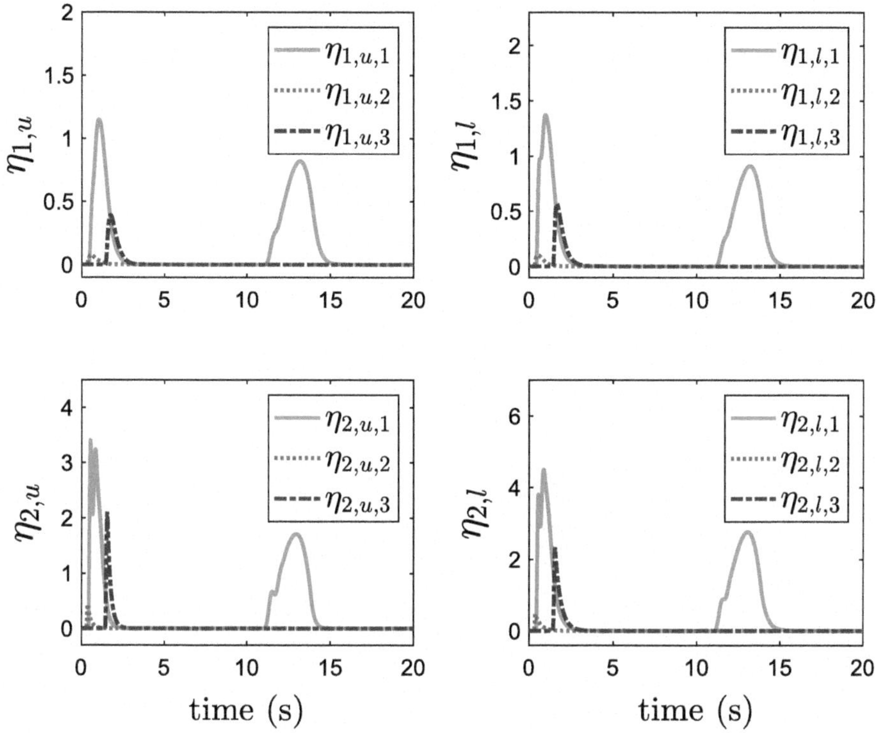

Fig. 6.14 The auxiliary system states

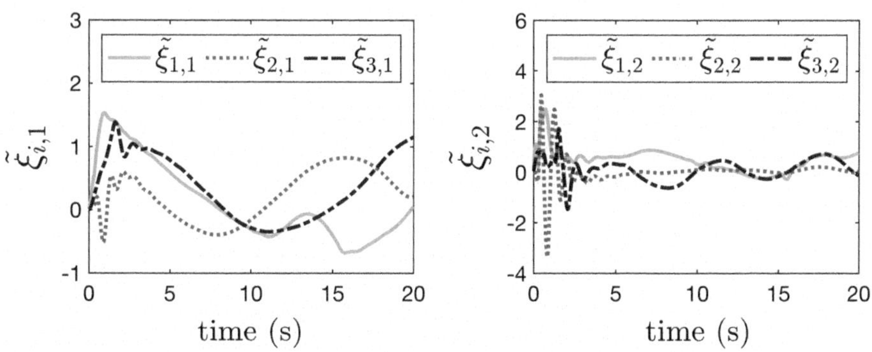

Fig. 6.15 The estimation errors of the fuzzy state observer

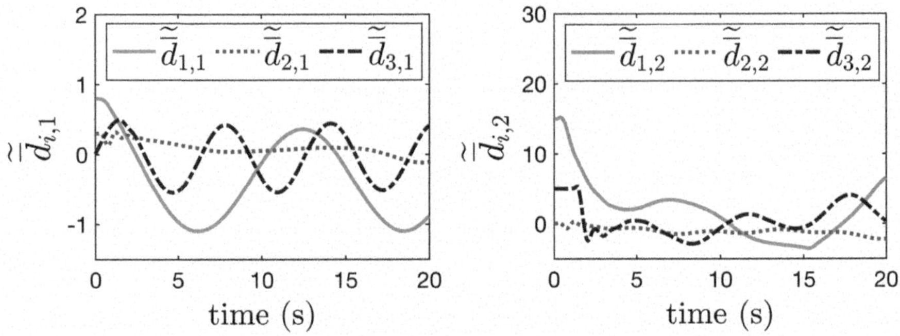

Fig. 6.16 The estimation errors of disturbance observer

References

1. Bechlioulis, C.P., Rovithakis, G.A.: Robust adaptive control of feedback linearizable mimo nonlinear systems with prescribed performance. IEEE Transactions on Automatic Control **53**(9), 2090–2099 (2008)
2. Bechlioulis, C.P., Rovithakis, G.A.: Robust partial-state feedback prescribed performance control of cascade systems with unknown nonlinearities. IEEE Trans. Autom. Control **56**(9), 2224–2230 (2011)
3. Cao, L., Li, H., Dong, G., Lu, R.: Event-triggered control for multiagent systems with sensor faults and input saturation. IEEE Trans. Syst., Man, Cybern.: Syst. (2019)
4. Chen, B., Liu, X.P., Ge, S.S., Lin, C.: Adaptive fuzzy control of a class of nonlinear systems by fuzzy approximation approach. IEEE Transactions on Fuzzy Systems **20**(6), 1012–1021 (2012)
5. Chen, M., Ge, S.S., Ren, B.: Adaptive tracking control of uncertain mimo nonlinear systems with input constraints. Automatica **47**(3), 452–465 (2011)
6. Chen, M., Shao, S., Jiang, B.: Adaptive neural control of uncertain nonlinear systems using disturbance observer. IEEE Transactions on Cybernetics **47**(10), 3110–3123 (2017)
7. Chen, M., Tao, G., Jiang, B.: Dynamic surface control using neural networks for a class of uncertain nonlinear systems with input saturation. IEEE transactions on neural networks and learning systems **26**(9), 2086–2097 (2014)
8. Chen, Z., Li, Z., Chen, C.P.: Adaptive neural control of uncertain mimo nonlinear systems with state and input constraints. IEEE Trans. Neural Netw. Learning Syst. **28**(6), 1318–1330 (2016)
9. Cheng, C., Zhang, Y., Liu, S.: Neural observer-based adaptive prescribed performance control for uncertain nonlinear systems with input saturation. Neurocomputing **370**, 94–103 (2019)
10. Cui, G., Yu, J., Shi, P.: Observer-based finite-time adaptive fuzzy control with prescribed performance for nonstrict-feedback nonlinear systems. IEEE Transactions on Fuzzy Systems (2020)
11. Dimanidis, I.S., Bechlioulis, C.P., Rovithakis, G.A.: Output feedback approximation-free prescribed performance tracking control for uncertain mimo nonlinear systems. IEEE Trans. Autom. Control **65**(12), 5058–5069 (2020)
12. Ding, S., Li, S., Zheng, W.X.: Nonsmooth stabilization of a class of nonlinear cascaded systems. Automatica **48**(10), 2597–2606 (2012)
13. Du, J., Hu, X., Krstić, M., Sun, Y.: Robust dynamic positioning of ships with disturbances under input saturation. Automatica **73**, 207–214 (2016)
14. Guo, Q., Zhang, Y., Celler, B.G., Su, S.W.: Neural adaptive backstepping control of a robotic manipulator with prescribed performance constraint. IEEE Trans. Neural Netw. Learning Syst. **30**(12), 3572–3583 (2018)

15. Hopfe, N., Ilchmann, A., Ryan, E.P.: Funnel control with saturation: Linear mimo systems. IEEE Transactions on Automatic Control **55**(2), 532–538 (2010)
16. Hu, C., Wang, Z., Qin, Y., Huang, Y., Wang, J., Wang, R.: Lane keeping control of autonomous vehicles with prescribed performance considering the rollover prevention and input saturation. IEEE transactions on intelligent transportation systems **21**(7), 3091–3103 (2019)
17. Huang, H., He, W., Li, J., Xu, B., Yang, C., Zhang, W.: Disturbance observer-based fault-tolerant control for robotic systems with guaranteed prescribed performance. IEEE Transactions on Cybernetics (2020)
18. Ji, R., Li, D., Ma, J., Ge, S.S.: Saturation-tolerant prescribed control of mimo systems with unknown control directions. IEEE Transactions on Fuzzy Systems pp. 1–1 (2022)
19. Ji, R., Ma, J., Li, D., Ge, S.S.: Finite-time adaptive output feedback control for mimo nonlinear systems with actuator faults and saturations. IEEE Transactions on Fuzzy Systems **29**(8), 2256 – 2270 (2021)
20. Ji, R., Ma, J., Sam Ge, S.: Modeling and control of a tilting quadcopter. IEEE Trans. Aerosp. Electro. Syst. **56**(4), 2823–2834 (2020). 10.1109/TAES.2019.2955525
21. Ji, R., Yang, B., Ma, J., Ge, S.S.: Saturation-tolerant prescribed control for a class of mimo nonlinear systems. IEEE Transactions on Cybernetics (2021)
22. Jing, Y.H., Yang, G.H.: Adaptive fuzzy output feedback fault-tolerant compensation for uncertain nonlinear systems with infinite number of time-varying actuator failures and full-state constraints. IEEE Transactions on Cybernetics (2019)
23. Li, H., Zhang, X., Feng, G.: Event-triggered output feedback control of switched nonlinear systems with input saturation. IEEE Trans. Cybern. (2020)
24. Li, L., Luo, H., Ding, S.X., Yang, Y., Peng, K.: Performance-based fault detection and fault-tolerant control for automatic control systems. Automatica **99**, 308–316 (2019)
25. Li, M., Liu, C., Li, T., Chen, C.P.: New second order sliding mode control design for course-keeping control of ship with input saturation. In: 2017 International Conference on Advanced Mechatronic Systems (ICAMechS), pp. 462–466. IEEE (2017)
26. Li, Y., Min, X., Tong, S.: Observer-based fuzzy adaptive inverse optimal output feedback control for uncertain nonlinear systems. IEEE Trans. Fuzzy Syst. (2020)
27. Li, Y., Tong, S.: Adaptive neural networks prescribed performance control design for switched interconnected uncertain nonlinear systems. IEEE transactions on neural networks and learning systems **29**(7), 3059–3068 (2017)
28. Liu, L., Liu, Y.J., Tong, S.: Neural networks-based adaptive finite-time fault-tolerant control for a class of strict-feedback switched nonlinear systems. IEEE Trans. Cybern. **49**(7), 2536–2545 (2018)
29. Lu, K., Xia, Y., Fu, M.: Controller design for rigid spacecraft attitude tracking with actuator saturation. Inf. Sci. **220**, 343–366 (2013)
30. Mehdifar, F., Bechlioulis, C.P., Hashemzadeh, F., Baradarannia, M.: Prescribed performance distance-based formation control of multi-agent systems. Automatica **119**, 109086 (2020)
31. Oliveira, T.R., Peixoto, A.J., Hsu, L.: Sliding mode control of uncertain multivariable nonlinear systems with unknown control direction via switching and monitoring function. IEEE Transactions on Automatic Control **55**(4), 1028–1034 (2010)
32. Salmanpour, Y., Arefi, M.M., Khayatian, A., Kaynak, O.: Event-triggered fuzzy adaptive leader-following tracking control of non-affine multi-agent systems with finite-time output constraint and input saturation. IEEE Transactions on Fuzzy Systems (2021)
33. Shao, X., Hu, Q., Shi, Y.: Adaptive pose control for spacecraft proximity operations with prescribed performance under spatial motion constraints. IEEE Trans. Contr. Syst. Technol. pp. 1–15 (2020)
34. Shao, X., Hu, Q., Shi, Y., Jiang, B.: Fault-tolerant prescribed performance attitude tracking control for spacecraft under input saturation. IEEE Transactions on Control Systems Technology **28**(2), 574–582 (2018)
35. Sui, S., Chen, C.P., Tong, S.: Finite-time adaptive fuzzy prescribed performance control for high-order stochastic nonlinear systems. IEEE Transactions on Fuzzy Systems (2021)

36. Sun, K., Liu, L., Qiu, J., Feng, G.: Fuzzy adaptive finite-time fault-tolerant control for strict-feedback nonlinear systems. IEEE Trans. Fuzzy Syst. (2020)
37. Sun, K., Qiu, J., Karimi, H.R., Fu, Y.: Event-triggered robust fuzzy adaptive finite-time control of nonlinear systems with prescribed performance. IEEE Transactions on Fuzzy Systems **29**(6), 1460–1471 (2020)
38. Sun, K., Qiu, J., Karimi, H.R., Gao, H.: A novel finite-time control for nonstrict feedback saturated nonlinear systems with tracking error constraint. IEEE Transactions on Systems, Man, and Cybernetics: Systems **51**(6), 3968–3979 (2019)
39. Tong, S., Huo, B., Li, Y.: Observer-based adaptive decentralized fuzzy fault-tolerant control of nonlinear large-scale systems with actuator failures. IEEE Trans. on Fuzzy Syst. **22**(1), 1–15 (2013)
40. Tong, S., Li, Y., Shi, P.: Observer-based adaptive fuzzy backstepping output feedback control of uncertain mimo pure-feedback nonlinear systems. IEEE Trans. Fuzzy Syst. **20**(4), 771–785 (2012)
41. Verginis, C.K., Nikou, A., Dimarogonas, D.V.: Robust formation control in se (3) for tree-graph structures with prescribed transient and steady state performance. Automatica **103**, 538–548 (2019)
42. Wang, C.C., Yang, G.H.: Neural network-based adaptive output feedback fault-tolerant control for nonlinear systems with prescribed performance. Neurocomputing **329**, 457–467 (2019)
43. Wang, F., Liu, Z., Li, X., Zhang, Y., Chen, C.P.: Observer-based finite time control of nonlinear systems with actuator failures. In. Sci. **500**, 1–14 (2019)
44. Wang, F., Zhang, X.: Adaptive finite time control of nonlinear systems under time-varying actuator failures. IEEE Trans. Syst., Man, Cybern.: Syst. (2018)
45. Wang, H., Liu, K., Liu, X., Chen, B., Lin, C.: Neural-based adaptive output-feedback control for a class of nonstrict-feedback stochastic nonlinear systems. IEEE Trans. Cybern. **45**(9), 1977–1987 (2014)
46. Wang, H., Liu, P.X., Zhao, X., Liu, X.: Adaptive fuzzy finite-time control of nonlinear systems with actuator faults. IEEE Trans. Cybern. (2019)
47. Wang, P., Zhang, X., Zhu, J.: Online performance-based adaptive fuzzy dynamic surface control for nonlinear uncertain systems under input saturation. IEEE Transactions on Fuzzy System **27**(2), 209–220 (2018)
48. Wang, T., Li, Y.: Neural-network adaptive output-feedback saturation control for uncertain active suspension systems. IEEE Trans. Cybern. (2020)
49. Wang, W., Liang, H., Pan, Y., Li, T.: Prescribed performance adaptive fuzzy containment control for nonlinear multiagent systems using disturbance observer. IEEE Trans. Cybern. (2020)
50. Wei, C., Luo, J., Dai, H., Duan, G.: Learning-based adaptive attitude control of spacecraft formation with guaranteed prescribed performance. IEEE transactions on cybernetics **49**(11), 4004–4016 (2018)
51. Wen, C., Zhou, J., Liu, Z., Su, H.: Robust adaptive control of uncertain nonlinear systems in the presence of input saturation and external disturbance. IEEE Transactions on Automatic Control **56**(7), 1672–1678 (2011)
52. Wu, C., Liu, J., Xiong, Y., Wu, L.: Observer-based adaptive fault-tolerant tracking control of nonlinear nonstrict-feedback systems. IEEE Trans. Neural Netw. Learning Syst. **29**(7), 3022–3033 (2017)
53. Wu, J., Chen, W., Li, J.: Global finite-time adaptive stabilization for nonlinear systems with multiple unknown control directions. Automatica **69**, 298–307 (2016)
54. Xia, G., Sun, C., Zhao, B., Xue, J.: Prescribed performance adaptive fuzzy containment control for with input saturation based on finite-time disturbance observer. International J. Contr., Autom. Syst. **17**(2), 370–379 (2019)
55. Xiao, B., Hu, Q., Zhang, Y.: Adaptive sliding mode fault tolerant attitude tracking control for flexible spacecraft under actuator saturation. IEEE Transactions on Control Systems Technology **20**(6), 1605–1612 (2011)
56. Xu, P., Li, Y., Tong, S.: Fuzzy adaptive finite time fault-tolerant control for multi-input and multi-output nonlinear systems with actuator faults. International J. Contr., Autom. Syst. **17**(7), 1655–1665 (2019)

57. Yang, C., Jiang, Y., Na, J., Li, Z., Cheng, L., Su, C.Y.: Finite-time convergence adaptive fuzzy control for dual-arm robot with unknown kinematics and dynamics. IEEE Trans. Fuzzy Syst. **27**(3), 574–588 (2018)
58. Yang, Y., Hua, C., Guan, X.: Adaptive fuzzy finite-time coordination control for networked nonlinear bilateral teleoperation system. IEEE Trans. Fuzzy Syst. **22**(3), 631–641 (2013)
59. Yang, Y., Tan, J., Yue, D.: Prescribed performance tracking control of a class of uncertain pure-feedback nonlinear systems with input saturation. IEEE Transactions on Systems, Man, and Cybernetics: Systems **50**(5), 1733–1745 (2018)
60. Yong, K., Chen, M., Shi, Y., Wu, Q.: Flexible performance-based robust control for a class of nonlinear systems with input saturation. Automatica **122**, 109268 (2020)
61. Yu, J., Shi, P., Lin, C., Yu, H.: Adaptive neural command filtering control for nonlinear mimo systems with saturation input and unknown control direction. IEEE Trans. Cybern. **50**(6), 2536–2545 (2019)
62. Yu, J., Zhao, L., Yu, H., Lin, C.: Barrier lyapunov functions-based command filtered output feedback control for full-state constrained nonlinear systems. Automatica **105**, 71–79 (2019)
63. Yu, J., Zhao, L., Yu, H., Lin, C., Dong, W.: Fuzzy finite-time command filtered control of nonlinear systems with input saturation. IEEE Trans. Cybern. **48**(8), 2378–2387 (2017)
64. Zhang, B., Jiang, L., Bai, J.: Velocity-free prescribed performance control for spacecraft hovering over an asteroid with input saturation. J. Franklin Institute (2020)
65. Zhang, C., Ma, G., Sun, Y., Li, C.: Observer-based prescribed performance attitude control for flexible spacecraft with actuator saturation. ISA Trans. **89**, 84–95 (2019)
66. Zhao, K., Song, Y., Chen, C.P., Chen, L.: Adaptive asymptotic tracking with global performance for nonlinear systems with unknown control directions. IEEE Transactions on Automatic Control **67**(3), 1566–1573 (2021)
67. Zheng, Z., Xie, L.: Finite-time path following control for a stratospheric airship with input saturation and error constraint. International J. Contr. **92**(2), 368–393 (2019)
68. Zhou, Q., Li, H., Wang, L., Lu, R.: Prescribed performance observer-based adaptive fuzzy control for nonstrict-feedback stochastic nonlinear systems. IEEE Transactions on Systems, Man, and Cybernetics: Systems **48**(10), 1747–1758 (2017)
69. Zhou, Q., Shi, P., Tian, Y., Wang, M.: Approximation-based adaptive tracking control for mimo nonlinear systems with input saturation. IEEE transactions on cybernetics **45**(10), 2119–2128 (2014)

Chapter 7
Tunnel Prescribed Control of Nonlinear Systems with Communication Constraint

7.1 Introduction

Practical systems often operate under limited computational loads but need to maintain certain performance constraints, arising from hardware capability and task requirements, for example, the target tracking problem of aerial robots or the manipulator grasping. Nowadays, event-based control has enticed sustained interest due to its resource efficiency. As systems grow in complexity, how to save communication resources and guarantee system performance at the same time deserve more in-depth studies. In this chapter, we propose a Coded Event-triggered Tunnel Prescribed Control and Self-adjustable Prescribed Performance to reduce the communication resources between control and actuator while addressing the Entry Capture Problem for a class of nonlinear systems. To solve the Entry Capture Problem, two strategies are proposed, that is, error self-tuning function-based and self-adjustable prescribed performance-based strategies. With the aid of these two strategies, the resulted control schemes can handle the Entry Capture Problem for either symmetric or asymmetric performance constraints, by which the tracking error is regulated into an allowable set before a given time. The initial-condition constraints are removed, and better transient performance is achieved. Moreover, we design two actuator update protocols, constant and coded event-triggered schemes such that only 1 bit or m-length string is transmitted between controller and actuator boxes for each communication. Compared with the existing event-triggered results, our communication protocols can further reduce the communication burden specifically from the encoding-decoding viewpoint.

R. Ji et al., *Constrained Nonlinear Control: Performance and Robustness*,
https://doi.org/10.1007/978-981-96-2971-8_7

7.2 Tunnel Prescribed Control with Constant Event-Triggered Scheme

7.2.1 Problem Formulation

Consider the following uncertain nonlinear control systems:

$$\begin{cases} \dot{x}_i = f_i(\bar{x}_i) + g_i x_{i+1} + d_i(t), \ i = 1, \ldots, n-1 \\ \dot{x}_n = f_n(\bar{x}_n) + g_n u + d_n(t) \\ y = x_1 \end{cases} \tag{7.1}$$

where $\bar{x}_i = [x_1, \ldots, x_i]^T \in \mathbb{R}^i, \ i = 1, \ldots n; \ f_i(\bar{x}_i) \in \mathbb{R}$ represents unknown nonlinear function; $d_i(t) \in \mathbb{R}, i = 1 \ldots, n$ is external disturbance; $g_i, i = 1, \ldots, n$ is an unknown constant control coefficient; u and y are the system input and output, respectively. Remarkably, this study is based on the fact that only y is measurable. For the subsequent control design, some basic assumptions are presented.

Assumption 7.1 The desired tracking trajectory $y_d(t)$ and its first-order time derivative $\dot{y}_d(t)$ are bounded.

Assumption 7.2 The sign of control coefficient g_i is unknown and bounded yet away from zero, namely, there exist positive constants $\underline{g}_0$ and $\bar{g}_0$ such that $0 < \underline{g}_0 \leq |g_i| \leq \bar{g}_0$ holds with $i = 1, \ldots, n$.

Assumption 7.3 The external disturbance is bounded by $|d_i(t)| < \bar{d}, \ i = 1, \ldots, n$, where $\bar{d}$ is a positive constant.

Remark 7.1 Assumption 7.1 is widely adopted in tracking control of nonlinear systems [16]. Assumption 7.2 is necessary to ensure that the nonlinear system (7.1) is controllable, which is commonly made in [10, 15, 36]. Assumption 7.3 indicates that the external disturbances are bounded, which is reasonable in practice [3, 6].

Before the control development, the definition of Entry Capture Problem (ECP) is given as:

Definition 7.1 Define the tracking error $e_1(t) = y(t) - y_d(t)$. The tracking error performance is said to be ECP if the initial tracking error is released from any constraints during some time interval, then it is regulated into the prescribed performance before a user given finite time, that is

$$-e_l(t) < e_1(t) < e_u(t), \ \forall t \geq \tau, \tag{7.2}$$

where $-e_l(t)$ and $e_u(t)$ are the lower and upper boundaries of Tunnel Prescribed Performance, and $\tau > 0$ denotes the specified settling time.

Control Objective: This work focuses on developing a low-complexity event-triggered Tunnel Prescribed Control for the nonlinear system with unknown control directions such that:

(i) All signals in the closed-loop system are bounded;
(ii) For any initial condition (although it may violate the performance constraints initially), $e_1(t)$ can be regulated into the prescribed allowable set right after a user-given settling time as depicted in (7.2); and
(iii) Only 1-bit string is required for each communication between control and actuator when the event-triggered scheme is satisfied.

Remark 7.2 In most existing PPC methods [4, 18, 20, 23, 34], the performance constraints should be satisfied from the beginning of system operation which limits the selection of system initial condition. However, the ECP describes a more general tracking scenario which involves no constraints initially but symmetric or asymmetric performance constraints right after some time instant. In this case, if the initial-constraint violation occurs, the classical PPC methods become invalid and suffer from singularity problem since they are not defined outside the allowable set. Regarding the control objectives of this section, there are two main difficulties in the control design. (i) How to remove such initial-constraint restriction to make the control scheme applicable to any initial condition; and (ii) How to develop an event-triggered mechanism in the ECP from an encoding-decoding viewpoint to achieve 1-bit string transmission for each communication. Either or both challenges are still left open, although many excellent studies have been reported on the PPC and event-triggered control.

7.2.2 *Tunnel Prescribed Performance*

Considering the progressively rigorous tracking requirements, we propose a Tunnel Prescribed Performance:

$$-e_l(t) < \xi(t) < e_u(t), \tag{7.3}$$

with

$$\begin{aligned} e_l(t) &= (\delta - \mathrm{sign}(e_0))\rho(t) + \rho_\infty \mathrm{sign}(e_0), \\ e_u(t) &= (\delta + \mathrm{sign}(e_0))\rho(t) - \rho_\infty \mathrm{sign}(e_0), \end{aligned} \tag{7.4}$$

where $\xi(t)$ denotes the system tracking signal, $e_0 = e_1(0)$, $0 < \delta < 1$, $\rho(t) = (\rho_0 - \rho_\infty)e^{-\ell t} + \rho_\infty$ with $0 < \rho_\infty < (1-\delta)\rho_0$ and $\ell > 0$.

The Tunnel Prescribed Performance is shown in Fig. 7.1 in the cases of $e_0 > 0$ and $e_0 = 0$. Compared with the classical performance functions, three salient features can be observed:

(i) By using sign($\cdot$) function in (7.4), we unify different initial condition into one simple expression. This makes the control design and stability analysis straightforward;

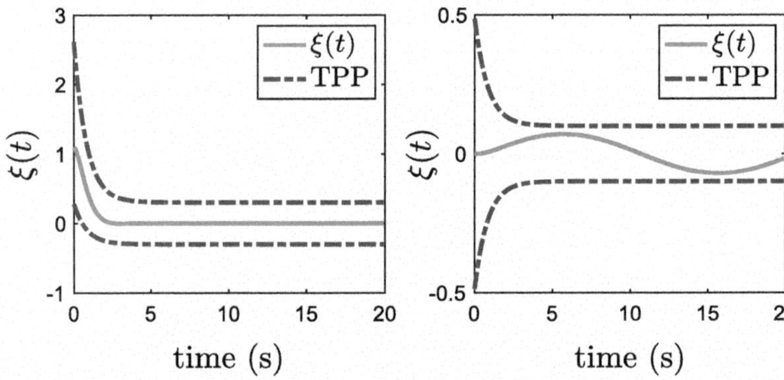

Fig. 7.1 The tracking state $\xi(t)$ under TPP. The left one in the case of $e_0 > 0$ and the right one in the case of $e_0 = 0$

(ii) From Fig. 7.1, when $e_0 > 0$, such Tunnel Prescribed Performance can provide more tight allowable set, which enables better transient performance; and
(iii) When $e_0 = 0$ in Fig. 7.1, Tunnel Prescribed Performance degenerates into the classical performance functions, which indicates that the classical ones are a special case of the proposed Tunnel Prescribed Performance.

The intrinsic mechanism of TPP is through the use of sign$(\cdot)$ function to unify different initial conditions into one concise performance description as (7.4). Subsequently, the TPP has the ability of adaptively adjusting its orientation with respect to the initial tracking condition. We further introduce a positive constant δ into $e_l(t)$ and $e_u(t)$, which serves as a scale function such that $-e_l(t) < e_u(t)$ always holds. Moreover, $-e_l(t)$ and $e_u(t)$ are always located on the same side of the origin in the case of $e_0 > 0$ or $e_0 < 0$, forming a tunnel-shape allowable set as shown in the left figure of Fig. 7.1.

Remark 7.3 More rigorous requirements are imperatively needed in practical engineering for achieving better tracking performance. For example, in missile or spacecraft mission, it is often required or desired the tracking errors evolving within a tight allowable set to avoid missing targets and improve mission efficiency. Unfortunately, most previous PPFs in [1, 20, 26] are symmetrically distributed on both sides, which cannot impose stricter constraints since they have loose performance boundaries. The proposed TPP addresses these problems, ascribed to its asymmetrical performance boundaries and concise expressions. Moreover, such TPP as a basic foundation can be easily combined with many excellent obtained results in [9, 21, 27, 31], i.e., finite-time convergence and flexible performance.

7.2.3 Fuzzy State Observer

Recently, considerable efforts have been made on the output feedback control since the states are rarely completely measurable, or are observable with requiring high observation costs in practice. As only the system output is measurable, a fuzzy state observer is therefore introduced. Upon (7.1), we define the following new variables

$$s_i = \frac{x_i}{\bar{g}_i},\ \bar{g}_i = \prod_{j=i}^{n} g_j \tag{7.5}$$

$$\tilde{s}_i = s_i - \hat{s}_i,\ i = 1, \ldots, n, \tag{7.6}$$

with $\hat{s}_i$ as the estimation of s_i. Hereby, the nonlinear system (7.1) can be rephrased as

$$\begin{cases} \dot{s}_i = \bar{f}_i(\bar{s}_i) + s_{i+1} + \bar{d}_i(t) \\ \dot{s}_n = \bar{f}_n(\bar{s}_n) + u + \bar{d}_n(t) \end{cases} \tag{7.7}$$

where $\bar{f}_i(\bar{s}_i) = \frac{f_i(\bar{x}_i)}{\bar{g}_i}$ and $\bar{d}_i(t) = \frac{d_i(t)}{\bar{g}_i}$. Owing to the universal approximation ability of FLSs in [28], we have

$$\bar{f}_i(\bar{s}_i) = \theta_i^T \varphi_i(\bar{s}_i) + \varepsilon_i,\ i = 1, \ldots, n, \tag{7.8}$$

where $\theta_i \in \mathbb{R}^l$ represents the optimal unknown parameter vector, $\varphi_i(\bar{s}_i)$ is the basic function, and ε_i is bounded by $|\varepsilon_i| \leq \bar{\varepsilon}$ with $\bar{\varepsilon} > 0$. We then develop the fuzzy state observer as

$$\begin{cases} \dot{\hat{s}}_i = -k_i \hat{s}_1 + \hat{s}_{i+1} + \hat{\theta}_i^T \varphi_i(\hat{\bar{s}}_i) \\ \dot{\hat{s}}_n = -k_n \hat{s}_1 + u + \hat{\theta}_n^T \varphi_n(\hat{\bar{s}}_n) \end{cases} \tag{7.9}$$

where $\hat{\theta}_i \in \mathbb{R}^l$ is the estimated vector of θ_i, and the adaptive law follows

$$\dot{\hat{\theta}}_i = -\sigma_i \hat{\theta}_i + r_i z_i \varphi_i(\hat{\bar{s}}_i), \tag{7.10}$$

where σ_i and r_i are positive constants, and $z_i, i = 1, \ldots, n$ is defined in (7.22) and (7.30), respectively. Let $\tilde{s}_i = s_i - \hat{s}_i$ and $\tilde{s} = [\tilde{s}_1, \ldots, \tilde{s}_n]^T$. Invoking (7.7) and (7.9), it can be derived that

$$\dot{\tilde{s}} = A\tilde{s} + \sum_{i=1}^{n} B_i(\tilde{\theta}_i^T \hat{\varphi}_i(\hat{\bar{s}}_i) + k_i s_1 + \theta_i^T \Delta\varphi_i + \varepsilon_i + \bar{d}_i),$$

where $\tilde{\theta}_i = \theta_i - \hat{\theta}_i$, $\Delta\varphi_i = \varphi_i(\bar{s}_i) - \varphi_i(\hat{\bar{s}}_i)$, and the matrices A and B are

$$A = \begin{bmatrix} -k_1 & & & \\ \vdots & & I_{n-1} & \\ -k_n & 0 & \cdots & 0 \end{bmatrix}, \quad B_i = [\underbrace{0, \ldots, 1}_{i}, \ldots, 0]^T \tag{7.11}$$

where $k_i, i = 1, \ldots, n$ are positive constants such that A becomes a Hurwitz matrix. Therefore, for a positive definite matrix $Q^T = Q$, there exist a positive definite matrix $P^T = P$ such that $A^T P + PA = -Q$ holds.

Subsequently, for the Lyapunov function candidate $V_0 = \tilde{s}^T P \tilde{s}$, its time derivative is

$$\begin{aligned} \dot{V}_0 = & -\tilde{s}^T Q \tilde{s} + 2\tilde{s}^T P \sum_{i=1}^{n} B_i(\tilde{\theta}_i^T \varphi_i(\hat{\bar{s}}_i) + k_i s_1) \\ & + 2\tilde{s}^T P \sum_{i=1}^{n} B_i(\theta_i^T \Delta\varphi_i + \bar{d}_i(t) + \varepsilon_i). \end{aligned} \tag{7.12}$$

In view of Assumptions 7.2–7.3, there is a positive constant D such that $|\bar{d}_i(t)| \leq D$ due to the bounded $d_i(t)$ and g_i. Then, we can have

$$\begin{aligned} & 2\tilde{s}^T P \sum_{i=1}^{n} B_i(\bar{d}_i(t) + \varepsilon_i) \leq \frac{2n}{\kappa_1} \|\tilde{s}\|^2 + n\kappa_1 \|P\|^2 (D^2 + \bar{\varepsilon}^2), \\ & 2\tilde{s}^T P \sum_{i=1}^{n} B_i(\tilde{\theta}_i^T \varphi_i(\hat{\bar{s}}_i) + \tilde{\theta}_i^T \Delta\varphi_i + k_i s_1) \leq \frac{4n}{\kappa_2} \|\tilde{s}\|^2 \\ & + \kappa_2 \|P\|^2 \sum_{i=1}^{n} (\|\tilde{\theta}_i\|^2 + 2\|\theta_i\|^2) + \kappa_2 \|P\|^2 \sum_{i=1}^{n} k_i^2 \|s_1\|^2, \end{aligned}$$

where κ_1 and κ_2 are positive constants. Substituting these inequalities into (7.12) yields

$$\begin{aligned} \dot{V}_0 \leq & -c_0 \|\tilde{s}\|^2 + \kappa_2 \|P\|^2 \sum_{i=1}^{n} \|\tilde{\theta}_i\|^2 + \kappa_2 \|P\|^2 \sum_{i=1}^{n} k_i^2 s_1^2 \\ & + n\kappa_1 \|P\|^2 (D^2 + \bar{\varepsilon}^2) + 2\kappa_2 \|P\|^2 \sum_{i=1}^{n} \|\theta_i\|^2, \end{aligned} \tag{7.13}$$

where $c_0 = \lambda_{\min}(Q) - \frac{2n}{\kappa_1} - \frac{4n}{\kappa_2}$ and $\lambda_{\min}(\cdot)$ denotes the corresponding minimum eigenvalue. We present the following sufficient condition for the boundedness of the observer errors.

Lemma 7.1 *If the tracking errors* $z_1, \ldots, z_n$ *defined in* (7.22) *and* (7.30) *are bounded, so are the observer errors* $\tilde{s}_1, \ldots, \tilde{s}_n$.

Proof According to the definition of V_0 and the inequality (7.13), given that $\tilde{\theta}_i \in L^\infty$ and $s_1 \in L^\infty$, we can get $\tilde{s}_i \in L^\infty, i = 1, \ldots, n$.

First, owing to $\tilde{\theta}_i = \theta_i - \hat{\theta}_i$, we have $\tilde{\theta}_i \in L^\infty$ if $\hat{\theta}_i \in L^\infty$. Based on the bounded-input and bounded-output stability and $\varphi(\hat{\tilde{s}}_i)^T \varphi(\hat{\tilde{s}}_i) < 1$ for (7.10), the boundedness of $\hat{\theta}_i$ can be guaranteed if $z_i \in L^\infty$. Obviously, it indicates $\tilde{\theta}_i \in L^\infty$ as long as $z_i \in L^\infty$. Next, Lemma 7.2 ensures the boundedness of the tracking error e_1 if $z_1 \in L^\infty$. This together with Assumption 7.1 and (7.5) implies that $s_1 \in L^\infty$ since x_1 and y_d are bounded. Hence, there is $s_1 \in L^\infty$ if $z_1 \in L^\infty$. This completes the proof. □

7.2.4 Error Self-tuning Function

To address ECP for nonlinear systems under any initial condition, an Error Self-tuning Function (ESF) is proposed on the tracking error $e_1(t) = y(t) - y_d(t)$ as

$$e_s(t) = w(t)e_1(t) + (1 - w(t))h(t), \tag{7.14}$$

where $w(t)$ and $h(t)$ are given by

$$w(t) = \begin{cases} \sin \frac{\pi t}{2\tau}, & \text{if } t < \tau, \\ 1, & \text{otherwise.} \end{cases} \tag{7.15}$$

$$h(t) = \frac{e_u(t) - e_l(t)}{2}, \tag{7.16}$$

where τ denotes the given settling time in ECP (7.2), and $e_u(t)$ and $-e_l(t)$ are Tunnel Prescribed Performance defined in (7.4). Remarkably, given any initial value of $e_1(0)$, there always have

$$e_s(0) = w(0)e_1(0) + (1 - w(0))h(0) = h(0), \tag{7.17}$$

$$e_s(t) = e_1(t), \ t \geq \tau. \tag{7.18}$$

To better illustrate the design mechanism of ESF (7.14), we take $e_1(0) < 0$ as an example, and simulation results are given in Fig. 7.2. Although the initial value of $e_1(0)$ may violate constraints as depicted in Fig. 7.2, the result of $w(t)e_1(t)$ offers a unified solution to handle different initial conditions due to $w(0) = 0$ which leads to $w(0)e_1(0) = 0$. Then, with the definitions of $w(t)$ and $h(t)$, the initial value of ESF $e_s(0)$ is always guaranteed within the allowable set, namely, $-e_l(0) < e_s(0) = h(0) < e_u(0)$ as depicted in Fig. 7.2. Such ESF makes the control design applicable to any initial condition. Considering the ECP where the tracking error $e_1(t)$ should satisfy the performance constraints right after a time instant $t = \tau$, the ESF hence becomes $e_s(t) = e_1(t)$ for $t \geq \tau$ due to $w(t) = 1$ for $t \geq \tau$ as described in (7.18). As discussed above, such ESF could allow the ECP to be addressed collectively, as

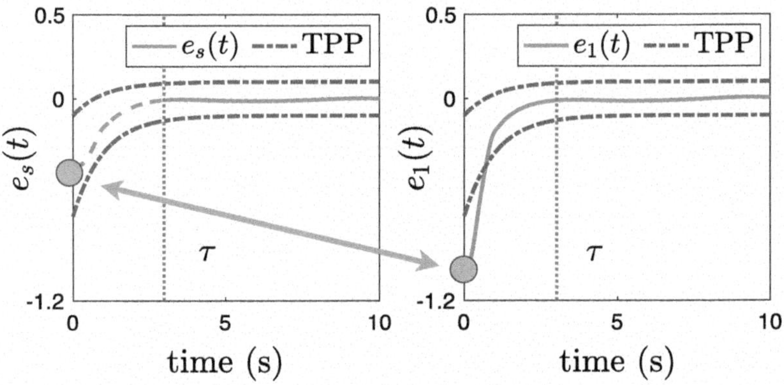

Fig. 7.2 The mechanism behind the time-varying scaling function

seen later. The following lemma demonstrates that the introduced ESF preserves the boundedness of $e_1(t)$.

Lemma 7.2 *If* $-e_l(t) < e_s(t) < e_u(t)$ *always holds for* $t \in [0, \tau)$, *the boundedness of* $e_1(t)$ *can be guaranteed during this interval. If* $-e_l(t) < e_s(t) < e_u(t)$ *holds for* $\forall t \geq 0$, *the ECP in* (7.2) *is addressed.*

Proof With aid of contradiction, we can verify the first claim. If $-e_l(t) < e_s(t) < e_u(t)$ for $t \in [0, \tau)$, we suppose a time instant t' such that

$$\lim_{t \to t'} |e_1(t)| = \infty. \tag{7.19}$$

Upon (7.14), the following inequality is obtained with $t \in [0, \tau)$

$$-e_l(t) < w(t)e_1(t) + (1 - w(t))h(t) < e_u(t).$$

By (7.19), if the above inequalities hold, one has $\lim_{t \to t'} w(t) = 0$. Following (7.16), it indicates that such equality holds if and only if $t' = 0$. Hereby, (7.19) is rephrased as

$$\lim_{t \to 0} |e_1(t)| = |y(0) - y_d(0)| = \infty. \tag{7.20}$$

From Assumption 7.1, the initial settings should be bounded, that is, $|y(0) - y_d(0)|$ is bounded, contradicting (7.20). Therefore, the assumption (7.19) is false and the first claim holds.

Then, for the second claim and based on (7.18), there have

$$-e_l(t) < e_1(t) < e_u(t), \ t \geq \tau, \tag{7.21}$$

since $w(t) = 1$ when $t \geq \tau$. Obviously, (7.21) satisfies the ECP in (7.2). This complete the proof. □

Based on Lemma 7.2, the objective is transformed to ensure the boundedness of $e_s(t)$ so that the ECP can be solved. To facilitate the extension of ESF to other approaches, some properties are summarized for developing the tuning functions $w(t)$ and $h(t)$:

(i) $w(0) = 0$; $w(t) = 1$ for $t \geq \tau$;
(ii) $0 \leq \dot{w}(t) < \infty$, for $t \geq 0$; and
(iii) $0 \leq \dot{h}(t) < \infty$, for $t \geq 0$, and $h(0) = \frac{e_u(0)-e_l(0)}{2}$.

From the definition of $w(t)$ and $h(t)$ in (7.15)–(7.16), the first property is clear established. The time derivative of $w(t)$ is $\dot{w}(t) = \frac{\pi}{2\tau}\cos\frac{\pi t}{2\tau}$ for $t < \tau$ and $\dot{w}(t) = 0$ for $t \geq \tau$, which satisfies the second property. As $h(t)$ is related to the tunnel prescribed performance, the smoothness of $h(t)$ can be guaranteed such that the third property holds. As discussed above, there will be many ESF candidates satisfying these properties, which can lead to fruitful studies and expressions in the future.

Remark 7.4 In the classical PPC [4, 18, 20, 23, 34], the control design becomes invalid if the initial value of $e_1(t)$ violates the performance constraints since there is no definition for the PPC outside the allowable set. The proposed ESF, therefore, provides an effective way to make the control design applicable to any initial condition. Some distinct features can be obtained for such ESF. (i) Different from the funnel control and global prescribed performance control in [2, 12, 36] that rely on some specific performance functions, we introduce a simple ESF on the tracking error signal that can be easily extended to other methods. (ii) Both funnel boundaries and global prescribed performance are only applicable to symmetric performance conditions. It indicates that these methods cannot be extended to asymmetric conditions, straightforwardly. However, our ESF can be applied to performance constraints that are either symmetric or asymmetric, thereby increasing its scalability to more general cases.

7.2.5 *Orientation Function-Based Tunnel Prescribed Control*

Considering the ECP defined in (7.2), we adopt an error transformation function

$$z_1(e_s) = \ln\left(\frac{e_l + e_s}{e_u - e_s}\right), \tag{7.22}$$

in which the constrained error $e_s(t)$ is transformed to an unconstrained one $z_1(e_s)$. For the nonlinear systems (7.1), we design an orientation function $\phi(z_1)$ to tackle the unknown control coefficients problem, where $\phi(z_1)$ should satisfy the following properties:

$$|\phi(z_1)| < \infty,\ |d\phi/dz_1| < \infty,\ \forall |z_1| < \infty, \tag{7.23}$$

and

$$\begin{cases} \lim_{|z_1|\to\infty} \sup z_1(e_s)\phi(z_1) = +\infty, \\ \lim_{|z_1|\to\infty} \inf z_1(e_s)\phi(z_1) = -\infty. \end{cases} \tag{7.24}$$

Some typical examples of $\phi(z_1)$ are $\cos(\cdot)$ and $\sin(\cdot)$. Although we introduce $w(t)$ in (7.14), the following lemma indicates that the above infinity property (7.24) is still satisfied.

Lemma 7.3 *Given any nonzero constant ζ, there have:*

$$\begin{cases} \lim_{|z_1|\to\infty} \sup \zeta w(t) z_1(e_s)\phi(z_1) = +\infty, \\ \lim_{|z_1|\to\infty} \inf \zeta w(t) z_1(e_s)\phi(z_1) = -\infty. \end{cases} \tag{7.25}$$

Proof Suppose a constant m such that

$$\lim_{|z_1|\to\infty} \sup \zeta w(t) z_1(e_s)\phi(z_1) = m, \tag{7.26}$$

or

$$\lim_{|z_1|\to\infty} \inf \zeta w(t) z_1(e_s)\phi(z_1) = m. \tag{7.27}$$

By (7.24), there is

$$\begin{cases} \lim_{|z_1|\to\infty} \sup \zeta z_1(e_s)\phi(z_1) = +\infty, \\ \lim_{|z_1|\to\infty} \inf \zeta z_1(e_s)\phi(z_1) = -\infty. \end{cases} \tag{7.28}$$

Considering either (7.26) or (7.27), we have $\lim_{|z_1|\to\infty} w(t) = 0$ due to (7.28). Upon the definition of $w(t)$ in (7.16), it yields that $\lim_{|z_1|\to\infty} t = 0$, namely, $|z_1(0)| = \infty$. However, at the initial time, there always has $z_1(0) = 0$ as presented in Lemma 7.4, which is paradoxical with $|z_1(0)| = \infty$. Consequently, the assumptions (7.26) and (7.27) are false, yet (7.25) is right. The proof is completed. □

From (7.22), a virtual control input α_1 is designed, where an orientation function $\phi(z_1)$ is introduced to tackle the unknown control direction

$$\alpha_1 = \lambda_1 z_1 \phi(z_1), \tag{7.29}$$

where λ_1 is a positive constant. Therefore, we have

$$z_i = \hat{s}_i - \alpha_{i-1}, \tag{7.30}$$

$$\alpha_i = -\lambda_i z_i, \tag{7.31}$$

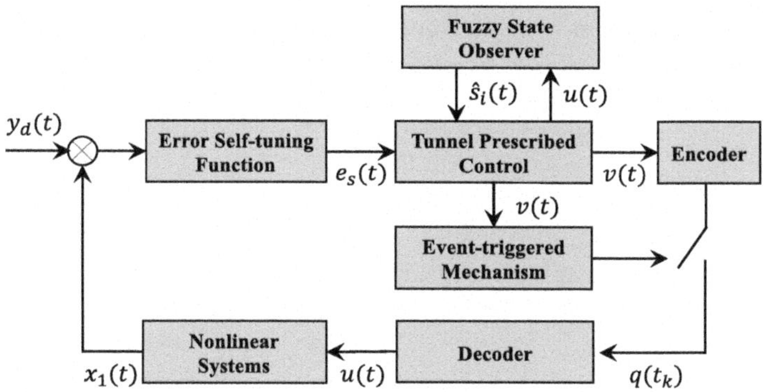

Fig. 7.3 The control scheme of the TPC

where $i = 2, \ldots, n$, λ_i represents a positive constant, α_i is a virtual control input, and in the case of $i = n$, let $v(t) = \alpha_n$ denote the actual control input. The control scheme is depicted in Fig. 7.3 to clearly show the design details. Upon the control design (7.29)–(7.31), such ESF (7.14) provides the following significant property.

Lemma 7.4 *For any initial conditions: $y(0)$ and $y_d(0)$, there hold:*

$$z_i = 0, \ i = 1, \ldots, n. \tag{7.32}$$

Proof Substituting (7.16) and (7.17) into (7.22) yields

$$z_1(e_1(0)) = \ln\left(\frac{e_l(0) + h(0)}{e_u(0) - h(0)}\right) = 0. \tag{7.33}$$

Then, (7.29) indicates that $\alpha_1(0) = 0$. Due to the initial value of the observer state $\hat{s}_i(0) = 0$ and by (7.30)–(7.31), it can be easily derived $z_i(0) = 0$, recursively. □

Remark 7.5 By introducing a novel ESF (7.14) together with an orientation function $\phi(z_1)$ into the control design, Lemma 7.4 indicates that $v(0) = a_n(0) = 0$ since $z_n(0) = 0$, which leads to two significant characteristics. (i) The initial control signal $v(0) = 0$ holds for any initial conditions, i.e., $y(0)$ and $y_d(0)$. Hence, a large initial control input, which may be introduced by the Nussbaum-gain technique, can be avoided or significantly alleviated. (ii) In lieu of Nussbaum-type functions, switching rules and parameter estimation methods, a simple but effective orientation function $\phi(z_1)$ is explored to tackle the unknown control directions. This reveals the scalability of our control design.

Remark 7.6 By inspection of the control design in (7.29)–(7.31), it does not require any prior knowledge of the nonlinear functions, control directions and external disturbances. Moreover, compared with the existing results, neither dynamic surface control nor command filter is required in the control design, that is, the intermediate derivatives are no longer needed. In this way, it leads to a low-complexity, less demanding and simple-structure control scheme, which not only facilitates the control implementation but also saves computation cost.

7.2.6 Actuator Update Protocol

To effectively lower the number and the bit of data transmission, we employ the following event-trigger rule on the control input $v(t)$ defined in (7.31):

$$\Delta v(t) = v(t) - v(t_k),\ t_k < t < t_{k+1}, \tag{7.34}$$

$$t_{k+1} = \inf\{t > t_k : |\Delta v(t)| = c\}, \tag{7.35}$$

$$q(t_k) = \begin{cases} 1, \text{ if } \Delta v(t_k) = c, \\ 0, \text{ if } \Delta v(t_k) = -c. \end{cases} \tag{7.36}$$

where $t_0 = 0$ as the initial time, $k = 0, 1, \ldots$, and c is a user specified positive constant. $q(t_k)$ denotes the only transmission signal between the channel of controller-to-actuator. Hereby, the control strategy is updated by

$$u(t) = 0,\ t_0 \le t < t_1, \tag{7.37}$$

$$u(t) = u(t_{k-1}) + c(2q(t_k) - 1),\ t_k \le t < t_{k+1}, \tag{7.38}$$

where $k = 1, 2, \ldots$. This protocol mechanism (7.34)–(7.38) is explained as follows.

At the initial stage of system, the control signal remains $u(t) = 0$ as shown in (7.37) until $|\Delta v(t)| = c$ is triggered for the first time. Hereafter, $v(t_{k-1})$ is updated as $v(t_k)$ and the signal $q(t_k)$ is transmitted to the actuator. Specifically, the actuator output is derived as $u(t_k) = u(t_{k-1}) + c$ if $q(t_k) = 1$ and $u(t_k) = u(t_{k-1}) - c$ if $q(t_k) = 0$ based on (7.38). Then, $u(t_k)$ remains unchanged on $[t_k, t_{k+1})$ until next event-trigger occurs. We summarize the following property of the proposed update protocol.

Lemma 7.5 *Upon the actuator update protocol* (7.34)–(7.38)*, we can have*

$$u(t) = v(t_k),\ t_k \le t < t_{k+1},\ k = 0, 1, 2, \ldots \tag{7.39}$$

Proof Upon (7.34)–(7.35), when the event-trigger condition is satisfied at t_{k+1}, we have

$$v(t_{k+1}) = \begin{cases} v(t_k) + c, \text{ if } \Delta v(t_{k+1}) = c, \\ v(t_k) - c, \text{ if } \Delta v(t_{k+1}) = -c. \end{cases} \tag{7.40}$$

where $k = 0, 1, 2, \ldots$. By (7.36), $v(t_{k+1})$ is rephrased as:

$$v(t_{k+1}) = v(t_k) + c(2q(t_k) - 1),\ k = 0, 1, 2, \ldots \tag{7.41}$$

According to Lemma 7.4, there has $v(0) = 0$, resulting in

$$v(t_{k+1}) = \sum_{i=1}^{k+1} c(2q(t_i) - 1), \ k = 0, 1, 2, \ldots \tag{7.42}$$

Considering (7.37)–(7.38), one gets

$$u(t) = \sum_{i=1}^{k+1} c(2q(t_i) - 1), \ t_{k+1} \leq t < t_{k+2}, \tag{7.43}$$

where $k = 0, 1, 2, \ldots$. From (7.42) and (7.43), there is $u(t) = v(t_{k+1})$ for $t_{k+1} \leq t < t_{k+2}$ and then Lemma 7.5 is established. □

Remark 7.7 Note that the event-triggered update protocol (7.35) relies on continuous and precise sensor measurements, which means that the occurrences of network attacks [7] and packet losses [8] are not considered in this work. In practice, it is well known that the system is sampled periodically, which could result in bounded error, that is, the event-triggered update protocol can be modified as:

$$t_{k+1} = \inf\{t > t_k : |\Delta v(t)| \geq c\}.$$

In this way, (7.39) is rewritten as:

$$u(t) = v(t_k) + \varepsilon(t_k), \ t_k \leq t < t_{k+1}, \ k = 0, 1, 2, \ldots$$

where $\varepsilon(t_k)$ represents the bounded error. Further, it can be derived that the effect of such bounded error can be referred to the external disturbance effect $d_n(t)$ in (7.1). Therefore, we, in this section, still employ (7.35) as the actuator update protocol for the clarity consequent analysis.

Remark 7.8 Different from most event-triggered control methods in [11, 19, 22, 25, 29, 33] that usually focus on reducing the frequency of signal update to save communication resources, the proposed event-triggered mechanism is designed from an encoding-decoding viewpoint. As shown in (7.34)–(7.36), only 1-bit string is required for each communication between control and actuator rather than a finite-length string, i.e., 16-bit string, in these existing results. In this way, the proposed mechanism can be more efficient in practical scenarios, especially for limited communication bandwidth since both the number and the bit of data transmission have been decreased. Moreover, the communication security is improved at the same time because the transmitted 1-bit string does not contain any valuable system information. Although some excellent works on 1-bit or 2-bit mechanisms have been investigated in [5, 30, 32], the initial value of control input should be transmitted over the network. In this article, with the aid of ESF, the initial control input no longer needs to be transmitted since it is always equal to zero as discussed in Remark 7.5.

7.2.7 *Performance Analysis*

Before the performance analysis, we give the following sufficient condition of the boundedness of the time derivative of control input signal α_i, $i = 1, \ldots, n$.

Lemma 7.6 *Given that* $-e_l(t) < e_s(t) < e_u(t)$, $\dot{e}_s(t)$, $z_1(t)$, *and* $\dot{z}_i, i = 2, \ldots, n$ *are bounded, it holds for* $\dot{\alpha}_i$.

Proof Suppose that $-e_l(t) < e_s(t) < e_u(t)$, and based on (7.22), the time derivative of z_1 is given by

$$\dot{z}_1 = \gamma_1 + \gamma_2 \dot{e}_s(t), \tag{7.44}$$

$$\gamma_1 = \frac{\dot{e}_l}{e_l + e_s} - \frac{\dot{e}_u}{e_u - e_s}, \tag{7.45}$$

$$\gamma_2 = \frac{e_l + e_u}{(e_l + e_s)(e_u - e_s)}. \tag{7.46}$$

The boundedness of z_1 implies that γ_1 and γ_2 are also bounded. Thereby, $\dot{z}_1$ is bounded if $\dot{e}_s(t)$ and $z_1(t)$ are bounded. From (7.29), it yields

$$\dot{\alpha}_1 = c_1 \dot{z}_1 (\phi(z_1) + z_1 \frac{d\phi}{dz_1}). \tag{7.47}$$

Then, based on the properties of $\phi(z_1)$ in (7.23) and the above condition, the boundedness of $\dot{\alpha}_1$ is guaranteed if $\dot{z}_1$ and z_1 are bounded. Further, differentiating (7.31), there have

$$\dot{\alpha}_i = -\lambda_i \dot{z}_i, \ i = 2, \ldots, n. \tag{7.48}$$

Obviously, the given conditions in Lemma 7.6 guarantee the boundedness of $\dot{\alpha}_i$. □

Hereby, we are at the position to present our main results as the following theorem.

Theorem 7.1 *Consider the nonlinear system* (7.1) *under Assumptions 7.1–7.3. The fuzzy state observer* (7.9) *and the proposed control scheme* (7.29)–(7.31) *guarantee the following: (i) First, all the closed-loop system states remain bounded and the tuned tracking error* $e_s(t)$ *is always within the TPP, i.e.,* $-e_l(t) < e_s(t) < e_u(t)$. *(ii) Second, the tracking error* $e_1(t)$ *meets ECP in* (7.2). *(iii) Zeno behavior does not occur during the whole system operation process.*

Proof Hereafter, the arguments in variables or functions sometimes are omitted if no confusion is caused. According to (7.9), (7.14) and (7.30), the time derivative of e_s and $z_i, i = 2, \ldots, n$ are derived as

$$\dot{e}_s = \dot{w}(y - y_d) + (1 - \dot{w})h + (1 - w)\dot{h} + w(f_1 + \bar{g}_1 \tilde{s}_2 + \bar{g}_1 \hat{s}_2 + d_1 - \dot{y}_d), \tag{7.49}$$

$$\dot{z}_i = -k_i\hat{s}_1 + \hat{s}_{i+1} + \hat{\theta}_i^T\varphi_i - \dot{\alpha}_{i-1}, i = 2, \ldots, n-1, \tag{7.50}$$

$$\dot{z}_n = -k_n\hat{s}_1 + u(t) + \hat{\theta}_n^T\varphi_n - \dot{\alpha}_{n-1}. \tag{7.51}$$

By recalling (7.34) and (7.39), there has

$$u(t) = v(t_k) = v(t) - \Delta v(t), \ |\Delta v(t)| \leq c. \tag{7.52}$$

Considering (7.29)–(7.31) and (7.52), there are

$$\hat{s}_2 = z_2 + \alpha_1 = z_2 - \lambda_1 z_1 \phi(z_1), \tag{7.53}$$

$$\hat{s}_{i+1} = z_{i+1} + \alpha_i = z_{i+1} - \lambda_i z_i, \ i = 2, \ldots, n-1, \tag{7.54}$$

$$u(t) = v(t) - \Delta v(t) = -\lambda_n z_n - \Delta v(t). \tag{7.55}$$

We substitute (7.53)–(7.55) into (7.49)–(7.51), respectively and there have

$$\dot{e}_s = \delta_1 - w\lambda_1\bar{g}_1 z_1\phi(z_1), \tag{7.56}$$

$$\dot{z}_i = \delta_i - \lambda_i z_i, \ i = 2, \ldots, n-1, \tag{7.57}$$

$$\dot{z}_n = \delta_n - \lambda_n z_n, \tag{7.58}$$

where $\delta_1 = \dot{w}(y - y_d) + (1 - \dot{w})h + (1 - w)\dot{h} + w(f_1(x_1) + \bar{g}_1\tilde{s}_2 + \bar{g}_1 z_2 + d_1 - \dot{y}_d)$, $\delta_i = -k_i\hat{s}_1 + z_{i+1} + \hat{\theta}_i^T\varphi_i - \dot{\alpha}_{i-1}$ for $i = 2, \ldots, n-1$, and $\delta_n = -k_n\hat{s}_1 - \Delta v(t) + \hat{\theta}_n^T\varphi_n - \dot{\alpha}_{n-1}$.

We first claim that the tuned tracking error $e_s(t)$ is always evolved within the allowable set as

$$-e_l(t) < e_s(t) < e_u(t), \ t \geq 0, \tag{7.59}$$

and

$$z_i \in L^\infty, i = 2, \ldots, n, \ t \geq 0, \tag{7.60}$$

which will be verified by seeking contradiction. From (7.17) and Lemma 7.4, it indicates that (7.59) and (7.60) are satisfied at the initial time $t = 0$. As $e_s(t)$ is continuous, so does z_1 defined in (7.22) if $-e_l(t) < e_s(t) < e_u(t)$ holds and $z_1 \in L^\infty$. Then, α_1 is continuous in time based on (7.23) and (7.29). Under (7.59) and (7.60), it can be derived that $z_i, i = 2, \ldots, n$ is continuous by (7.9), (7.30), (7.31) and Lemma 7.1 as long as $z_{i-1} \in L^\infty$.

Suppose that there exists the first time instant t_1 such that

$$-e_l(t) < e_s(t) < e_u(t), \ z_i \in L^\infty, \ 0 \leq t < t_1, \tag{7.61}$$

$$\lim_{t\to t_1} e_s = e_u, \text{ or } \lim_{t\to t_1} e_s = -e_l, \text{ or } \lim_{t\to t_1} |z_i| = \infty, \tag{7.62}$$

where $i = 2, \ldots, n$. The following analysis will verify the above scenarios do not exist by enumerating the case.

Step 1: At the outset, we suppose

$$\lim_{t \to t_1} e_s = e_u \text{ or } \lim_{t \to t_1} e_s = -e_l. \tag{7.63}$$

Following the definition of z_1 in (7.22), the above condition indicates $\lim_{t \to t_1} |z_1| = \infty$. Recalling the TPP (7.3), the tuned tracking error e_s either increases to the upper boundary or decreases to the lower boundary, irrespective of the sign of the initial tracking error e_0. Therefore, there should exist two constants p_1 and p_2 such that the following inequalities hold respectively:

$$\liminf_{t \to t_1} \dot{e}_s \geq p_1, \tag{7.64}$$

$$\limsup_{t \to t_1} \dot{e}_s \leq p_2. \tag{7.65}$$

We will give the analysis about $\dot{e}_s(t)$ as follows. Upon (7.16), we have $w(t), \dot{w}(t), h(t), \dot{h}(t) \in L^\infty$. By Assumptions 7.1–7.3, it yields $d_1, y_d, \dot{y}_d \in L^\infty$. Based on (7.61) and Lemmas 7.1–7.2, it can be deduced that $\tilde{s}_2, z_2, x_1, f(x_1) \in L^\infty$ on $t \in [0, t_1)$. The above derivations imply the boundedness of δ_1 in (7.56) when $t \in [0, t_1)$. Subsequently, we adopt Lemma 7.3 to (7.56) under $\lim_{t \to t_1} |z_1| = \infty$, which leads to

$$\liminf_{t \to t_1} \dot{e}_s = \liminf_{t \to t_1} \delta_1 + \liminf_{t \to t_1} \zeta w z_1 \phi_1 = -\infty, \tag{7.66}$$

$$\limsup_{t \to t_1} \dot{e}_s = \limsup_{t \to t_1} \delta_1 + \limsup_{t \to t_1} \zeta w z_1 \phi_1 = +\infty, \tag{7.67}$$

with $\zeta = \lambda_1 \bar{g}_1$. Obviously, (7.66) and (7.67) contradict (7.64) and (7.65), respectively. Therefore, the assumption (7.63) is false, which yields

$$-e_l(t) + b_1 < e_s(t) < e_u(t) - b_1, \ 0 \leq t < t_1, \tag{7.68}$$

where b_1 is a positive constant. The boundedness of z_1 can be guaranteed over $t \in [0, t_1)$ according to (7.22). By (7.56), $\dot{e}_s$ is also bounded there. Under Lemma 7.6, we have $\dot{\alpha}_1$ is bounded for $0 \leq t < t_1$. This enables the analysis for $z_i, i = 2, \ldots, n-1$.

Step i ($i = 2, \ldots, n-1$)**:** Based on (7.57), the boundedness of δ_i on $t \in [0, t_1)$ is discussed as follows. As the definition in (7.5) and (7.5), it yields that $\hat{s}_1 = \frac{x_1}{\bar{g}_1} - \tilde{s}_1$. Thus, δ_i can be rephrased as $\delta_i = -k_i(\frac{x_1}{\bar{g}_1} - \tilde{s}_1) + z_{i+1} + \hat{\theta}_i^T \varphi_i - \dot{\alpha}_{i-1}$. From the last step analysis, $x_1, \dot{\alpha}_{i-1} \in L^\infty$ holds when $0 \leq t < t_1$. By (7.61) and Lemma 7.1, it implies that $\tilde{s}_1, z_{i+1} \in L^\infty$ there. Also, we have $\hat{\theta}_i^T \varphi_i \in L^\infty$ from (7.10) since $z_i \in L^\infty$ on $t \in [0, t_1)$. Therefore, there exists a positive constant b_i such that $|\delta_i| \leq b_i$ holds for $0 \leq t < t_1$.

Then, we construct the Lyapunov function candidate as $V_i = \frac{1}{2} z_i^2$. Its time derivative along (7.57) is given by

$$\dot{V}_i = z_i(\delta_i - \lambda_i z_i) \leq |z_i|(b_i - \lambda_i|z_i|). \tag{7.69}$$

Apparently, $\dot{V}_i < 0$ if $|z_i| > \frac{b_i}{\lambda_i}$ on $t \in [0, t_1)$, which means that $|z_i|$ decreases if $|z_i| > \frac{b_i}{\lambda_i}$ based on the definition of V_i. Namely, $|z_i| \leq \frac{b_i}{\lambda_i}$ when $t \in [0, t_1)$. Hence, the assumed condition $\lim_{t\to t_1} |z_i| = \infty$ with $i = 2, \ldots, n-1$ in (7.62) does not exist. Instead, $z_i \in L^\infty$, $i = 2, \ldots, n-1$ holds for $t \geq 0$ as depicted in (7.60). Moreover, we can further guarantee the boundedness of $\dot{z}_i$ there from (7.57), so does $\dot{\alpha}_i$ based on Lemma 7.6.

Step n: Identically, the boundedness of δ_n in (7.58) when $0 \leq t < t_1$ can be established as follows. As discussed in Step $n-1$, it has been shown that $\hat{s}_1, \dot{\alpha}_{n-1} \in L^\infty$ on $t \in [0, t_1)$. In accordance with (7.10), we can obtain $\hat{\theta}_n^T \varphi_n \in L^\infty$ during this period since $z_n \in L^\infty$ as depicted in (7.61). It is not difficult to find that $\Delta v(t)$ is bounded on $t \in [0, t_1)$ based on (7.52). Hence, there exists a positive constant b_n such that $|\delta_n| \leq b_n$ when $0 \leq t < t_1$.

Define the Lyapunov function $V_n = \frac{1}{2}z_n^2$. Its time derivative along (7.58) is given by

$$\dot{V}_n \leq z_n(\delta_n - \lambda_n z_n) \leq |z_n|(b_n - \lambda_n|z_n|). \tag{7.70}$$

Careful inspection of the above inequality reveals that $\dot{V}_n < 0$ if $|z_n| > \frac{b_n}{\lambda_n}$, that is, $z_n \leq \frac{b_n}{\lambda_n}$ holds for $t \in [0, t_1)$. Therefore, the condition of $\lim_{t\to t_1} |z_n| = \infty$ is excluded, which indicates that $z_n \in L^\infty$ on $t \geq 0$ and (7.60) is true.

Consequently, (7.59) and (7.60) are established according to the above performance analysis. Next, the boundedness of other closed-loop system states are proved. The boundedness of z_i ensures $\alpha_i, v(t), u(t) \in L^\infty$ with $i = 1, \ldots, n$ from (7.29), (7.31) and (7.55). Also, there are $\tilde{s}_i \in L^\infty$, $i = 1, \ldots, n$ from Lemma 7.1. By (7.9), (7.10) and (7.11), there have $\hat{s}_i \in L^\infty$, $i = 1, \ldots, n$ since the Hurwitz matrix A and $\hat{\theta}_i \in L^\infty$, $i = 1, \ldots, n$. As $x_i = \bar{g}_i s_i = \bar{g}_i(\tilde{s}_i + \hat{s}_i)$, $i = 1, \ldots, n$, it thereby indicates that x_i, $i = 1, \ldots, n$ are bounded. This completes the proof of the boundedness of all the closed-loop system states. According to (7.59) and Lemma 7.2, the ECP in (7.2) is achieved.

Now, we need to confirm that the actuator update protocol is free of Zeno phenomenon. During each data transmission interval, the time derivative of $\Delta v(t)$ along (7.34) is given by

$$\Delta\dot{v}(t) = \dot{v}(t) - \dot{v}(t_k) = \dot{v}(t),\ t_k \leq t < t_lemmak + 1, \tag{7.71}$$

where $k = 0, 1, \ldots$. Noting $v(t) = \alpha_n$ as defined in (7.31) and differentiating $v(t)$, there has

$$\dot{v}(t) = -\lambda_n \dot{z}_n. \tag{7.72}$$

From (7.58) and the analysis in Step n, one sees that δ_n and z_n are bounded, leading to $\dot{z}_n \in L^\infty$. Therefore, $\dot{v}(t) \in L^\infty$. In order to ease the derivation, $\dot{v}(t)$ is bounded by $|\dot{v}(t)| \le b_{n+1}$ with b_{n+1} being a positive constant. The equation (7.71) can be rewritten as

$$\Delta\dot{v}(t) \le b_{n+1},\ t_k \le t < t_{k+1},\ k = 0, 1, 2\ldots \tag{7.73}$$

Given that $\lim_{t\to t_k^+} |\Delta v(t)| = 0$ and $\lim_{t\to t_{k+1}^-} |\Delta v(t)| = c$, it is certain that there exists a positive constant $t^* = \frac{c}{b_{n+1}}$ such that

$$t_{k+1} - t_k \ge t^*,\ k = 0, 1, 2\ldots \tag{7.74}$$

where t^* denotes the minimal interval for the data transmission. It implies that the Zeno behavior does not occur. This completes the proof. □

Remark 7.9 Here, we pause to highlight several additional and attractive features behind the control scheme. (i) We fully consider the case of the initial condition out of the allowable set, while the control designs in [1, 20, 26, 27, 34] are unworkable under this case. We therefore define a novel Entry Capture Problem in (7.2), where the tracking error is steered into the prescribed performance set before the given time. (ii) Many previous results can only guarantee the uniform ultimate boundedness of the tracking error, such as [29, 30, 33]. Contrary to these results, the transient and steady-state performance of this section is preset by users, instead of depending on the control parameters and system states.

Remark 7.10 The guidelines for selecting the control parameters are provided here. For the settling time τ in (7.2), it determines the time instant after which the performance constraints are strictly imposed on the tracking error. Its value should satisfy $0 < T_m < \tau$ where T_m denotes the minimum time necessary for state observer and signal processing. Such assumption is widely used in many existing works [24, 36]. Although a small value of τ leads to fast convergence, it results in a large control input at the same time, especially during the initial time interval. For the control parameter $\lambda_i, i = 1, \ldots, n$, a large λ_i leads to fast convergence rate and high tracking accuracy. However, if the control gain is too high, it will make the system vulnerable to external disturbances, even leading to instability and undesirable behaviors. Therefore, the selections of these control parameters are trade-off products between achieving good tracking performance and ensuring system stability.

7.2.8 Numerical Simulations

7.2.8.1 Validity Verification of the Proposed Control

In this section, simulation studies are conducted to demonstrate our theoretical finding. As a typical example of nonlinear system, a jet engine compressor is considered

$$\begin{aligned}\dot{x}_1 &= -\frac{3}{2}x_1^2 - \frac{1}{2}x_1^3 - x_2\\ \dot{x}_2 &= u + d_2(t)\\ y &= x_1\end{aligned} \tag{7.75}$$

where the initial system states are set as $x_1(0) = x_2(0) = 0$ and $d_2(t) = 0.2\sin 0.2t$. The specifications of TPP (7.4) are given by $\delta = 0.5$, $\rho_\infty = 0.2$, $\rho_0 = 0.6$, and $\ell = 1$.

The control objective is to address the ECP with $-e_l(t) < e_1(t) < e_u(t)$, $\forall t \geq \tau$, where $y_d = \cos(0.5t)$ and $\tau = 3.0, 5.0, 7.0$, respectively. Since $f_2(\bar{x}_2) = 0$, then $\hat{\theta}_2$ in state observer is not required. The membership functions in FLSs can be described as

$$\mu_{1,j} = \exp\left(-\frac{(\hat{s}_1 - 1.5 + 0.5j)^2}{2}\right), \quad j = 1, \ldots, 5. \tag{7.76}$$

The other parameters related to the observer (7.9) and adaptive law (7.10) are selected as $k_1 = 10$, $k_2 = 5$, $\sigma_1 = 2$ and $r_1 = 1$. To fulfill the control objective, a global TPC as well as an actuator update protocol are developed in the light of (7.29)–(7.38). Therefore, their parameters are set as $\lambda_1 = 2$, $\lambda_2 = 15$, and $c = 0.3$.

The simulation results are shown in Figs. 7.4, 7.5, 7.6, 7.7 and 7.8. Although the initial value of $e_1(0)$ violates the initial performance constraints as shown in Fig. 7.4, it can be regulated into the allowable set under the proposed TPC within different settling times τ whose values are $\tau = 3.0, 5.0, 7.0$, respectively, to illustrate the effectiveness of the TPC. In this case, the traditional PPC becomes invalid since it

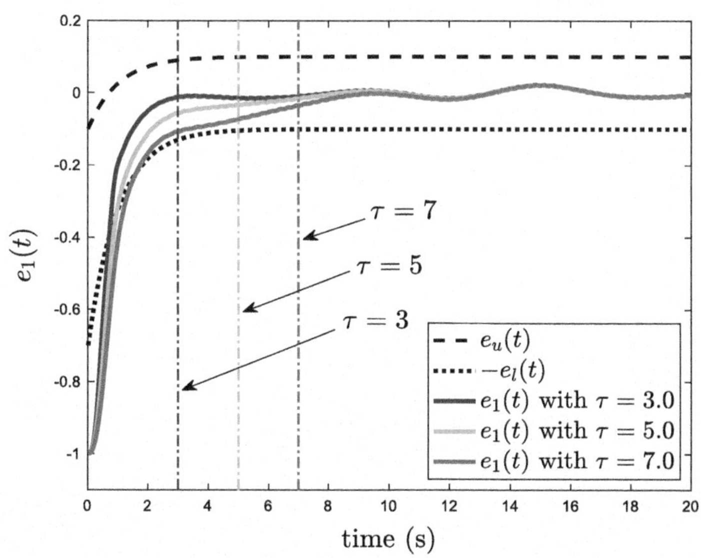

Fig. 7.4 The trajectory of the tracking error $e_1(t)$

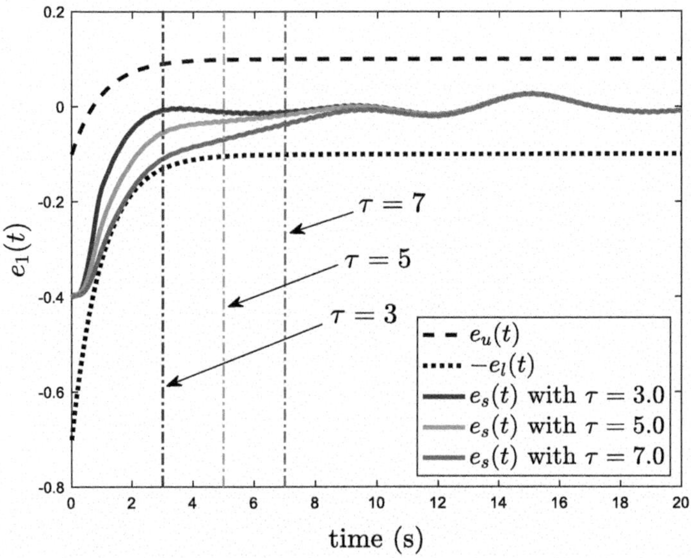

Fig. 7.5 The trajectory of the tuned tracking error $e_s(t)$

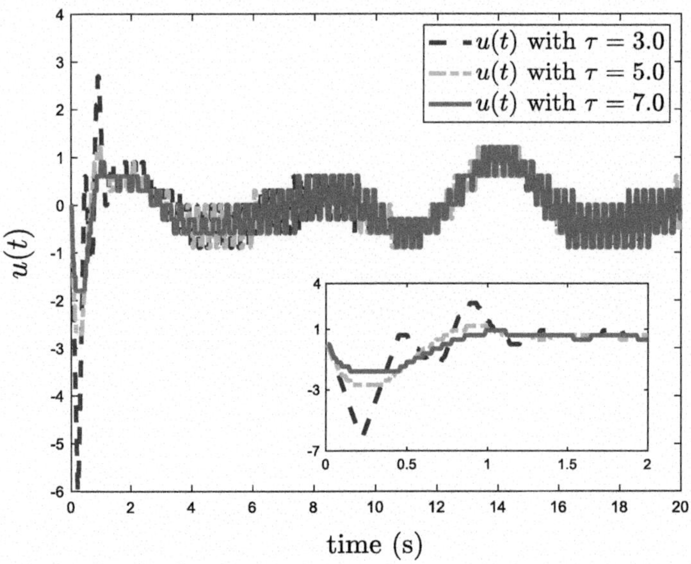

Fig. 7.6 The event-triggered control input

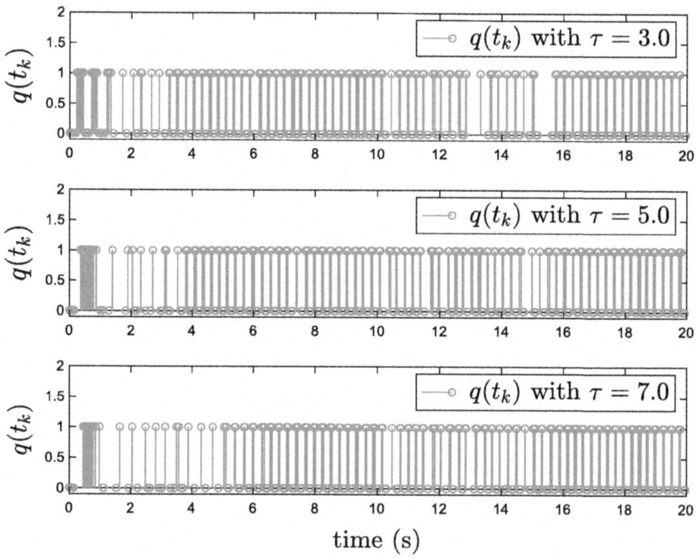

Fig. 7.7 The transmission of 1-bit string $q(t_k)$

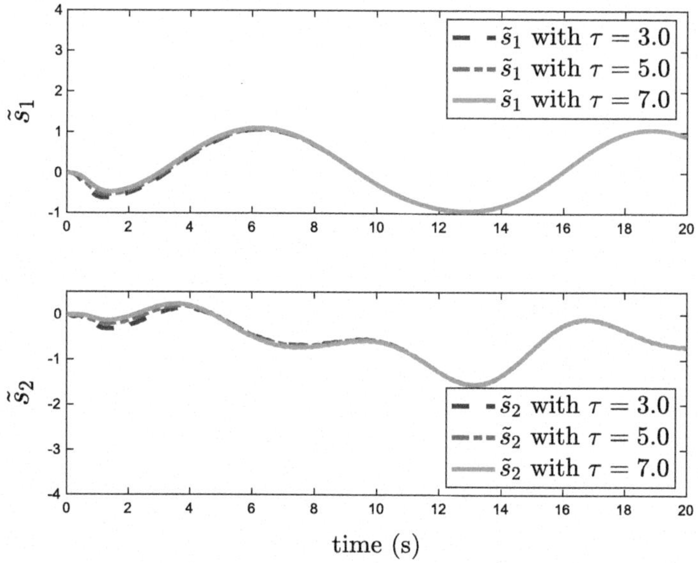

Fig. 7.8 The estimation error $\tilde{s}_i$

suffers from singularity problem if the initial condition is outside constraints. With the aid of ESF, the initial-condition restriction is removed as the ESF $e_s(t)$ is always ensured within the allowable set as depicted in Fig. 7.5, regardless of initial value of $e_1(0)$. Therefore, the ECP is well addressed. In Fig. 7.6, the control input $u(t)$ under different settling times is plotted, whose initial value is equal to zero alleviating the control peaking phenomenon. Moreover, when the event-triggered mechanism is satisfied, only 1-bit string $q(t_k)$ is required to be transmitted over the communication channel between control and actuator. Such event-triggered mechanism can dramatically reduce the communication burden. Specifically, the total numbers of event-triggered time are 411, 365, and 326 for different cases as shown in Fig. 7.7. Note that only 1 binary string $q(t_k)$, (either 1 or 0), is needed for each communication rather than finite-length string (i.e., 16-bit string), which saves communication consumption and enhances the communication security at the same time since no system information is transmitted over the network. The estimation error $\tilde{s}_i, i = 1, 2$ is depicted in Fig. 7.8, where the boundedness of $\tilde{s}_i$ is guaranteed.

7.2.8.2 Comparison With the Existing Work

We conduct a comparison simulation between our proposed TPC and Funnel Control in [13] to further illustrate the merits of the developed method. The following nonlinear system is considered:

$$\begin{cases} \dot{x}_1 = f_1(x_1) + g_1 x_2 + d_1(t) \\ \dot{x}_2 = f_2(\bar{x}_2) + g_2 u + d_2(t) \end{cases} \tag{7.77}$$

where $f_1(x_1) = 0$, $g_1 = 1$, $f_2 = -0.98 \sin x_1 - 0.2x_2$, $g_2 = 2$, $d_1 = 0.02 \sin t$, and $d_2(t) = 0.01 \sin 0.5t$. The simulation parameters of the observer and TPC are $k_1 = 1$, $k_2 = 4$, $\sigma_2 = 0.2$, $r_2 = 0.1$, $\lambda_1 = 1$, $\lambda_2 = 7$, and $c = 0.3$. The parameters of the tunnel prescribed performance are $\delta = 0.7$, $\rho_\infty = 0.6$, $\rho_0 = 0.9$, and $\ell = 0.8$. Let $\tau = 3$ and the reference trajectory is $y_d = \cos 0.5t$.

For comparison, the funnel control in [13] is given by

$$\begin{aligned} u &= -\bar{\mu}[(\cos(\chi_2) - \chi_2 \sin(\chi_2))^2 e_1^2]\chi_2^4(1 + \bar{\omega}^2)\vartheta \\ &\quad - \bar{\mu} N_2 e_1 - \bar{\mu} N_2^2 \chi_2^4 (1 + \bar{\omega}^2)\vartheta \\ N_2 &= \chi_2 \cos(\chi_2), \ \chi_2 = \frac{1}{1 - \mathcal{F}^2 e_1^2}, \\ \vartheta &= \bar{\omega} - N_2 e_1, \ \dot{\bar{\omega}} = -\bar{\mu}\bar{\omega} + u, \ e_1 = x_1 - y_d, \end{aligned} \tag{7.78}$$

where $\bar{\mu} = 8.0$ and $\mathcal{F} = \frac{\sqrt{1-\psi^2}}{\sqrt{\ell}\psi}$ with $\psi = (1 - 0.4)e^{-0.5t} + 0.4$ and $\ell = 1 - 0.4^2$. For the funnel control, the initial conditions are selected as $x_1(0) = 1.0$, and $x_2(0) = 0$.

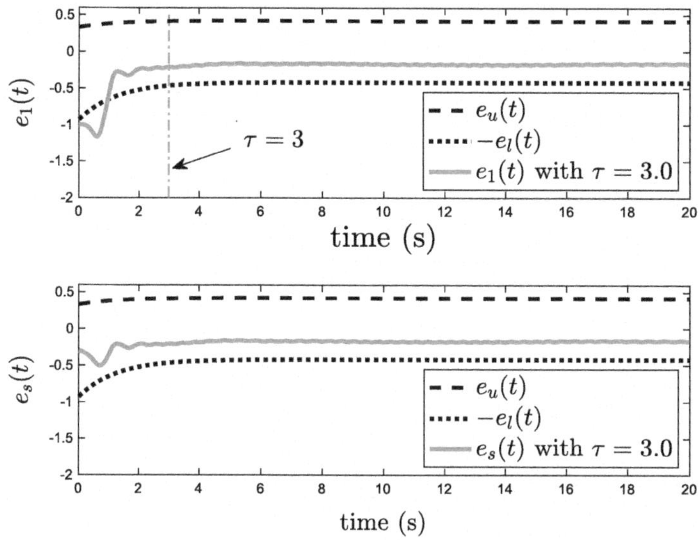

Fig. 7.9 The tracking error $e_1(t)$ and the ESF $e_s(t)$ by the TPC

The tracking error $e_1(t)$ under the proposed TPC is depicted in Fig. 7.9, whose initial condition violates the performance constraints. With the aid of TPC, $e_1(t)$ can converge into the allowable set within a given time τ, and the ECP is, therefore, addressed. By the ESF, $e_s(t)$ in Fig. 7.9 is guaranteed within the performance constraints from the beginning of system operation. The initial control input of the TPC is equal to zero as shown in Fig. 7.10, which can avoid a large initial control input. Moreover, when the event-triggered mechanism is satisfied, only 1-bit string $q(t_k)$, either 1 or 0, is needed for each transmission as depicted in Fig. 7.10. Therefore, in a 20-second simulation, the number of event triggers is 339, indicating that only 339 binary strings are transmitted over the network, which reduces the communication burden and enhances the communication security at the same time. For the funnel control in Fig. 7.11, although it can be employed to any initial condition, its design relies on specific funnel boundaries and a large initial control input is inevitable. In addition, the funnel control is executed periodically, leading to 32000-bit strings to be transmitted in a 20-second simulation with the sample time being 0.01. By comparison simulation results, the merits of the proposed TPC are illustrated.

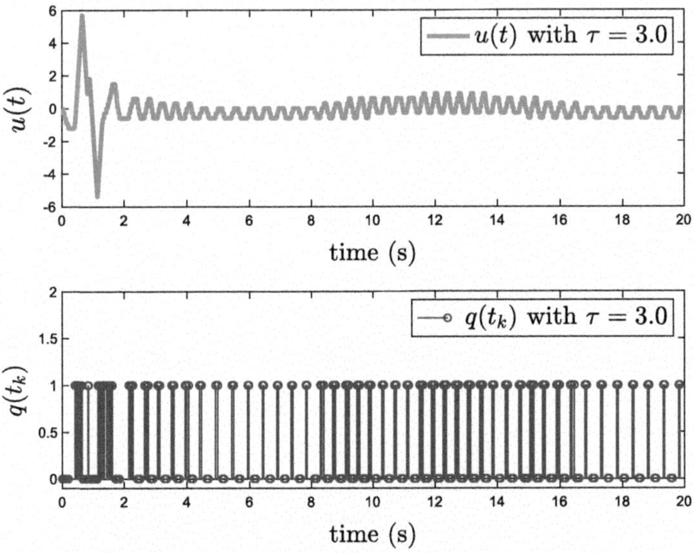

Fig. 7.10 The control input $u(t)$ and the transmission of 1-bit string $q(t_k)$

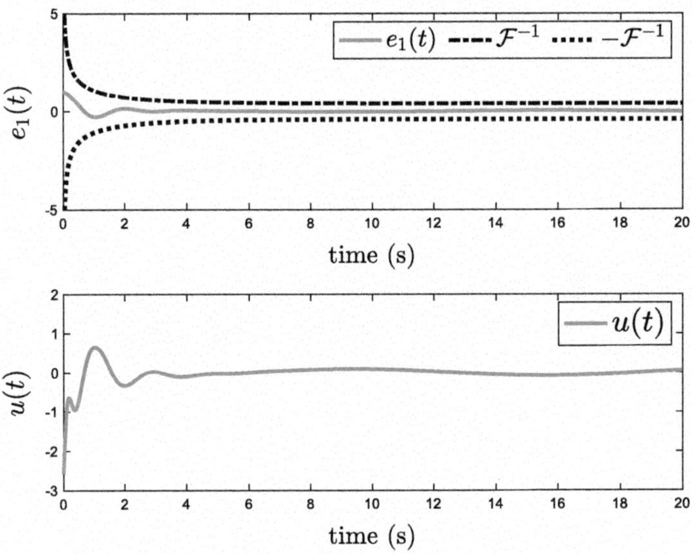

Fig. 7.11 The tracking error $e_1(t)$ and the control input $u(t)$ by the funnel control

7.3 Self-adjustable Tunnel Prescribed Control with Coded Event-Triggered Scheme

7.3.1 Problem Formulation

Consider a class of strict-feedback nonlinear systems

$$\begin{cases} \dot{x}_i = f_i(\bar{x}_i) + g_i(\bar{x}_i)x_{i+1}, \; i = 1, \ldots, n-1 \\ \dot{x}_n = f_n(\bar{x}_n) + g_n(\bar{x}_n)u \\ y = x_1 \end{cases} \tag{7.79}$$

where $x_i \in \mathbb{R}$, $i = 1, \ldots, n$, is the system state with $\bar{x}_i = [x_1, \ldots, x_i]^T \in \mathbb{R}^i$, $f_i(\cdot) : \mathbb{R}^i \to \mathbb{R}$ is an unknown but continuous function, $g_i(\cdot) : \mathbb{R}^i \to \mathbb{R}$ denotes an unknown time-varying control coefficient, $u \in \mathbb{R}$ and $y \in \mathbb{R}$ are system input and output, respectively.

Define the tracking error $e_1(t) = y(t) - y_d(t)$ with $y_d(t)$ being the desired trajectory. The control objective is to design a CEC for the nonlinear systems (7.79) under any initial condition such that:

(i) All signals in the closed-loop systems are bounded;
(ii) Only a coded m-length string is required for each communication between CEC and actuator when the event-trigger condition is satisfied; and
(iii) For any initial condition, the tracking error $e_1(t)$ can be regulated into the prescribed allowable set in a customizable manner before a given time and then be constrained within there.

Definition 7.2 For any initial tracking condition (including initial constraint violation), if the tracking error $e_1(t)$ is fully constrained by the prescribed performance boundaries right after a user-given settling time $T > 0$:

$$-e_l(t) < e_1(t) < e_u(t), \; \forall t \geq T, \tag{7.80}$$

then, such error tracking performance is said to be Entry Capture Problem, where $-e_l(t)$ and $e_u(t)$ denote lower and upper prescribed boundaries, respectively.

Assumption 7.4 The desired trajectory $y_d(t)$ and its time derivatives up to $(n+1)$th-order are known, bounded, and piece-wise continuous.

Assumption 7.5 The nonlinear function $f_i(\bar{x}_i)$, $i = 1, \ldots, n$, is unknown but certain crude information is available so that $|f_i(\bar{x}_i)| \leq b_i\phi_i(\bar{x}_i)$, $\forall t \geq 0$, where $b_i \geq 0$ is an unknown constant and $\phi_i(\bar{x}_i) \geq 0$ is a known smooth function.

Assumption 7.6 The control coefficient $g_i(\bar{x}_i)$, $i = 1, \ldots, n$, is unknown and time-varying, but away from zero, that is, there exist positive constants $\underline{g}_i$ and $\bar{g}_i$ such that $0 < \underline{g}_i \leq |g_i(\bar{x}_i)| \leq \bar{g}_i$. Without loss of generality, we assume that the signs of $g_i(\bar{x}_i)$ are known and all positive.

7.3.2 Coded Event-Triggered Scheme

With respect to the second objective of CEC, we introduce a Coded Event-triggered Scheme (CES) as follows:

$$u(t) = v(t_k),\ \forall t \in [t_k, t_{k+1}),\ k = 0, 1, \ldots \tag{7.81}$$

$$t_{k+1} = \inf\{t > t_k \mid |\Delta v(t)| \geq \omega p^{\beta}\}, \tag{7.82}$$

where $v(t)$ is the actual control to be developed, $\Delta v(t) = v(t) - u(t)$ denotes the control signal error, t_k represents the update time, p is an even number leading to base-p system (i.e., Binary, Octal number system), the selections of β and ω adopt the following rules:

$$b = q,\ \text{if } |u(t)| \in [\pi_q, \pi_{q+1}), \tag{7.83}$$

$$\beta = b = \begin{cases} \sum_{j=1}^{m} s_{j,k} p^{j-1} & \text{if } s_{m,k} < \frac{p}{2}, \\ \sum_{j=1}^{m} s_{j,k} p^{j-1} - s_c & \text{if } s_{m,k} \geq \frac{p}{2}, \end{cases} \tag{7.84}$$

$$\omega = \omega_b, \tag{7.85}$$

where $q \in \{0, 1, \ldots, s_c - 1\}$, $s_c = \frac{p^m}{2}$ with m being the length of encoded string to be transmitted, as such as there are s_c variations, we specify $0 = \pi_0 < \pi_1 < \cdots < \pi_{s_c-1} < \pi_{s_c} = +\infty$ to measure the level of control input, $\omega_0, \ldots, \omega_{s_c-1}$ are positive constants chosen by designers, and $s_{j,k} \in \{0, 1, \ldots, p-1\}$ denotes each digit value of base-p number system at k-th updated time. From the triggering condition in (7.82), it becomes essential to determine the sign of $\Delta v(t)$ for accurate encoding-decoding procedures. As shown in the encoding procedure (7.84) and the latter decoding procedure (7.87), it indicates that if $\Delta v(t) > 0$, let $s_{m,k} < \frac{p}{2}$, otherwise, $s_{m,k} \geq \frac{p}{2}$. Therefore, supposing $t \in [t_k, t_{k+1})$, $k = 0, 1, \ldots$, the encoded string S_k, used for the communication between control and actuator, is constructed by

$$S_k = s_{m,k} s_{m-1,k} \ldots s_{2,k} s_{1,k}, \tag{7.86}$$

where S_k is an m-length string to be transmitted whenever the criteria in (7.82) is satisfied. Based on (7.82) and (7.84), S_k can be seen as a sign-and-magnitude representation of the threshold for the control signal error $\Delta v(t)$. More specifically, $s_{m,k}$ denotes a sign digit, whose value determines the sign of $\Delta v(t)$ (i.e., if $s_{m,k} < \frac{p}{2}$, $\Delta v(t) \geq 0$ and if $s_{m,k} \geq \frac{p}{2}$, $\Delta v(t) < 0$); and the magnitude of the threshold for $\Delta v(t)$ is derived by the remaining digital numbers, i.e., $s_{m-1,k}, \ldots, s_{1,k}$.

If CES in (7.82) is satisfied at t_k, S_k is expected to be broadcasted to the actuator. When the actuator receives S_k, it adopts the following decoder process to update $u_d(t)$ with the aid of the last control input and parameters stored:

$$u_d(t) = v(t_k) = \begin{cases} v(t_{k-1}) + \omega p^{\beta} & \text{if } s_{m,k} < \frac{p}{2} \\ v(t_{k-1}) - \omega p^{\beta} & \text{if } s_{m,k} \geq \frac{p}{2} \end{cases} \tag{7.87}$$

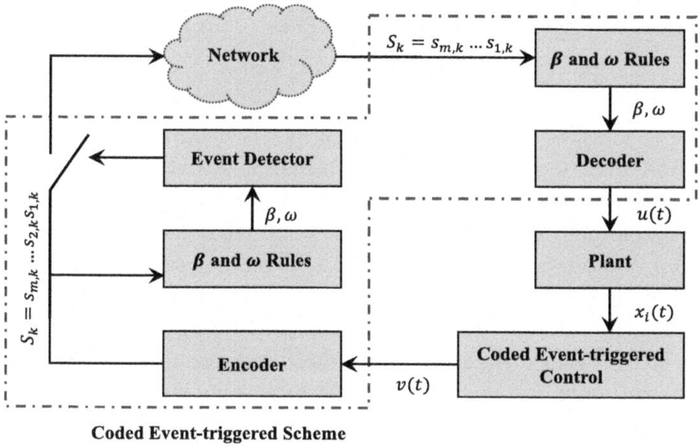

Fig. 7.12 Coded Event-triggered Scheme

where $u_d(0) = v(t_0)$, and the selections of β and ω follow the same rules in (7.83)–(7.85), which are also stored in the actuator. Based on the triggering condition in (7.82) and the decoder rule in (7.87), it can be derived that $u(t) = u_d(t)$. Throughout this section, we only use $u(t)$ to simplify the control design. In summary, the CES is illustratively described in Fig. 7.12.

To clearly illustrate the CES, we give a simple case with $p = 2$ and $m = 3$. This results in a Binary number system and 3-bit binary string is used for each communication between CEC and actuator, where each digit value of Binary number system has a value of either 1 or 0. The encoding and decoding processes in (7.86) and (7.87) are executed as shown in Table 7.1 whenever the condition in (7.82) is triggered.

Table 7.1 Encoder and decoder design under $p = 2$ and $m = 3$

Encoder	Decoder
$S_k = 000$	$u(t) = v(t_k) + \omega p^{\beta},\ \beta = 0,\ \omega = \omega_0,\ p = 2$
$S_k = 001$	$u(t) = v(t_k) + \omega p^{\beta},\ \beta = 1,\ \omega = \omega_1,\ p = 2$
$S_k = 010$	$u(t) = v(t_k) + \omega p^{\beta},\ \beta = 2,\ \omega = \omega_2,\ p = 2$
$S_k = 011$	$u(t) = v(t_k) + \omega p^{\beta},\ \beta = 3,\ \omega = \omega_3,\ p = 2$
$S_k = 100$	$u(t) = v(t_k) - \omega p^{\beta},\ \beta = 0,\ \omega = \omega_0,\ p = 2$
$S_k = 101$	$u(t) = v(t_k) - \omega p^{\beta},\ \beta = 1,\ \omega = \omega_1,\ p = 2$
$S_k = 110$	$u(t) = v(t_k) - \omega p^{\beta},\ \beta = 2,\ \omega = \omega_2,\ p = 2$
$S_k = 111$	$u(t) = v(t_k) - \omega p^{\beta},\ \beta = 3,\ \omega = \omega_3,\ p = 2$

Some salient features of such CES can be observed from the following aspects.

(i) Considering the relative threshold in [30] with $t_{k+1} = \inf\{t > t_k \mid |\Delta v(t)| \geq \bar{\delta}|u(t)| + d, 0 < \bar{\delta} < 1, d > 0\}$, it is unable to promptly respond to signal changes due to the high threshold when $|u(t)|$ is extremely large. This signal distortion often degrades tracking performance. A switching threshold scheme is proposed in [29] which switches between fixed and relative thresholds based on the magnitude of control input. However, it indicates that the transmission of raw control input between control and actuator is necessary, which aggravates the communication burden, especially when the bandwidth is limited. Therefore, our CES is designed from an encoding-decoding viewpoint by which only a concise m-length string is transmitted for each communication instead of the raw control input and m can be chosen by designers. It not only reduces the bandwidth required but also provides a balanced strategy between system performance and network constraints. Please find the following critical point for more details.

(ii) Different from the fixed threshold in [14, 17, 29, 35] and the above relative threshold where thresholds are either constants or increase monotonically with $|u(t)|$, the proposed CES adopts a piecewise increasing threshold related to $|u(t)|$ and tends to a constant threshold when $|u(t)|$ is excessively large. This dynamic threshold facilitates more accurate tracking performance due to the lower threshold when $|u(t)|$ is small, and ensures a rapid response to the signal's change when $|u(t)|$ is large with the aid of the constant threshold. Therefore, the signal distortion problem commonly observed in fixed and relative thresholds is effectively alleviated since our CES has a good balance between system performance and resource constraints. Comparative simulations are conducted to illustrate the effectiveness of our CES. Moreover, the fixed threshold can be seen as a special case of our CES if we set $p = 2$ and $m = 1$ in (7.82).

(iii) The excellent work in [30] also studies a 1-bit communication protocol for the relative threshold scheme. However, the original control input needs to be transmitted sometimes to re-calibrate the decoder due to the severe signal distortion as the aforementioned discussed. In this section, our CES not only alleviates the signal distortion problem but also addresses the challenge of co-designing an event-trigger scheme with encoding-decoding rules. As a result, the CES provides a leaner communication protocol, which is friendly for practical systems with limited communication bit resources. On the other hand, by encrypting the control signal into an encoded string S_k, we inherently improve communication security against potential cyber threats without the knowledge of encoding-decoding rules.

For the Self-adjustable Prescribed Performance design, please refer to Sects. 5.3.2 and 5.3.3 for more details, and we omit here for clarity.

7.3.3 Coded Event-Triggered Control for Second-Order Systems

To clearly illustrate our design methodology, we first consider the following second-order nonlinear system:

$$\begin{cases} \dot{x}_1 = f_1(\bar{x}_1) + g_1(\bar{x}_1)x_2, \\ \dot{x}_2 = f_2(\bar{x}_2) + g_2(\bar{x}_2)u, \end{cases} \tag{7.88}$$

where x_1 and x_2 are system states, $f_i(\bar{x}_i)$ and $g_i(\bar{x}_i), i = 1, 2$, are unknown yet smooth nonlinear functions satisfying Assumptions 7.5–7.6, and u denotes the system control input.

Step 1: We first define the tracking errors:

$$e_1 = x_1 - y_d \tag{7.89}$$

$$e_2 = x_2 - \alpha_1, \tag{7.90}$$

where α_1 represents the virtual control input to be defined shortly. In order to guarantee e_1 satisfying SPP (5.157) for all $t \geq 0$, we introduce the following error transformation function

$$z_1 = \ln\left(\frac{E_l + e_1}{E_u - e_1}\right) \tag{7.91}$$

then, by invoking (5.158)–(5.159) and (7.90), its time derivative is

$$\dot{z}_1 = \mu_1 \dot{e}_1 + \mu_2 = \mu_1(f_1 + g_1 e_2 - \dot{y}_d + g_1\alpha_1) + \mu_2 \tag{7.92}$$

where $\mu_1 = \ell(E_u + E_l)$ and $\mu_2 = \ell((\dot{e}_l - \dot{\eta}_l)(E_u - e_1) - (\dot{e}_u + \dot{\eta}_u)(E_l + e_1))$ and $\ell = \frac{1}{(e_1+E_l)(E_u-e_1)}$. From (5.158)–(5.161) and the property (iv) in Definition 5.5, it can be derived that $E_u(t) + E_l(t) > 0$ and $(E_u(t) - e_1(t))(E_l(t) + e_l(t)) > 0$ are bounded functions for $e_1(t)$ in the compact set $\Omega_{e_1} = \{e_1(t) \in \mathbb{R} : -E_l(t) < e_1(t) < E_u(t)\}$. Therefore, we have $\mu_1 \neq 0$ and $\mu_1 \in L_\infty$. Then, the time derivative of $\frac{1}{2}z_1^2$ along (7.92) is:

$$\begin{aligned} z_1\dot{z}_1 &= z_1\mu_1(f_1 + g_1 e_2 - \dot{y}_d) + z_1\mu_1 g_1\alpha_1 + z_1 u_2 \\ &= z_1\mu_1 g_1\alpha_1 + \Xi_1, \end{aligned} \tag{7.93}$$

where $\Xi_1 = z_1\mu_1(f_1 + g_1 e_2 - \dot{y}_d) + z_1\mu_2$. Upon using Assumptions 7.4–7.6 and Young's inequality, we have

$$-z_1\mu_1\dot{y}_d \leq \underline{g}_1 z_1^2\mu_1^2\dot{y}_d^2 + \frac{1}{4\underline{g}_1}, \tag{7.94}$$

$$z_1\mu_1 f_1 \le \underline{g}_1 z_1^2\mu_1^2 b_1^2\phi_1^2 + \frac{1}{4\underline{g}_1}, \tag{7.95}$$

$$z_1\mu_1 g_1 e_2 \le \underline{g}_2 z_1^2\mu_1^2 e_2^2 + \frac{\bar{g}_1^2}{4\underline{g}_2}, \tag{7.96}$$

$$z_1\mu_2 \le \underline{g}_1 z_1^2\mu_2^2 + \frac{1}{4\underline{g}_1}. \tag{7.97}$$

Therefore, Ξ_1 in (7.93) is bounded by

$$\Xi_1 \le \underline{g}_1\theta_1 z_1^2\Phi_1 + \underline{g}_2 z_1^2\mu_1^2 e_2^2 + \frac{3}{4\underline{g}_1} + \frac{\bar{g}_1^2}{4\underline{g}_2} \tag{7.98}$$

with

$$\theta_1 = \max\left\{1, b_1^2\right\}, \tag{7.99}$$

$$\Phi_1 = \mu_1^2\dot{y}_d^2 + \mu_1^2\phi_1^2 + \mu_2^2. \tag{7.100}$$

We develop the virtual control input α_1 as

$$\alpha_1 = -\frac{1}{\mu_1}(c_1 z_1 + z_1\hat{\theta}_1\Phi_1), \tag{7.101}$$

$$\dot{\hat{\theta}}_1 = r_1 z_1^2\Phi_1 - \sigma_1\hat{\theta}_1, \;\; \hat{\theta}_1(0) \ge 0 \tag{7.102}$$

where c_1, r_1, and σ_1 are positive constants, and $\hat{\theta}_1$ is the estimation of θ_1.

Consider the following Lyapunov function

$$V_1 = \frac{1}{2}z_1^2 + \frac{\underline{g}_1}{2r_1}\tilde{\theta}_1^2,$$

where $\tilde{\theta}_1 = \theta_1 - \hat{\theta}_1$. Then, from (7.93), (7.98), (7.101) and (7.102), its time derivative can be derived as

$$\begin{aligned}\dot{V}_1 =& z_1\mu_1 g_1\alpha_1 + \Xi_1 - \frac{\underline{g}_1}{r_1}\tilde{\theta}_1\dot{\hat{\theta}}_1 \\ \le& -c_1\underline{g}_1 z_1^2 + \underline{g}_2 z_1^2\mu_1^2 e_2^2 + \frac{3}{4\underline{g}_1} + \frac{\bar{g}_1^2}{4\underline{g}_2} + \frac{\underline{g}_1\sigma_1}{r_1}\tilde{\theta}_1\hat{\theta}_1.\end{aligned}$$

Based on the definition of $\tilde{\theta}_1$, it yields

$$\tilde{\theta}_1\hat{\theta}_1 = \tilde{\theta}_1(\theta_1 - \tilde{\theta}_1) \le -\frac{1}{2}\tilde{\theta}_1^2 + \frac{1}{2}\theta_1^2.$$

We then have

$$\dot{V}_1 \leq -c_1\underline{g}_1 z_1^2 - \bar{\sigma}_1\tilde{\theta}_1^2 + \underline{g}_2 z_1^2\mu_1^2 e_2^2 + \varepsilon_1, \tag{7.103}$$

where $\bar{\sigma}_1 = \frac{\underline{g}_1\sigma_1}{2r_1}$ and $\varepsilon_1 = \frac{3}{4\underline{g}_1} + \frac{\bar{g}_1^2}{4\underline{g}_2} + \frac{\underline{g}_1\sigma_i}{2r_1}\theta_1^2$ is a bounded signal. Notice that the item $\underline{g}_2 z_1^2\mu_1^2 e_2^2$ will be tackled in the next step.

Step 2: From (7.88) and (7.90), the time derivative of e_2 is

$$\dot{e}_2 = \dot{x}_2 - \dot{\alpha}_1 = f_2 + g_2 u - \dot{\alpha}_1, \tag{7.104}$$

with

$$\dot{\alpha}_1 = \frac{\partial \alpha_1}{\partial x_1}(f_1 + g_1 x_2) + \Delta\alpha_1, \tag{7.105}$$

where $\Delta\alpha_1 = \sum_{k=0}^{1} \frac{\partial \alpha_1}{\partial y_d^{(k)}} y_d^{(k+1)} + \sum_{k=0}^{1} \frac{\partial \alpha_1}{\partial \rho^{(k)}} \rho^{(k+1)} + \sum_{k=0}^{1} \frac{\partial \alpha_1}{\partial \eta_u^{(k)}} \dot{\eta}_u^{(k+1)} + \sum_{k=0}^{1} \frac{\partial \alpha_1}{\partial \eta_l^{(k)}} \dot{\eta}_l^{(k+1)} + \frac{\partial \alpha_1}{\partial \hat{\theta}_1}\dot{\hat{\theta}}_1$, which is computable. According to the definition of $\Delta v(t)$, we can obtain

$$u(t) = v(t) - \Delta v(t),\ |\Delta v(t)| \leq \bar{p}, \tag{7.106}$$

where $\bar{p} = \max\{\omega_0 p^0, \omega_1 p, \ldots, \omega_{s_c} p^{s_c}\}$ is a positive constant. We then consider the Lyapunov function V_2:

$$V_2 = V_1 + \frac{1}{2}e_2^2 + \frac{\underline{g}_2}{2r_2}\tilde{\theta}_2^2, \tag{7.107}$$

where $\tilde{\theta}_2 = \theta_2 - \hat{\theta}_2$, $\hat{\theta}_2$ is the estimation of θ_2 in (7.111), and r_2 is a positive constant. The time derivative of V_2, along (7.104)–(7.106), is derived

$$\begin{aligned} \dot{V}_2 =& \dot{V}_1 + e_2\dot{e}_2 - \frac{\underline{g}_2}{r_2}\tilde{\theta}_2\dot{\hat{\theta}}_2 \\ \leq& -c_1\underline{g}_1 z_1^2 - \bar{\sigma}_1\tilde{\theta}_1^2 + e_2 g_2 v - \frac{\underline{g}_2}{r_2}\tilde{\theta}_2\dot{\hat{\theta}}_2 + \Xi_2 + \varepsilon_1 \end{aligned} \tag{7.108}$$

where $\Xi_2 = e_2(f_2 - \frac{\partial \alpha_1}{\partial x_1}(f_1 + g_1 x_2) - \Delta\alpha_1) - e_2 g_2 \Delta v + \underline{g}_2 z_1^2\mu_1^2 e_2^2$. Using Young's inequality, we can also expand Ξ_2 as inequalities in (7.94)–(7.97). Note that the control input signal will be updated whenever the coded event-triggered scheme (7.82) is triggered, which indicates that $|\Delta v| \leq \omega p^\beta$ holds. As ω, β and p are all bounded numbers, from (7.106), it can be derived that $\bar{p}$ is a bounded constant, leading to:

$$-e_2 g_2 \Delta v \leq \underline{g}_2 \bar{e}_2^2 + \frac{\bar{g}_2^2}{4\underline{g}_2} \bar{p}^2. \tag{7.109}$$

It yields that

$$\Xi_2 \leq \underline{g}_2 \theta_2 e_2^2 \Phi_2 + \frac{3}{4\underline{g}_2} + \frac{\bar{g}_1^2}{4\underline{g}_2} + \frac{\bar{g}_2^2}{4\underline{g}_2} \bar{p}^2, \tag{7.110}$$

where

$$\theta_2 = \max\left\{1, b_1^2, b_2^2\right\}, \tag{7.111}$$

$$\Phi_2 = \left(\frac{\partial \alpha_1}{\partial x_1}\phi_1\right)^2 + \left(\frac{\partial \alpha_1}{\partial x_1}x_2\right)^2 + (\Delta\alpha_1)^2 + \mu_1^2 z_1^2 + \phi_2^2 + 1. \tag{7.112}$$

To move forward, the actual control is developed as

$$v(t) = -(c_2 e_2 + \hat{\theta}_2 e_2 \Phi_2), \tag{7.113}$$

$$\dot{\hat{\theta}}_2 = r_2 e_2^2 \Phi_2 - \sigma_2 \hat{\theta}_2, \ \ \hat{\theta}_2(0) \geq 0 \tag{7.114}$$

where $c_2, r_2, \sigma_2 > 0$ and $\hat{\theta}_2$ is the estimation of θ_2. By (7.110)–(7.114), the inequality (7.108) can be rewritten as

$$\begin{aligned} \dot{V}_2 \leq & - c_1 \underline{g}_1 z_1^2 - \bar{\sigma}_1 \tilde{\theta}_1^2 - \underline{g}_2 c_2 e_2^2 + \frac{\sigma_2 \underline{g}_2}{r_2} \tilde{\theta}_2 \hat{\theta}_2 \\ & + \frac{3}{4\underline{g}_2} + \frac{\bar{g}_1^2}{4\underline{g}_2} + \frac{\bar{g}_2^2}{4\underline{g}_2} \bar{p}^2 + \varepsilon_1. \end{aligned} \tag{7.115}$$

Based on the definition of $\tilde{\theta}_2$, we obtain

$$\tilde{\theta}_2 \hat{\theta}_2 = \tilde{\theta}_2(\theta_2 - \tilde{\theta}_2) \leq -\frac{1}{2}\tilde{\theta}_2^2 + \frac{1}{2}\theta_2^2. \tag{7.116}$$

By substituting (7.116) into (7.115) yields

$$\dot{V}_2 \leq - c_1 \underline{g}_1 z_1^2 - \bar{\sigma}_1 \tilde{\theta}_1^2 - \underline{g}_2 c_2 e_2^2 - \bar{\sigma}_2 \tilde{\theta}_2^2 + \varepsilon_2, \tag{7.117}$$

where $\bar{\sigma}_2 = \frac{\underline{g}_2 \sigma_2}{2r_2}$ and $\varepsilon_2 = \frac{\underline{g}_2 \sigma_2}{2r_2}\theta_2^2 + \frac{3}{4\underline{g}_2} + \frac{\bar{g}_1^2}{4\underline{g}_2} + \frac{\bar{g}_2^2}{4\underline{g}_2}\bar{p}^2 + \varepsilon_1$.

To summarize, we establish the following theorem.

Theorem 7.2 *Consider the second-order nonlinear system* (7.88) *under Assumptions 7.4–7.6. The virtual control* α_1 *and the actual control input* $v(t)$ *are developed* *in* (7.101) *and* (7.113). *The adaptive laws are given by* (7.102) *and* (7.114). *Following the Coded Event-triggered Scheme* (7.81)–(7.82), *the proposed control method guarantees the following.*

(i) The tracking error e_1 *satisfies Entry Capture Problem property since it is strictly constrained by the prescribed performance:* $-e_l(t) < e_1(t) < e_u(t)$ *right after a user-given finite time* T *as presented in Definition 7.2;*
(ii) All signals in the closed-loop system are guaranteed to be bounded regardless of initial conditions; and
(iii) Zeno behavior is excluded.

Proof By revisiting (7.117), there is

$$\begin{aligned}\dot{V}_2 \leq &- c_1 \underline{g}_1 z_1^2 - \bar{\sigma}_1 \tilde{\theta}_1^2 - \underline{g}_2 c_2 e_2^2 - \bar{\sigma}_2 \tilde{\theta}_2^2 + \varepsilon_2 \\ \leq &- \pi V_2 + \varepsilon_2,\end{aligned}$$

where $\pi = \min\{2c_1\underline{g}_1, 2c_2\underline{g}_2, 2\frac{r_1\bar{\sigma}_1}{\underline{g}_1}, 2\frac{r_2\bar{\sigma}_2}{\underline{g}_2}\} > 0$. Subsequently, we can have

$$0 \leq V_2(t) \leq \frac{\varepsilon_2}{\pi} + (V_2(0) - \frac{\varepsilon_2}{\pi})e^{-\pi t}.$$

It can be concluded that V_2 is bounded which guarantees the boundedness of z_1, e_2, $\tilde{\theta}_1$, and $\tilde{\theta}_2$. From the error transformation function in (7.91), it is certain that $-E_l(t) < e_1(t) < E_u(t)$ holds due to $z_1 \in L_\infty$. According to Lemma 5.8, e_1 satisfies $-e_l(t) < e_1(t) < e_u(t)$ for $t \geq T$, which indicates that e_1 follows the Entry Capture Problem in Definition 7.2.

Next, other tracking signals in the closed-loop system are proved to be bounded. Considering the adaptive laws (7.102) and (7.114), θ_1, θ_2, $\dot{\hat{\theta}}_1$ and $\dot{\hat{\theta}}_2$ remain bounded. The virtual control input α_1 defined in (7.101) belong to L_∞. Moreover, by (7.113), the boundedness of actual control signal $v(t)$ is guaranteed. From (7.89) and (7.90), it can be derived that x_1 and x_2 also keep bounded. All the closed-loop signals are therefore bounded for the second-order nonlinear system under any initial condition.

To illustrate the exclusion of Zeno behavior, the time derivative of $\Delta v(t)$ during the inter-execution interval is given

$$\Delta \dot{v}(t) = \dot{v}(t) - \dot{v}(t_k) = \dot{v}(t), \ t_k \leq t < t_{k+1},$$

where $k = 1, 2, \ldots$ By (7.113), it yields $\dot{v}(t) = \sum_{k=1}^{2} \frac{\partial v}{\partial x_k}\dot{x}_k + \sum_{k=0}^{2} \frac{\partial v}{\partial y_d^{(k)}} y_d^{(k+1)} + \sum_{k=1}^{2} \frac{\partial v}{\partial \hat{\theta}_k}\dot{\hat{\theta}}_k + \sum_{k=0}^{2} \frac{\partial v}{\partial \rho^{(k)}} \rho^{(k+1)} + \sum_{k=0}^{2} \frac{\partial v}{\partial \eta_u^{(k)}} \dot{\eta}_u^{(k+1)} + \sum_{k=0}^{2} \frac{\partial v}{\partial \eta_l^{(k)}} \dot{\eta}_l^{(k+1)}$. In accordance with Assumptions 7.4–7.6, Definition 5.5, prescribed performance functions $e_u(t)$ and $e_l(t)$ defined in (5.160)–(5.161), and the fact that $\dot{e}_2$ governed by (7.104) and other closed-loop signals remain bounded, the boundedness of $\dot{v}$ is ensured.

For convenience, the upper bound of $|\dot{v}|$ can be specified by a positive constant κ. Therefore, we can obtain

$$|\Delta\dot{v}(t)| \leq \kappa,\ t_k \leq t < t_{k+1},$$

where $k = 1, 2, \ldots$ Given that $\lim_{t\to t_k^+} |\Delta v| = 0$ and $\lim_{t\to t_{k+1}^-} |\Delta v| = \omega p^{\beta}$, it indicates that there exists a strictly positive constant $t^* = \frac{\omega p^{\beta}}{\kappa}$ such that

$$t_{k+1} - t_k \geq t^*,\ k = 1, 2, \ldots$$

with t^* denoting minimal inter-transmission interval. Thereby, the Zeno behavior is excluded, and the proof is completed. □

7.3.4 *Coded Event-Triggered Control for High-Order Systems*

In this section, the proposed CEC will be extended to nth-order systems (7.79). Instead of step-by-step procedure, its control design is omitted here for conciseness since its stability analysis is identical to the second-order system (7.88).

We first define the tracking error variables:

$$e_1 = x_1 - y_d \tag{7.118}$$

$$e_i = x_i - \alpha_{i-1},\ i = 2, \ldots, n, \tag{7.119}$$

where α_{i-1} represent virtual control inputs. In order to guarantee e_1 fulfilling the Self-adjustable Prescribed Performance (5.157), we introduce the following error transformation function:

$$z_1 = \ln\left(\frac{E_l + e_1}{E_u - e_1}\right) \tag{7.120}$$

Then, the virtual control inputs are designed as

$$\alpha_1 = -\frac{1}{\mu_1}(c_1 z_1 + z_1\hat{\theta}_1\Phi_1), \tag{7.121}$$

$$\alpha_i = -(c_i e_i + \hat{\theta}_i e_i \Phi_i),\ i = 2, \ldots, n-1, \tag{7.122}$$

and the adaptive laws are updated by

$$\dot{\hat{\theta}}_1 = r_1 z_1^2 \Phi_1 - \sigma_1\hat{\theta}_1,\ \ \hat{\theta}_1(0) \geq 0, \tag{7.123}$$

$$\dot{\hat{\theta}}_i = r_i e_i^2 \Phi_i - \sigma_i\hat{\theta}_i,\ \ \hat{\theta}_i(0) \geq 0, \tag{7.124}$$

for $i = 2, \ldots, n-1$, where c_i, r_i and σ_i, $i = 1, \ldots, n-1$, are positive constants. $\hat{\theta}_i$ is the estimation of θ_i, and Φ_i is a computable function, where θ_i and Φ_i are

$$\theta_i = \max\{1, b_1^2, b_2^2, \ldots, b_i^2\},\ i = 1, \ldots, n-1, \tag{7.125}$$

$$\Phi_1 = \mu_1^2 \dot{y}_d^2 + \mu_1^2 \phi_1^2 + \mu_2^2, \tag{7.126}$$

$$\Phi_2 = \left(\frac{\partial \alpha_1}{\partial x_1}\phi_1\right)^2 + \left(\frac{\partial \alpha_1}{\partial x_1}x_2\right)^2 + (\Delta\alpha_1)^2 + \mu_1^2 z_1^2 + \phi_2^2, \tag{7.127}$$

$$\Phi_i = \sum_{k=1}^{i-1}\left(\frac{\partial \alpha_{i-1}}{\partial x_k}\phi_k\right)^2 + \sum_{k=1}^{i-1}\left(\frac{\partial \alpha_{i-1}}{\partial x_k}x_{k+1}\right)^2 + (\Delta\alpha_{i-1})^2 + e_{i-1}^2 + \phi_i^2,\ i = 3, \ldots, n-1, \tag{7.128}$$

where $\mu_1 = \ell(E_u + E_l)$ and $\mu_2 = \ell(\dot{e}_l - \dot{\eta}_l)(E_u - e_1) - (\dot{e}_u + \dot{\eta}_u)(E_l + e_1)$ and $\ell = \frac{1}{(e_1+E_l)(E_u-e_1)}$. We can also derive that $E_u(t) + E_l(t) > 0$ and $(E_u(t) - e_1(t))(E_l(t) + e_l(t)) > 0$ are bounded functions if $e_1(t)$ belongs to a compact set $\Omega_{e_1} = \{e_1(t) \in \mathbb{R} : -E_l(t) < e_1(t) < E_u(t)\}$. Therefore, we have $\mu_1 \neq 0$ and $\mu_1 \in L_\infty$. Identically, the actual control input and the adaptive law are developed:

$$v(t) = -(c_n e_n + \hat{\theta}_n e_n \Phi_n), \tag{7.129}$$

$$\dot{\hat{\theta}}_n = r_n e_n^2 \Phi_n - \sigma_n \hat{\theta}_n,\ \ \hat{\theta}_n(0) \geq 0, \tag{7.130}$$

where $r_n, r_n, \sigma_n > 0, \hat{\theta}_n$ denotes the estimation of θ_n and Φ_n is a computable function. Similar to (7.125)–(7.128), we have

$$\theta_n = \max\{1, b_1^2, b_2^2, \ldots, b_n^2\}, \tag{7.131}$$

$$\Phi_n = \sum_{k=1}^{n-1}\left(\frac{\partial \alpha_{i-1}}{\partial x_k}\phi_k\right)^2 + \sum_{k=1}^{n-1}\left(\frac{\partial \alpha_{i-1}}{\partial x_k}x_{k+1}\right)^2 + (\Delta\alpha_{n-1})^2 + e_{n-1}^2 + \phi_n^2 + 1. \tag{7.132}$$

In the following theorem, we summarize the result on the Coded Event-triggered Control for the nth-order nonlinear systems.

Theorem 7.3 *Consider nth-order nonlinear systems* (7.79) *under Assumptions* 7.4–7.6. *If the control inputs and the adaptive laws are designed as* (7.123)–(7.130), *by adopting the Coded Event-triggered Scheme* (7.81)–(7.82), *we can guarantee the following properties:*

(i) Given any initial condition, the tracking error $e_1(t)$ is constrained by the prescribed performance: $-e_l(t) < e_1(t) < e_u(t)$, for $t \geq T$ with T being a preassigned finite time, that is, e_1 fulfills Entry Capture Problem properties in Definition 7.2;

(ii) All signals in the closed-loop system are guaranteed to be bounded regardless of initial conditions;
(iii) Zeno behavior is excluded.

Proof Consider the following Lyapunov candidates:

$$V_1 = \frac{1}{2}z_1^2 + \frac{\underline{g}_1}{2r_1}\tilde{\theta}_1^2,$$
$$V_i = V_{i-1} + \frac{1}{2}e_n^2 + \frac{\underline{g}_n}{2r_n}\tilde{\theta}_n^2.$$

where $\tilde{\theta}_i = \theta_i - \hat{\theta}_i, i = 1, \ldots, n$. Similar to the stability analysis in Theorem 7.2, it can be also derived:

$$\begin{aligned}\dot{V}_n \leq & -c_1\underline{g}_1 z_1^2 - \sum_{k=2}^{n} c_k\underline{g}_k e_k^2 - \sum_{k=1}^{n} \bar{\sigma}_k\tilde{\theta}_k^2 + \varepsilon_n \\ \leq & -\pi V_n + \varepsilon_n,\end{aligned}$$

where $\pi = \min\{2c_1\underline{g}_1, \ldots, 2c_n\underline{g}_n, 2\frac{r_1\bar{\sigma}_1}{\underline{g}_1}, \ldots, 2\frac{r_n\bar{\sigma}_n}{\underline{g}_n}\} > 0$, $\varepsilon_n = \frac{\underline{g}_n\sigma_n}{2r_n}\theta_n^2 + \frac{3}{4\underline{g}_n} + \frac{\bar{g}_{n-1}^2}{4\underline{g}_n} + \frac{\bar{g}_n^2}{4\underline{g}_n}\bar{p}^2 + \varepsilon_1$. In accordance with (7.81)–(7.82), we have $|\Delta v(t)| \leq \bar{p}$, where $\bar{p} = \max\{\omega_0 p^0, \omega_1 p, \ldots, \omega_{s_c} p^{s_c}\}$ is a bounded constant. Then, we obtain

$$0 \leq V_n(t) \leq \frac{\varepsilon_n}{\pi} + (V_n(0) - \frac{\varepsilon_n}{\pi})e^{-\pi t}.$$

It indicates that $V_n(t)$ is bounded, leading to $z_1, e_2, \ldots, e_n, \tilde{\theta}_1, \ldots, \tilde{\theta}_n \in L_\infty$. According to the error transformation function (7.120), we can derive that e_1 is evolving within the SPP, namely, $-E_l(t) < e_1(t) < E_u(t)$. Based on Lemma 5.8, the Entry Capture Problem is achieved, that is, $-e_l(t) < e_1(t) < e_u(t)$ for all $t \geq T$.

Next, the boundedness of other signals in the closed-loop system is proved. From (7.123)–(7.125) and (7.130)–(7.131), and the definition of $\tilde{\theta}_i$, it leads to $\hat{\theta}_i$ and $\dot{\hat{\theta}}_i \in L_\infty$, for $i = 1, \ldots, n$. By (7.121)–(7.122) and (7.129), the boundedness of control input signals $\alpha_1, \ldots, \alpha_{n-1}, v$ is guaranteed. Considering (7.118)–(7.119), $x_1, \ldots, x_n$ belong to L_∞. Therefore, all signals in the closed-loop system are bounded.

To verify the exclusion of Zeno phenomenon, the time derivative of $\Delta v(t)$ before next triggering is given

$$\Delta\dot{v}(t) = \dot{v}(t) - \dot{v}(t_k) = \dot{v}(t),\ t_k \leq t < t_{k+1}.$$

where $k = 1, 2, \ldots$ By invoking (7.129), there has $\dot{v}(t) = \sum_{k=1}^{n} \frac{\partial v}{\partial x_k} \dot{x}_k + \sum_{k=0}^{n} \frac{\partial v}{\partial y_d^{(k)}} y_d^{(k+1)} + \sum_{k=1}^{n} \frac{\partial v}{\partial \hat{\theta}_k} \dot{\hat{\theta}}_k + \sum_{k=0}^{n} \frac{\partial v}{\partial \rho^{(k)}} \rho^{(k+1)} + \sum_{k=0}^{n} \frac{\partial v}{\partial \eta_u^{(k)}} \dot{\eta}_u^{(k+1)} + \sum_{k=0}^{n} \frac{\partial v}{\partial \eta_l^{(k)}} \dot{\eta}_l^{(k+1)}$. Since all the signals in the closed-loop system are bounded, from Assumptions 7.4–7.6, Definition 5.5, prescribed performance functions $e_u(t)$ and $e_l(t)$ defined in (5.160)–(5.161), $\dot{v}(t)$ is, therefore, bounded by

$$|\Delta \dot{v}(t)| \leq \kappa, \; t_k \leq t < t_{k+1},$$

where $k = 1, 2, \ldots$ Given that $\lim_{t \to t_k^+} |\Delta v| = 0$ and $\lim_{t \to t_{k+1}^-} |\Delta v| = \omega p^{\beta}$, it indicates that there exists a strictly positive constant $t^* = \frac{\omega p^{\beta}}{\kappa}$ such that

$$t_{k+1} - t_k \geq t^*, \; k = 1, 2, \ldots$$

with t^* being the minimal inter-transmission interval. Thereby, the Zeno behavior is excluded and proof is completed. □

7.3.5 Numerical Simulations

To illustrate the effectiveness of our proposed CEC, some comparative simulations are conducted in this section. Consider the following second-order nonlinear system:

$$\begin{cases} \dot{x}_1 = f_1(\bar{x}_1) + g_1(\bar{x}_1) x_2, \\ \dot{x}_2 = f_2(\bar{x}_2) + g_2(\bar{x}_2) u, \end{cases} \tag{7.133}$$

where $f_1(\bar{x}_1) = x_1^2 + 0.1\cos(0.5x_1)$, $f_2(\bar{x}_2) = 4x_1x_2 + x_1 e^{-|x_2|} + 0.05\sin(x_1x_2)$, $g_1(\bar{x}_1) = 5 + 0.5\sin(x_1)$, and $g_2(\bar{x}_2) = 3 + 0.2\cos(x_1x_2)$. By Assumption 7.5, it can be derived that $|f_1(\bar{x}_1)| \leq x_1^2 + 0.1$ with $b_1 = 1$ and $\phi_1 = 1 + x_1^2$, and $|f_2(\bar{x}_2)| \leq 3x_1^2 + 2x_2^2 + 0.3$ with $b_2 = 3$ and $\phi_2 = x_1^2 + x_2^2 + 1$. The initial condition is set by $x_1(0) = 1.5$ and $x_2(0) = 0$. The desired tracking trajectory is $y_d = \sin(0.5t)$. The control parameters are selected as: $m = 3$ (i.e., $\beta = 0, 1, 2, 3$ and $\omega = 0.3, 0.4, 0.5, 0.6$), $T = 4$, $c_1 = 2$, $c_2 = 15$, $\sigma_1 = \sigma_2 = 0.01$, $r_1 = r_2 = 0.002$, $l = 0.6$, $\lambda = 2$, $\delta = 0.5$, $\rho_0 = 1$, $\rho_\infty = 0.4$, and $\varsigma = 1$. Moreover, our CEC is compared with our control design with relative threshold [30] and switching threshold [29].

(Case I) The simulation results of our CEC under different threshold strategies are shown in Figs. 7.13 and 7.14. From Fig. 7.13, the initial tracking error $e_1(0)$ violates the initial performance constraints colored in brown, leading to the classical prescribed control methods being inapplicable. By our CEC, we can regulate $e_1(0)$ into the allowable set within the given time T. With the aid of SPP, bounded yet customizable virtual performance boundaries are established during the initial interval as shown in blue color. Therefore, the severe overshoot or unconstrained behaviors over $[0, T)$ are effectively avoided. On the other hand, the triggering time of three

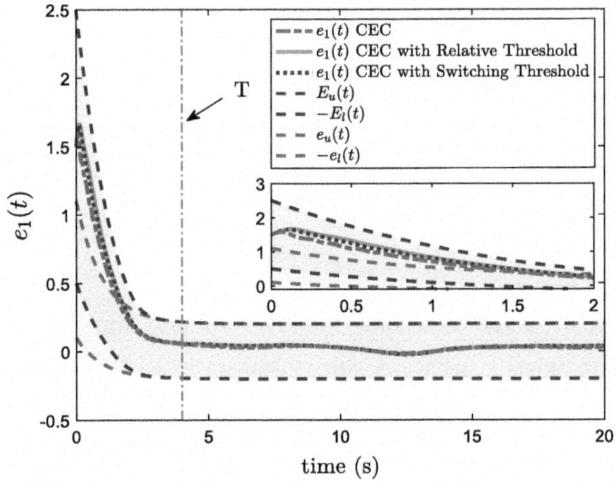

Fig. 7.13 The tracking error $e_1(t)$ for case I

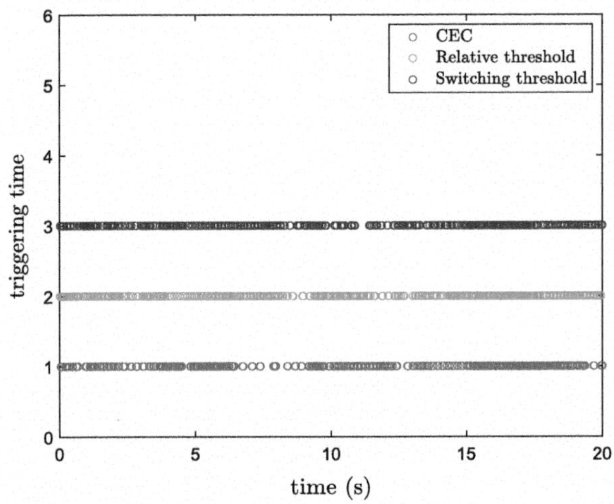

Fig. 7.14 The triggering time for case I

threshold strategies is shown in Fig. 7.14, and the number of triggering events and bit assumptions are summarized in the second and third columns of Table 7.2, in which we exclude the initial signal transmission. Due to $m = 3$ in the CES, it indicates that only a 3-bit string is transmitted for each communication when the event condition is satisfied. Compared with the relative threshold and the switching threshold where control inputs should be encoded by 8-bit strings before each transmission, our CES consumes fewer bits while maintaining the tracking performance, which is significant for the practical system with limited communication bandwidth.

Table 7.2 The number of triggering events for different strategies

Different strategies	Case I		Case II	
	Trigger number	Bit consumption	Trigger number	Bit consumption
CES	332	996	460	1380
Relative threshold	339	2712	X	X
Switching threshold	373	2984	403	3224

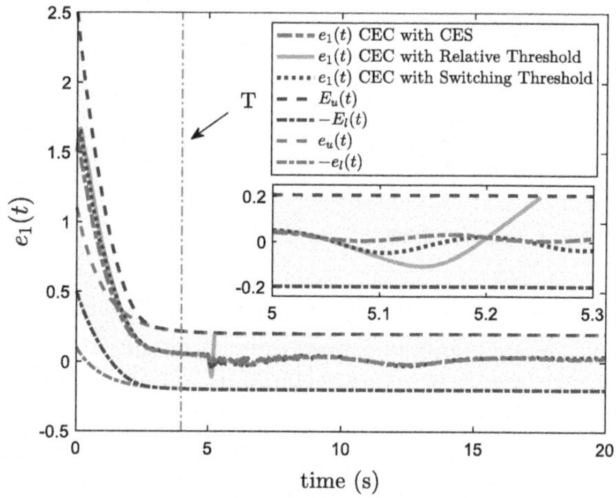

Fig. 7.15 The tracking error $e_1(t)$ for case II

(Case II) To further illustrate the effectiveness of our CES, we consider the introduction of external disturbance $d(t) = 2\cos 0.5t$ during 7–10s in the second row of (7.133). The comparative results under different threshold strategies are shown in Figs. 7.15 and 7.16. Due to the signal distortion problem associated with the relative threshold strategy as discussed above, the tracking performance is severely degraded with $e_1(t)$ depicted by the green line violating performance constraints at 5.24s, as illustrated in Fig. 7.15. It makes the control design suffer from the singularity problem. Note that our CES and switching threshold strategy accomplish a good balance between tracking performance and network constraints. Therefore, they can guarantee the tracking errors within the allowable set even the introduced external disturbances. However, from the fourth and fifth columns of Table 7.2 and the triggering time in Fig. 7.16, our CES strategy can consume fewer bit resources compared to the switching threshold one. On the other hand, for each communication case, we only transmit an encrypted 3-bit string through the public network rather than the sensitive real control input in the switching threshold strategy. Therefore, our control design reduces the communication burden and addresses the security concern at the same time.

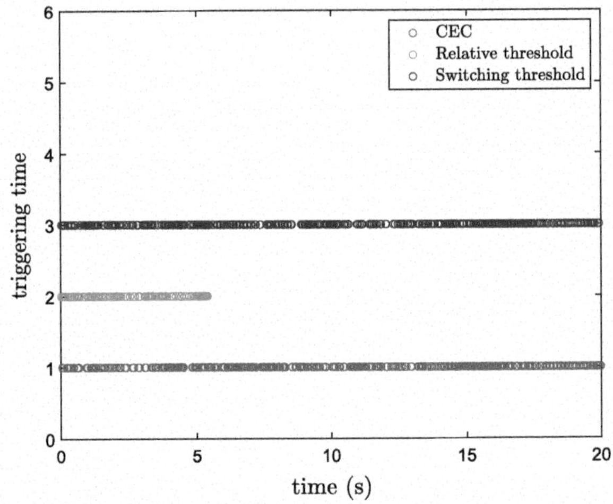

Fig. 7.16 The triggering time for case II

7.4 Conclusion

In this chapter, the control design for nonlinear system under prescribed performance and limited communication resources is investigated. With the aid of previous proposed Error Self-tuning Function-based Tunnel Prescribed Performance and Self-adjustable Tunnel Prescribed Performance, the Entry Capture Problem is addressed, and the tracking error is strictly constrained by the prescribed performance right after a finite time given by the designer. To reduce the communication burden, two actuator update protocols have been proposed, namely, constant event-triggered scheme and coded event-triggered scheme. In this way, our proposed control methods consume less communication bandwidth, and enhance secure communication capability, since only 1 bit or m-length string is encoded and transmitted for each communication between controller and actuator.

References

1. Bechlioulis, C.P., Rovithakis, G.A.: Robust adaptive control of feedback linearizable mimo nonlinear systems with prescribed performance. IEEE Transactions on Automatic Control **53**(9), 2090–2099 (2008)
2. Berger, T., Dennstädt, D., Ilchmann, A., Worthmann, K.: Funnel model predictive control for nonlinear systems with relative degree one. SIAM Journal on Control and Optimization **60**(6), 3358–3383 (2022)
3. Bu, X., He, G., Wei, D.: A new prescribed performance control approach for uncertain nonlinear dynamic systems via back-stepping. Journal of the Franklin Institute **355**(17), 8510–8536 (2018)

4. Bu, X., Qi, Q., Jiang, B.: A simplified finite-time fuzzy neural controller with prescribed performance applied to waverider aircraft. IEEE Transactions on Fuzzy systems **30**(7), 2529–2537 (2021)
5. Cao, L., Ren, H., Li, H., Lu, R.: Event-triggered output-feedback control for large-scale systems with unknown hysteresis. IEEE Transactions on Cybernetics **51**(11), 5236–5247 (2020)
6. Chen, M., Shao, S., Jiang, B.: Adaptive neural control of uncertain nonlinear systems using disturbance observer. IEEE Transactions on Cybernetics **47**(10), 3110–3123 (2017)
7. Cui, Y., Liu, Y., Zhang, W., Alsaadi, F.E.: Sampled-based consensus for nonlinear multiagent systems with deception attacks: The decoupled method. IEEE Transactions on Systems, Man, and Cybernetics: Systems **51**(1), 561–573 (2018)
8. Dolk, V., Heemels, M.: Event-triggered control systems under packet losses. Automatica **80**, 143–155 (2017)
9. Gao, S., Liu, X., Jing, Y., Dimirovski, G.M.: A novel finite-time prescribed performance control scheme for spacecraft attitude tracking. Aerospace Science and Technology **118**, 107044 (2021)
10. Ge, S.S., Wang, C.: Direct adaptive nn control of a class of nonlinear systems. IEEE Transactions on Neural Networks **13**(1), 214–221 (2002)
11. Girard, A.: Dynamic triggering mechanisms for event-triggered control. IEEE Transactions on Automatic Control **60**(7), 1992–1997 (2014)
12. Hua, C., Li, H., Li, K., Ding, W.: Low-computation tracking control of nonlinear systems with asymmetric full-state constraints and unknown control directions. IEEE Transactions on Systems, Man, and Cybernetics: Systems **53**(2), 1051–1059 (2023)
13. Ilchmann, A., Ryan, E.P., Townsend, P.: Tracking with prescribed transient behavior for nonlinear systems of known relative degree. SIAM Journal on Control and Optimization **46**(1), 210–230 (2007)
14. Ji, R., Ge, S.S.: Event-triggered tunnel prescribed control for nonlinear systems. IEEE Transactions on Fuzzy Systems (2023)
15. Jin, X.: Adaptive fixed-time control for mimo nonlinear systems with asymmetric output constraints using universal barrier functions. IEEE Transactions on Automatic Control **64**(7), 3046–3053 (2018)
16. Krstic, M., Kokotovic, P.V., Kanellakopoulos, I.: Nonlinear and adaptive control design. John Wiley & Sons, Inc. (1995)
17. Kumari, K., Bandyopadhyay, B., Reger, J., Behera, A.K.: Event-triggered sliding mode control for a high-order system via reduced-order model based design. Automatica **121**, 109163 (2020)
18. Li, Y., Tong, S.: Adaptive neural networks prescribed performance control design for switched interconnected uncertain nonlinear systems. IEEE transactions on neural networks and learning systems **29**(7), 3059–3068 (2017)
19. Liu, T., Jiang, Z.P.: Event-based control of nonlinear systems with partial state and output feedback. Automatica **53**, 10–22 (2015)
20. Mehdifar, F., Bechlioulis, C.P., Hashemzadeh, F., Baradarannia, M.: Prescribed performance distance-based formation control of multi-agent systems. Automatica **119**, 109086 (2020)
21. Qiu, J., Wang, T., Sun, K., Rudas, I.J., Gao, H.: Disturbance observer-based adaptive fuzzy control for strict-feedback nonlinear systems with finite-time prescribed performance. IEEE Transactions on Fuzzy Systems **30**(4), 1175–1184 (2021)
22. Sahoo, A., Xu, H., Jagannathan, S.: Adaptive neural network-based event-triggered control of single-input single-output nonlinear discrete-time systems. IEEE transactions on neural networks and learning systems **27**(1), 151–164 (2015)
23. Shao, X., Hu, Q., Shi, Y., Jiang, B.: Fault-tolerant prescribed performance attitude tracking control for spacecraft under input saturation. IEEE Transactions on Control Systems Technology **28**(2), 574–582 (2018)
24. Song, Y.D., Zhou, S.: Tracking control of uncertain nonlinear systems with deferred asymmetric time-varying full state constraints. Automatica **98**, 314–322 (2018)
25. Tabuada, P.: Event-triggered real-time scheduling of stabilizing control tasks. IEEE Transactions on Automatic Control **52**(9), 1680–1685 (2007)

26. Theodorakopoulos, A., Rovithakis, G.A.: Low-complexity prescribed performance control of uncertain mimo feedback linearizable systems. IEEE Transactions on Automatic Control **61**(7), 1946–1952 (2015)
27. Wang, H., Bai, W., Zhao, X., Liu, P.X.: Finite-time-prescribed performance-based adaptive fuzzy control for strict-feedback nonlinear systems with dynamic uncertainty and actuator faults. IEEE transactions on Cybernetics (2021)
28. Wang, L.X.: Adaptive fuzzy systems and control: design and stability analysis. Prentice-Hall, Inc. (1994)
29. Xing, L., Wen, C., Liu, Z., Su, H., Cai, J.: Event-triggered adaptive control for a class of uncertain nonlinear systems. IEEE Transactions on Automatic Control **62**(4), 2071–2076 (2016)
30. Xing, L., Wen, C., Liu, Z., Su, H., Cai, J.: Event-triggered output feedback control for a class of uncertain nonlinear systems. IEEE Transactions on Automatic Control **64**(1), 290–297 (2018)
31. Yong, K., Chen, M., Shi, Y., Wu, Q.: Flexible performance-based robust control for a class of nonlinear systems with input saturation. Automatica **122**, 109268 (2020)
32. Zhang, C., Chen, Z., Wang, J., Liu, Z., Chen, C.P.: Fuzzy adaptive two-bit-triggered control for a class of uncertain nonlinear systems with actuator failures and dead-zone constraint. IEEE Transactions on Cybernetics **51**(1), 210–221 (2020)
33. Zhang, C.H., Yang, G.H.: Event-triggered adaptive output feedback control for a class of uncertain nonlinear systems with actuator failures. IEEE transactions on cybernetics **50**(1), 201–210 (2018)
34. Zhang, J.X., Yang, G.H.: Low-complexity tracking control of strict-feedback systems with unknown control directions. IEEE Transactions on Automatic Control **64**(12), 5175–5182 (2019)
35. Zhang, Z., Wen, C., Zhao, K., Song, Y.: Decentralized adaptive control of uncertain interconnected systems with triggering state signals. Automatica **141**, 110283 (2022)
36. Zhao, K., Song, Y., Chen, C.P., Chen, L.: Adaptive asymptotic tracking with global performance for nonlinear systems with unknown control directions. IEEE Transactions on Automatic Control **67**(3), 1566–1573 (2021)

Chapter 8
Conclusions

This book systematically investigates and addresses constrained nonlinear control problem for systems under various conditions, such as input saturation, prescribed performance, communication constraints, actuator faults, time delay, unknown control directions, etc. Throughout the chapters, we propose a variety of control strategies tailored to specific single constraint or multiple constraints. In additional to ensuring the stability and robustness of the closed-loop systems, the proposed control strategies aim to further improve the tracking performance of the controlled systems especially under multiple constraints.

We first consider representative Euler–Lagrange Systems with single input saturation constraint. To compensate for the nonlinearities caused by input saturation, an auxiliary system is designed based on the input saturation error. In order to address the chattering phenomenon due to the discontinuous sign functions, we propose a second-order dynamic surface control based on the second-order sliding mode control. A new saturation mechanism is introduced for the time derivative of control input to ensure the control input signal always within the feasible region. Based on the corresponding control designs, all the closed-loop signals are proved to be bounded, and the tracking errors are regulated to a small neighborhood around equilibrium within finite time.

We then addresses the tracking control problem of more general nonlinear systems with input saturation and performance constraints at the same time. Novel conceptions of Tunnel Prescribed Performance and Finite-time Tunnel Prescribed Performance are proposed to not only simplify the control design but also improve the tracking performance in terms of overshoot performance and settling time. With the aid of the above proposed performance constraints, state/input-driven observers and Nussbaum technique, Tunnel Prescribed Control is proposed for nonlinear systems with known and unknown control directions, respectively. It has been proved the stability and finite-time convergence of the closed-loop system even in the presence of actuator faults and time delays.

R. Ji et al., *Constrained Nonlinear Control: Performance and Robustness*,
https://doi.org/10.1007/978-981-96-2971-8_8

A new Entry Capture Problem (ECP) is introduced, where the tunnel prescribed performance is satisfied after a certain period of system operation, as opposed to starting from the beginning, thereby, making the control design complex yet challenging. To this end, an error self-tuning function and an auxiliary function are designed to adjust the tracking error signal or the boundaries of prescribed performance based on system initial conditions. In this way, our proposed control schemes provide solutions for the underlying ECP, which also exhibits a low complexity level. Irrespective of initial conditions, including those outside the prescribed performance boundaries, the tracking errors are regulated to the tunnel prescribed performance within user-defined finite time.

To comprehensively tackle the conflict between the input saturation and prescribed performance, two auxiliary systems are designed for more complex Multi-input Multi-output (MIMO) nonlinear systems under known and unknown control directions. These two auxiliary systems establish interconnections between the above two constraints, instead of neglecting their contradictions. By introducing the generated non-negative signals into Tunnel Prescribed Performance, the resulted Saturation-tolerant Prescribed Performance possesses flexible performance. Namely, SPP can temporarily enlarge the performance boundaries to guarantee both constraints when input saturation occurs, and fast recover back to the prescribed boundaries when input saturation disappears. It can guarantee both input and performance constraints even under actuator faults and external disturbances.

Finally, communication resource constraints are investigated in addressing the Entry Capture Problem for a class of nonlinear systems. A novel coded event-triggered Scheme is proposed, by which only m-length string of base-p number system is transmitted for each communication between control box and actuator. Compared with other existing event-triggered results, our control design can further reduce the communication burden specifically from the encoding-decoding viewpoint. Moreover, as no sensitive information is transmitted over the vulnerable communication channel, the proposed coded event-triggered scheme can effectively enhance the communication security. Compared with constant and relative event-triggered schemes, our scheme adopts a piecewise increasing threshold, which can well balance between system performance and network constraints.

Despite the results introduced in this book, there exist more interesting yet challenging problems that remain open. We list some possible future directions as follows:

(i) Prescribed-time Entry Capture Problem: Although the methods in this book ensure finite-time convergence, their settling times are heavily dependent on initial conditions or control parameters such that settling times are not tunable by designers and inherently limit their applications in practice. It could be significant to investigate the Prescribed-time Entry Capture Problem, which can guarantee the system to reach the prescribed performance just at the user-defined time, regardless of initial conditions.

(ii) *Asymptotic Convergence:* All obtained results only guarantee finite-time stability, that is, the tracking error converges to a small neighborhood around zero. If we extend our methods to asymptotic convergence, it brings an immediate challenging in developing the control design since some auxiliary systems or fuzzy state observers inevitable involve some unexpected signals. However, it is of significance to investigate such control designs to further improve the tracking performance.

(iii) *Saturation-tolerant Prescribed Consensus Control:* Saturation-tolerant Prescribed Control has been proposed for MIMO nonlinear systems in this book. It is more interesting to extend our control schemes to the consensus control problem of multi-agent systems (MASs). The interconnections between agents make the control design more complex and how to guarantee the group of agents always within the prescribed performance boundaries is a challenging problem and worth further investigation due to the practical and theoretical significance.

(iv) *Secure Communication Protocol for MASs:* This book investigates a coded event-triggered scheme. If such a scheme is applied to heterogeneous MASs, the communication security among neighbors can be dramatically strengthened. However, the discontinuous triggering signals lead to non-differentiable virtual control inputs, making backstepping design inapplicable. Therefore, it is an interesting research direction to design a secure communication protocol for MASs under limited communication resources.

The authors hope this book will also catalyze efforts towards solving the above interesting problems.

Index

R. Ji et al., *Constrained Nonlinear Control: Performance and Robustness*,
https://doi.org/10.1007/978-981-96-2971-8

T

U

The manufacturer's authorised representative in the EU is Springer Nature Customer Service Centre GmbH, Europaplatz 3, 69115 Heidelberg, Germany. If you have any concerns regarding our products, please contact ProductSafety@springernature.com

Printed and bound by CPI Group (UK) Ltd, Croydon, CR0 4YY

30/09/2026

02225249-0002